Managing BR

Anees B. Chagpar
Editor

Managing BRCA Mutation Carriers

Springer

Editor
Anees B. Chagpar
Yale Comprehensive Cancer Center
Yale University School of Medicine
New Haven, CT
USA

ISBN 978-3-319-59197-1 ISBN 978-3-319-59198-8 (eBook)
DOI 10.1007/978-3-319-59198-8

Library of Congress Control Number: 2017940586

Printed on acid-free paper

This Springer imprint is published by Springer Nature
The registered company is Springer International Publishing AG
The registered company address is: Gewerbestrasse 11, 6330 Cham, Switzerland

Preface

Over the last several decades, there has been an explosion in our knowledge base regarding the genetic basis of breast cancer. With the discovery of novel genes, and with novel technology that has made genetic testing more easily available and affordable, there has been increasing interest on the part of patients and providers to access these services. Celebrities like Angelina Jolie have come forth to tell their story, and more patients are now aware of their family history and genetic risk. At the same time, with the Supreme Court's overturning of the Myriad patent and with the development of affordable panels for genetic testing for breast cancer risk, there has been a democratization of access to this technology. With this, however, has come a barrage of issues that bombard clinicians working in this space. How do we risk stratify our patients? Who should receive genetic testing? When should we use panels? How do we interpret a variant of uncertain significance? As more patients undergo genetic testing, we are increasingly faced with questions surrounding the value of prophylactic mastectomy, oophorectomy, or how best to surveil our patients who opt not to undergo mastectomy. In patients who are diagnosed with cancer and who harbor a genetic mutation, nuances exist in terms of their surgical, medical, and radiation management. Clearly, fertility preservation and preimplantation genetic assessment are also considerations. And of course, more men with genetic mutations are diagnosed with breast cancer, and they warrant specific considerations. Finally, throughout the journey—from genetic assessment through surgery and survivorship—patients with genetic mutations have a variety of psychosocial considerations and may have concerns about the legal implications of their diagnosis. This book is intended to help clinicians navigate this complex landscape.

We are incredibly grateful to the stellar authors who have participated in this project. Their expertise is unparalleled, and we appreciate their thoughtful consideration in preparing the chapters herein. In addition, we are grateful to Stephanie Frost and the rest of the team at Springer for their assistance in bringing this work to fruition.

We hope that you should find this volume of interest, and helpful to you in your practice. We look forward to your feedback!

New Haven, CT, USA Anees B. Chagpar

Contents

Contributors

Banu K. Arun Department of Breast Medical Oncology, The University of Texas MD Anderson Cancer Center, Houston, TX, USA

Soley Bayraktar Department of Medical Oncology, Memorial Cancer Center, Istanbul, Turkey

Patrick I. Borgen Department of General Surgery, Maimonides Medical Center, Brooklyn, NY, USA

Karina L. Brierley Smilow Cancer Genetics and Prevention Program, Yale New Haven Hospital, New Haven, CT, USA

Jessica Cintolo-Gonzalez Department of Surgical Oncology, Massachusetts General Hospital, Boston, MA, USA

Hiram S. Cody, III Breast Service, Department of Surgery, Memorial Sloan Kettering Cancer Center, Weill Cornell Medical College, New York, NY, USA

Patricia A. Cronin Memorial Sloan Kettering Cancer Center, Breast Service, Evelyn Lauder Breast and Imaging Center, New York, NY, USA

Don S. Dizon Department of Medicine, Massachusetts General Hospital Cancer Center, Boston, MA, USA

Elizabeth S. Ginsburg Division of Reproductive Endocrinology and Infertility, Department of Obstetrics and Gynecology, Brigham and Women's Hospital and Harvard Medical School, Boston, MA, USA

Ryland J. Gore Department of General Surgery, Maimonides Medical Center, Brooklyn, NY, USA

Claire L. Healy Smilow Cancer Genetics and Prevention Program, Yale New Haven Hospital, New Haven, CT, USA

Kevin S. Hughes Department of Surgical Oncology, Massachusetts General Hospital, Boston, MA, USA; Harvard Medical School, Boston, MA, USA; Bermuda Cancer Genetics and Risk Assessment Clinic, Hamilton, Bermuda; Avon Comprehensive Breast Evaluation Center, Massachusetts General Hospital, Boston, MA, USA

Erin W. Hofstatter Department of Medical Oncology, Smilow Cancer Genetics and Prevention Program, Yale New Haven Hospital/Yale School of Medicine, New Haven, CT, USA

Julio A. Ibarra Orange Coast Memorial Medical Center, Fountain Valley, CA, USA; Department of Pathology, University of California at Irvine, Irvine, CA, USA

Iris G. Insogna Department of Obstetrics and Gynecology, Brigham and Women's Hospital and Harvard Medical School, Boston, MA, USA

Christine Laronga Department of Breast Oncology, H. Lee Moffitt Cancer Center & Research Institute, Tampa, FL, USA

Jessica M. Long Department of Medicine, Division of Hematology-Oncology and Basser Center for BRCA, University of Pennsylvania, Philadelphia, PA, USA

Donna-Marie Manasseh Department of General Surgery, Maimonides Medical Center, Brooklyn, NY, USA

Willonie Mendonca GeneDx Laboratories, Gaithersburg, MD, USA

Meena S. Moran Therapeutic Radiology, Yale University School of Medicine, New Haven, CT, USA; Yale New Haven Hospital, New Haven, CT, USA

Anvy Nguyen Department of Surgical Oncology, Massachusetts General Hospital, Boston, MA, USA

Cristina O'Donoghue Department of Breast Oncology, H. Lee Moffitt Cancer Center & Research Institute, Tampa, FL, USA

Sonia Orcutt Department of Breast Oncology, H. Lee Moffitt Cancer Center & Research Institute, Tampa, FL, USA

Tuya Pal Department of Cancer Epidemiology, H. Lee Moffitt Cancer Center & Research Institute, Tampa, FL, USA

Ann H. Partridge Department of Medical Oncology, Dana-Farber Cancer Institute, Brookline, MA, USA; Department of Medicine, Brigham and Women's Hospital and Harvard Medical School, Boston, USA

Jennifer K. Plichta Department of Surgery, Duke University Medical Center, Durham, NC, USA

Jacquelyn M. Powers Department of Medicine, Division of Hematology-Oncology and Basser Center for BRCA, University of Pennsylvania, Philadelphia, PA, USA

Dario R. Roque Department of Obstetrics and Gynecology, Program in Women's Oncology, Women and Infants Hospital/The Warren Alpert Medical School at Brown University, Providence, RI, USA

Ellen Warner Medical Oncologist, Sunnybrook Health Sciences Centre, Department of Medicine, Division of Medical Oncology, University of Toronto, Toronto, ON, Canada

Risk Assessment for Breast Cancer

1

Anvy Nguyen, Jessica Cintolo-Gonzalez, Jennifer K. Plichta and Kevin S. Hughes

It is critical to identify all women who have a *BRCA1* or *BRCA2* mutation before they develop their first cancer. Identification of mutation carriers enhances the accuracy of breast cancer prediction and allows treatment plans to be tailored to the individual's cancer. Preventive efforts, including increased screening intensity and frequency, prophylactic surgery, and the use of chemopreventive agents, have reduced the risk for breast cancer and increased survival among *BRCA1/2* mutation carriers [1–3]. Underestimating carrier risk has a negative impact on breast cancer prevention and may increase morbidity and mortality.

A. Nguyen (✉) · J. Cintolo-Gonzalez · K.S. Hughes
Department of Surgical Oncology, Massachusetts General Hospital,
55 Fruit Street, Yawkey 7, Boston, MA 02214, USA
e-mail: anvy_nguyen@dfci.harvard.edu

J. Cintolo-Gonzalez
e-mail: jcintolo-gonzalez@partners.org

K.S. Hughes
e-mail: kshughes@partners.org

J.K. Plichta
Department of Surgery, Duke University Medical Center, DUMC 3513,
Durham, NC 27705, USA
e-mail: jennifer.plichta@duke.edu

K.S. Hughes
Harvard Medical School, Boston, MA, USA

K.S. Hughes
Bermuda Cancer Genetics and Risk Assessment Clinic, Hamilton, Bermuda

K.S. Hughes
Avon Comprehensive Breast Evaluation Center, Massachusetts General Hospital,
55 Fruit Street, Boston 02114, MA, USA

A.B. Chagpar (ed.), *Managing BRCA Mutation Carriers*,
DOI 10.1007/978-3-319-59198-8_1

As Mary-Claire King has stated, "to identify a woman as a carrier only after she develops cancer is a failure of cancer prevention" [4]. We wholeheartedly agree, but unfortunately, this is the current state of affairs with regard to genetic risk assessment. While genetic testing has been available for the past 20 years, approximately 95% of unaffected *BRCA* mutation carriers remain unaware of their status. Far worse, many mutation carriers who develop cancer still remain unidentified [5].

While these numbers are disconcerting, of all the hereditary cancer syndromes, the rate of identification of *BRCA1/2* mutation carriers among individuals with hereditary breast and ovarian cancer syndrome is likely the highest. Lynch syndrome carriers (i.e., hereditary non-polyposis colorectal cancer) are even more likely to remain unaware. As many as 99% of these carriers remain untested, owing to incomplete documentation of family histories and poor awareness of genetic referral criteria [6–8]. Moreover, since new breast cancer susceptibility genes, including *PALB2*, *CHEK2*, and *ATM*, have been introduced over the past several years, awareness of these genes and application of multi-gene panel testing are also critical [9–11] and likely insufficient.

Risk Assessment

Hereditary cancer risk assessment is a multi-step process used to identify and counsel individuals at high risk for hereditary or familial cancer. The current method of identifying mutation carriers is flawed, is inefficient, and presents multiple barriers. To identify patients for genetic testing one must (1) take a family history; (2) ensure that the history is comprehensive; (3) analyze the family history data using risk models, guidelines, or intuition; (4) arrange access to resources for genetic testing; and (5) finally, persuade the patient to proceed with testing. At each of these barriers, a certain proportion of patients is lost.

After these barriers have been surmounted, the patient is ready for genetic testing. Unfortunately, many physicians lack sufficient time during their examination to take a simple, much less comprehensive, family history. As a result, the family history, considered the first genetic test, is often neglected or incomplete. Even oncologists may fail to take a directed family history from patients who already have cancer, leading to lower rates of referral for genetic counseling and testing [12]. The physicians taking the family history also may have insufficient expertise to characterize the conditions associated with increased risk for hereditary syndromes and to identify which patients are testable. In addition to hereditary breast and ovarian cancer (HBOC) syndrome, physicians should be familiar with Li–Fraumeni syndrome (mutations in p53), Cowden syndrome (*PTEN*), Peutz–Jeghers syndrome (*STK11*), the hereditary diffuse gastric cancer syndrome (*CDH1*), and Lynch syndrome (*MLH1*, *MSH2*, *MSH6*, *PMS2*, and EPCAM), to name but a few. Furthermore, some physicians have insufficient knowledge of genetics, may feel uncomfortable with genetic testing, or lack accessible guidance for determining risk stratification and management [13–16]. Finally, when a complete family history is taken, it requires vigilance on the part of patients and physician providers to update the history every time a previously unaffected family member develops disease; failure to do so, as so often occurs, can lead to missed opportunities for genetic testing.

Some physicians feel comfortable performing the genetic tests themselves and providing pre- and post-test counseling, whereas others prefer to refer their patients to a professional geneticist or counselor. The goal of pre-test counseling is to ensure that the patient is sufficiently prepared to understand the possible genetic basis of disease, the personal risk for developing certain cancers, the consequences of undergoing testing, and the available preventive and treatment options. Enhancing the patient's understanding of their cancer risk allows them to make informed decisions regarding testing, and it has been shown that patients who receive pre-test counseling have improved understanding and satisfaction [17]. Although genetic counselors are skilled in communicating complicated genetic information, there are not nearly enough counselors to meet the growing demand for genetic services. Travel time, transportation, and childcare also pose barriers to access, and a long delay between the time the appointment is made and the availability of a counselor can produce substantial attrition [18].

Unfortunately, owing to the presence of these many barriers, the majority of candidates are never tested. This is one of the reasons Mary-Claire King has recently espoused that all women should undergo *BRCA* testing by age 30, regardless of family history. While criticism has been levied by those who consider this viewpoint to be extreme, given the low cost of genetic testing and expanded appreciation for its value, it is hardly an unreasonable proposal. Indeed, testing has become more commonplace, causes less anxiety among physicians, and is much less expensive, making widespread testing a possibility worth serious consideration.

Nevertheless, as population-level screening for *BRCA* is still considered unacceptable, the alternative strategy is to identify high-risk individuals before they develop cancer. This responsibility falls to the primary care physicians (PCPs), obstetrician/gynecologists, and to breast imaging centers where risk assessment can be accomplished prior to cancer development. Assuming a complete cancer family history has been obtained, patients can be identified for testing either through the use of risk models, published guidelines, or based on the clinician's intuition.

Guidelines for Genetic Risk Assessment and Testing

The guidelines for genetic risk assessment and/or testing are generally similar, since most rely heavily on the individual's family history. One of the most extensive are the guidelines published by the National Comprehensive Cancer Network (NCCN), which have recently been updated (February 2016) [19]. These guidelines contain detailed criteria regarding who may qualify for additional genetic risk evaluation, and in certain cases, extensive knowledge of an individual's family history may be required to adequately perform the assessment. Individuals with breast cancer may be eligible for an additional genetic evaluation if they have a known mutation in a cancer susceptibility gene within the family, if their age at the time of diagnosis was under 50 (or under the age of 60 for triple-negative breast cancer), or if a male family member has been afflicted with breast cancer. Additional criteria focus on the family history, regardless of the individual's breast cancer history, such as family members with breast, ovarian, prostate, and/or pancreatic cancer and families of Ashkenazi Jewish descent. For individuals referred for genetic testing, additional criteria are used to determine who may be eligible, although determination of such eligibility typically requires similar knowledge of one's family history. Of note, the NCCN guidelines have lowered the threshold for testing individuals with limitations in the family tree (i.e., few female relatives) or those who were adopted.

Other authoritative entities have published guidelines related to genetic testing, including the American Society of Clinical Oncology (ASCO), the American College of Obstetricians and Gynecologists (ACOG), the US Preventive Services Task Force (USPSTF), and the American College of Medical Genetics and Genomics (ACMG) [20–24]. In particular, ASCO has created several webpages dedicated to helping providers collect a complete family history as well as decision support for when to refer patients for further evaluation [20, 25, 26]. In response to the quickly emerging technologies in genetics, ASCO also published a statement related to genetic testing and the challenges that have been created by new technologies in an effort to ensure appropriate deployment in clinical practice [25]. In determining who should receive further genetic risk assessment, ACOG's recommendations again rely on an individual's personal and family histories, including details such as a family history of breast cancer (with known age[s] at diagnosis), ovarian cancer, Ashkenazi Jewish ancestry, and male breast cancer. Although these guidelines identify those who may require further evaluation, they do not specifically detail those who should undergo genetic testing. The USPSTF guidelines are even less specific, but also suggest initial screening by a primary care provider based on personal and family histories (to be updated/reviewed every 5–10 years) and ultimately recommend additional evaluation for genetic testing [22]. Similar to the NCCN guidelines, the ACMG's guidelines suggest referral for genetic testing for those diagnosed with breast cancer under the age of 50, those with triple-negative breast cancer under the age of 60, those with 2 or more breast cancers (in the same person), and other select criteria [24]. Although these criteria are seemingly easier to follow, they are limited to those at risk for hereditary breast and/or ovarian cancer syndromes and are not ideal screening tools for those at risk for other hereditary conditions related to breast cancer, such as Li–Fraumeni Syndrome or Cowden Syndrome.

Mutation Risk Prediction Models

Mathematical risk models can assist providers to determine who should be referred for genetic testing. These models integrate data on risk factors, including family history, in order to calculate the estimated risk that a given patient may carry a deleterious *BRCA1* or *BRCA2* mutation. By identifying particular risk factors, these models can guide history taking to assure that providers collect appropriate data and then provide a concrete prediction of the patient's risk. While many such models have been developed, four of the most highly validated and clinically applied statistical models are the Myriad, BRCAPRO, BOADICEA, and Tyrer-Cuzick. The most widely used model in the USA is BRCAPRO, while the BOADICEA model is commonly used in Europe.

Myriad Model

The first Myriad model, known as Myriad I or the Shattuck-Eidens model [27], has since been updated as the Myriad II model, also known as the Frank model [28]. Myriad uses logistic regression to predict the risk for a deleterious *BRCA1/2* mutation using risk factors that are presented in tables that are accessible online and routinely updated (https://www.myriadpro.com/hereditary-cancer-testing/hereditary-breast-and-ovarian-cancer-hboc-syndrome/prevalence-tables/). Separate tables exist for those with and without Ashkenazi Jewish (AJ) ancestry. Risk factors incorporated into the Myriad II tables include personal breast cancer history (invasive or in situ) with age of diagnosis categorized as ≥ 50 or <50 years, history of ovarian cancer, history of male breast cancer, and the combination of both breast and ovarian cancers. Family history assessed includes breast cancer (diagnosed at age ≥ 50 or <50 years) and ovarian cancer diagnosed at any age in first- or second-degree relatives. Attributing the same degree of risk to all breast cancer patients under the age of 50 is a limitation of the Myriad II model. An online risk calculator also exists for application in the clinical setting (https://www.myriadpro.com/hereditary-cancer-testing/hereditary-breast-and-ovarian-cancer-hboc-syndrome/brca-risk-calculator/).

BRCAPRO Model

BRCAPRO predicts the likelihood that a woman may have a deleterious *BRCA1/2* germline mutation [29, 30] and also predicts the likelihood of developing invasive breast cancer or ovarian cancer within a set time period, including the risk for contralateral breast cancer in those already diagnosed with breast cancer [31]. Using Bayes' theorem, BRCAPRO calculates risk based on frequently updated estimates of the prevalence and penetrance of *BRCA1* and *BRCA2* mutations and baseline rates of breast cancer in the population. BRCAPRO considers relatives of any degree and incorporates into its calculations their relation to the counselee, current age, breast cancer and ovarian cancer status, and age at diagnosis if affected. BRCAPRO also incorporates pathologic markers (ER, PR, HER2, CK14, CK5/6) for known breast cancer cases as well as race and ethnicity. Not accounting for interventions undergone by family members may lead to misleading mutation carrier probabilities [32]. BRCAPRO addresses this issue by incorporating family members who have undergone mastectomy, including male mastectomy, bilateral mastectomy, and/or oophorectomy [31]. BRCAPRO has been found to be highly sensitive [33], and its application has been shown to predict breast cancer risk in high-risk populations. Estimation of breast and ovarian cancer risks by BRCAPRO and subsequent genetic testing have led to prophylactic oophorectomy in 89.5% and prophylactic mastectomy in 11.1% of Jewish *BRCA1/2* mutation carriers, in addition to reducing the risks of breast and ovarian cancers in this population of women [34]. BRCAPRO can be accessed for clinical use via its native R implementation as part of the open-source Bayes Mendel [35] package or via multiple web-based and commercial software packages [36–39].

BOADICEA Model

The Breast and Ovarian Analysis of Disease Incidence and Carrier Estimation Algorithm (BOADICEA model) also uses Bayes' theorem to calculate the risk for carrying a harmful *BRCA1/2* mutation and, unlike other models, addresses a polygenic component that reflects the multiplicative effects of lower-penetrance breast cancer susceptibility genes on breast cancer risk [40, 41]. Similar to the BRCAPRO and Tyrer-Cuzick models, BOADICEA predicts the risk for developing breast cancer over time [41]. Unlike other models discussed, BOADICEA incorporates third-degree family members and accounts for history of prostate and pancreatic cancers as well as breast and ovarian cancers [42]. Although including all relatives within a counselee's pedigree may permit more complete data, it also increases the risk for recall bias, which may affect the results [40]. The model also incorporates into its calculations tumor pathology characteristics, such as estrogen receptor (ER) status, triple-negative status, and expression of basal markers (CK5/6 and CK14), as well as updated breast cancer incidence [43]. BOADICEA can be accessed for clinical use via an online web application (http://ccge.medschl.cam.ac.uk/boadicea/boadicea-web-application/).

Tyrer-Cuzick Model

The Tyrer-Cuzick (TC) model, also known as the International Breast Cancer Study (IBIS) Breast Cancer Risk Evaluation Tool, predicts both risk for carrying a deleterious *BRCA1/2* mutation and risk for invasive breast cancer within 10 years. Hereditary information assessed in this model includes Ashkenazi inheritance and first- and second-degree relatives with breast cancer and ovarian cancer. Additionally, it incorporates the age of diagnosis and the presence of bilateral breast cancer [44]. Additional non-genetic risk factors include hormonal and pathologic factors such as age, body mass index (BMI), age at menarche, age at first live birth, parity, age at menopause, use of hormone replacement therapy, breast biopsies, hyperplasia, atypical hyperplasia (ductal or lobular), and LCIS [44]. Unlike the Myriad II, BRCAPRO, and BOADICEA models, the Tyrer-Cuzick model does not account for male breast cancer. Future iterations of this model are being tested and may incorporate mammographic density adjusted for age and BMI [45–47].

Studies comparing the accuracy of risk calculations performed by the Myriad II, BRCAPRO, BOADICEA, and Tyrer-Cuzick models have generally found similar accuracy between the Myriad II, BRCAPRO, and BOADICEA models [48, 49], although other studies found BOADICEA and BRCAPRO to demonstrate better accuracy and prediction [42, 50].

Table 1.1 summarizes the risk factors incorporated in the foregoing models.

The medical provider or genetics professional uses the individual's genetic risk assessment as the basis for determining whether genetic testing is indicated. After pre-test counseling, which includes a discussion of the risks, benefits, limitations, and possible results of genetic testing, a variety of tests may be recommended, including a test to determine whether there are mutations in the *BRCA1* and *BRCA2* genes. Traditional methods for genetic testing prior to 2006 failed to detect large rearrangement mutations, which account for 6–10% of all *BRCA1/2* mutations [51]. A more comprehensive test capable of full sequencing to detect large rearrangements in *BRCA1* and *BRCA2* is currently recommended and is supported in the NCCN guidelines [19].

Multi-gene panel testing is also recognized by the NCCN as an efficient and cost-effective method to interrogate multiple genes simultaneously. Panel testing is recommended when more than one gene may explain a heritable cancer syndrome or when a patient with a suggestive personal or family history has previously tested negatively. Some would even conjecture that panel testing should be recommended for everyone having testing. Given the introduction of over 25 breast cancer susceptibility genes (e.g., p53, *PTEN*, *CDH1*, *STK11*, *PALB2*, *ATM*, *CHEK2*, *BARD1*, *NBN*, *BRIP1*, *RAD50*, and *RAD51*), future genetic testing will likely involve multi-gene panel testing for a number of clinically actionable mutated genes [52, 53]. Limitations to panel testing include the higher likelihood of identifying variants of unknown significance (VUS) and the lack of uniform guidelines for screening of mutation carriers of these genes. This is discussed in further detail in Chap. 2.

To estimate the risk going forward, the most comprehensive approach is to use a risk model such as BRCAPRO. For patients without cancer and having all at-risk organs intact, the lifetime risk estimation for breast and ovarian cancers is determined by simply entering the type of mutation and the age of the patient. The BRCAPRO risk calculations, depicted in Table 1.2, reveal the impact of risk-reducing surgery, if performed, on age-specific breast and ovarian cancer risk in *BRCA1* and *BRCA2* mutation carriers. These data may help guide specific recommendations regarding the use of prophylactic surgery to reduce cancer risk in mutation carriers.

Table 1.1 Incorporation of risk factors into risk prediction models

Variable	Myriad II	BRCAPRO	BOADICEA	Tyrer-Cuzick
Personal information				
Body mass index	No	No	No	Yes
Age	Yes	Yes	Yes	Yes
Hormonal/reproductive factors				
Age at menarche	No	No	No	Yes
Age at first live birth	No	No	No	Yes
Age at menopause	No	No	No	Yes
Oral contraceptive use	No	No	No	No
Hormone replacement therapy	No	No	No	Yes
Breastfeeding	No	No	No	No
Personal history of breast disease				
Breast biopsies	No	No	No	Yes
Atypical ductal hyperplasia	No	No	No	Yes
Lobular carcinoma in situ	No	No	No	Yes
Family history of breast and/or ovarian cancer				
First-degree relatives with breast cancer	Yes	Yes	Yes	Yes
Second-degree relatives with breast cancer	Yes	Yes	Yes	Yes
Third-degree relatives with breast cancer	No	No	Yes	No
Age of onset of breast cancer in a relative	Yes	Yes	Yes	Yes
Ovarian cancer in a relative	Yes	Yes	Yes	Yes
Male breast cancer	Yes	Yes	Yes	No
Bilateral breast cancer in a relative	No	Yes	Yes	Yes
Personal history of risk-reducing surgery				
Risk-reducing mastectomy (RRM)	No	Yes	No	No
Risk-reducing salpingo-oophorectomy (RRSO)	No	Yes	No	No
RRM and RRSO	No	Yes	No	No

If the patient has already had breast or ovarian cancer, recommendations for follow-up will be determined more by the stage and prognosis of these cancers than by the mutation status. For the patient with breast cancer, BRCAPRO still provides the risk for ovarian cancer going forward, which is likely accurate and can be acted upon based on the prognosis and stage of the breast cancer. BRCAPRO also produces a breast cancer risk which purports to predict contralateral breast cancer [54]. This determination should be interpreted with caution, however, as mitigating circumstances such as the use of tamoxifen may render this prediction less accurate. Modifications of the model will likely be required before contralateral cancer risk can be used clinically. On the other hand, a patient with ovarian cancer will obtain a fairly accurate breast cancer risk going forward, which takes into account the impact of their oophorectomy.

Table 1.2 BRCAPRO calculations of lifetime risk for breast and ovarian cancers according to mutation, age, and risk-reducing surgery

Mutation	Age	Risk-reducing surgery	Lifetime breast cancer risk (%)	Lifetime ovarian cancer risk (%)
BRCA1	25	None	63.1	58
	65	None	27.9	43.4
	25	RRSO	41.2	17.8
	65	RRSO	19	10
	25	RRM	6.3	58
	65	RRM	1.4	43.4
	25	RRSO and RRM	6.3	17.8
	65	RRSO and RRM	1.4	10
BRCA2	25	None	55.4	30
	65	None	26.5	22.7
	25	RRSO	33	8.4
	65	RRSO	19	5.7
	25	RRM	5.5	30
	65	RRM	1.6	22.7
	25	RRSO and RRM	5.5	8.4
	65	RRSO and RRM	1.6	5.7

Genetic and Non-genetic Modifiers

Great variation exists in the risk for breast cancer arising from genetic and non-genetic modifiers that impact the penetrance of *BRCA* genes. Genetic polymorphisms known to modify the risk for breast and ovarian cancers in mutation carriers have been recently explored. Several single nucleotide polymorphisms (SNPs) have been found in association with increased breast cancer risk for *BRCA1/2* mutation carriers [55–57], while other polymorphisms are associated with decreased risk [58]. It has been suggested that incorporating validated SNPs will improve the predictive performance of risk models [59], but further studies are warranted regarding the clinical impact of SNPs on cancer risk.

Non-genetic modifiers have been extensively discussed but remain controversial. Age of the mother at first birth, which is associated with a reduction in breast cancer development in the general population, has not been shown to be protective in *BRCA1* mutation carriers [60]. Other hormonal factors, including age at menarche [61], the effect of pregnancy [62], and the protective effect of breastfeeding in *BRCA1* mutation carriers, have also been studied [63, 64]. Use of oral contraceptives before age 25 has been shown to increase the risk for early-onset breast cancer among *BRCA1* mutation carriers [65], and oophorectomy was found to be protective in pre-menopausal breast cancer in *BRCA2* carriers [66, 67]. The effects of weight loss as well as alcohol and caffeine consumption have been investigated as protective factors for breast cancer among mutation carriers, but the results on alcohol consumption have been mixed [68–72].

Clinical Decision Support

Although these observations are extremely interesting, they are difficult to implement clinically without clinical decision support (CDS) [73, 74]. CDS involves the use of computers and health information technologies designed to improve clinical decision-making by providing a platform to integrate evidence-based knowledge and clinical guidelines, to interpret medical data, and to make recommendations and predictions. The benefits of CDS include enhanced quality of care and outcomes, decreased adverse events, and improved efficiency.

Conclusion

For physicians who lack sufficient knowledge to perform their own genetic testing, it is important to refer individuals with increased risk for hereditary breast cancer for genetic counseling. Genetic counselors are uniquely trained to identify individuals with mutations in cancer susceptibility genes and to provide cancer risk assessment. Medical providers, particularly PCPs, are also responsible for identifying and referring individuals deemed appropriate for genetic testing. An accurate qualitative risk assessment includes a detailed and complete personal and family history. Quantitative risk assessment involves the use of risk prediction models to determine the individual's risk for carrying a deleterious mutation in a breast cancer susceptibility gene. Clinical guidelines for genetic high-risk assessment, such as those published by the NCCN, are useful, easily accessible resources that help direct clinicians through the difficult process of deciding who is appropriate for genetic testing. The expanding role of genetics in the management of breast cancer, coupled with rapid advances in genomic sequencing technology, requires health professionals to obtain critical patient information and to integrate current guidelines and evidence-based knowledge with health information technology in order to identify mutation carriers through risk assessment and genetic testing. Thus, the need for continuing medical education regarding the emergence of new tests and technologies cannot be understated. Identifying *BRCA1/2* mutation carriers before they develop cancer is the only sure path to cancer prevention for individuals with hereditary or familiar risk for breast cancer and is essential to improving survival.

References

1. Ingham SL, Sperrin M, Baildam A, Ross GL, et al. Risk-reducing surgery increases survival in BRCA1/2 mutation carriers unaffected at time of family referral. Breast Cancer Res Treat. 2013;142(3):611–8.
2. Calderon-Margalit R, Paltiel O. Prevention of breast cancer in women who carry BRCA1 or BRCA2 mutations: a critical review of the literature. Int J Cancer. 2004;112(3):357–64.
3. Grann VR, Jacobson JS, Whang W, et al. Prevention with tamoxifen or other hormones versus prophylactic surgery in BRCA1/2-positive women: a decision analysis. Cancer J Sci Am. 2000;6(1):13–20.
4. Levy-Lahad E, Lahad A, King MC. Precision medicine meets public health: population screening for BRCA1 and BRCA2. J Natl Cancer Inst. 2015;107(1):420.
5. Drohan B, Roche CA, Cusack JC Jr, Hughes KS. Hereditary breast and ovarian cancer and other hereditary syndromes: using technology to identify carriers. Ann Surg Oncol. 2012;19 (6):1732–7.
6. van Dijk DA, Oostindiër MJ, Kloosterman-Boele WM, Krijnen P, Vasen HF, Hereditary Tumor Study Group of the Comprehensive Cancer Centre West. Family history is neglected in the work-up of patients with colorectal cancer: a quality assessment using cancer registry data. Fam Cancer. 2007;6(1):131–4.
7. Overbeek LI, Hoogerbrugge N, van Krieken JHJM, Nagengast FM et al, on behalf of the MIPA Study Group. Most patients with colorectal tumors at young age do not visit a cancer genetics clinic. Dis Colon Rectum. 2008;51(8):1249–54.
8. Singh H, Schiesser R, Anand G, et al. Underdiagnosis of Lynch syndrome involves more than family history criteria. Clin Gastroenterol Hepatol. 2010;8(6):523–9.
9. Rahman N, Seal S, Thompson D, et al, Breast Cancer Susceptibility Collaboration (UK), Easton DF, Stratton MR. PALB2, which encodes a BRCA2-interacting protein, is a breast cancer susceptibility gene. Nat Genet. 2007;39(2):165–7.
10. Meijers-Heijboer H, van den Ouweland A, Klijn J, et al, CHEK2-Breast Cancer Consortium. Low-penetrance susceptibility to breast cancer due to CHEK2(*)1100delC in noncarriers of BRCA1 or BRCA2 mutations. Nat Genet. 2002;31(1):55–9.
11. Renwick A, Thompson D, Seal S, et al. ATM mutations that cause ataxia-telangiectasia are breast cancer susceptibility alleles. Nat Genet. 2006;38(8):873–5.
12. Wood ME, Kadlubek P, Pham TH, et al. Quality of cancer family history and referral for genetic counseling and testing among oncology practices: a pilot test of quality measures as part of the American Society of Clinical Oncology Quality Oncology Practice Initiative. J Clin Oncol. 2014;32(8):824–9.
13. Wood ME, Flynn BS, Stockdale A. Primary care physician management, referral, and relations with specialists concerning patients at risk for cancer due to family history. Public Health Genom. 2013;16(3):75–82.
14. Flynn BS, Wood ME, Ashikaga T, et al. Primary care physicians' use of family history for cancer risk assessment. BMC Fam Pract. 2010;11:45.
15. Wood ME, Stockdale A, Flynn BS. Interviews with primary care physicians regarding taking and interpreting the cancer family history. Fam Pract. 2008;25(5):334–40.
16. Wideroff L, Vadaparampil S, Greene M, et al. Hereditary breast/ovarian and colorectal cancer genetics knowledge in a national sample of US physicians. J Med Genet. 2005;42(10): 749–55.
17. Armstrong J, Toscano M, Kotchko N, et al. Utilization and outcomes of BRCA genetic testing and counseling in a national commercially insured population: the ABOUT study. JAMA Oncol. 2015;1(9):1251–60.
18. Anderson B, McLosky J, Wasilevich E, et al. Barriers and facilitators for utilization of genetic counseling and risk assessment services in young female breast cancer survivors. J Cancer Epidemiol. 2012;2012:298745, 11 p.

19. NCCN Clinical Practice Guidelines in Oncology: Genetic/Familial High Risk Assessment: Breast and Ovarian. Cited 2016; Available from: https://www.nccn.org.
20. ASCO Genetics Toolkit. Cited 2016; Available from: http://www.asco.org/practice-guidelines/practice-management-issues/genetics-toolkit/.
21. American College of Obstetricians and Gynecologists, ACOG Committee on Practice Bulletins–Gynecology, ACOG Committee on Genetics, Society of Gynecologic Oncologists. ACOG Practice Bulletin No. 103: Hereditary breast and ovarian cancer syndrome. Obstet Gynecol. 2009;113(4):957–66.
22. Nelson HD, Pappas M, Zakher B, et al. Risk assessment, genetic counseling, and genetic testing for BRCA-related cancer in women: a systematic review to update the U.S. Preventive Services Task Force recommendation. Ann Int Med. 2014;160(4):255–66.
23. Routine Screening for Hereditary Breast and Ovarian Cancer Recommended. 2009 March 23, 2009. Cited 2016; Available from: http://www.acog.org/About-ACOG/News-Room/News-Releases/2009/Routine-Screening-for-Hereditary-Breast-and-Ovarian-Cancer-Recommended.
24. Hampel H, Bennett RL, Buchanan A, et al, Guideline Development Group, American College of Medical Genetics and Genomics Professional Practice and Guidelines Committee and National Society of Genetic Counselors Practice Guidelines Committee. A practice guideline from the American College of Medical Genetics and Genomics and the National Society of Genetic Counselors: referral indications for cancer predisposition assessment. Genet Med. 2015;17(1):70–87.
25. Robson ME, Bradbury AR, Arun B, et al. American Society of Clinical Oncology Policy Statement update: genetic and genomic testing for cancer susceptibility. J Clin Oncol. 2015;33(31):3660–7.
26. Lu KH, Wood ME, Daniels M, et al. American Society of Clinical Oncology Expert Statement: collection and use of a cancer family history for oncology providers. J Clin Oncol. 2014;32(8):833–40.
27. Shattuck-Eidens D, Oliphant A, McClure M, et al. BRCA1 sequence analysis in women at high risk for susceptibility mutations. Risk factor analysis and implications for genetic testing. JAMA. 1997;278(15):1242–50.
28. Frank TS, Manley SA, Olopade OI, et al. Sequence analysis of BRCA1 and BRCA2: correlation of mutations with family history and ovarian cancer risk. J Clin Oncol. 1998; 16(7):2417–25.
29. Berry DA, Parmigiani G, Sanchez J, et al. Probability of carrying a mutation of breast-ovarian cancer gene BRCA1 based on family history. J Natl Cancer Inst. 1997;89(3):227–38.
30. ParmigianiG, Berry D, Aguilar O. Determining carrier probabilities for breast cancer-susceptibility genes BRCA1 and BRCA2. Am J Hum Genet. 1998;62(1):145–58.
31. Mazzola E, Blackford A, Parmigiani G, Biswas S. Recent enhancements to the genetic risk prediction model BRCAPRO. Cancer Inform. 2015;14(Suppl 2):147–57.
32. Katki HA. Incorporating medical interventions into carrier probability estimation for genetic counseling. BMC Med Genet. 2007;8:13.
33. Berry DA, Iversen ES Jr, Gudbjartsson DF, et al. BRCAPRO validation, sensitivity of genetic testing of BRCA1/BRCA2, and prevalence of other breast cancer susceptibility genes. J Clin Oncol. 2002;20(11):2701–12.
34. Metcalfe KA, Mian N, Enmore M, et al. Long-term follow-up of Jewish women with a BRCA1 and BRCA2 mutation who underwent population genetic screening. Breast Cancer Res Treat. 2012;133(2):735–40.
35. http://www.la-press.com/recent-enhancements-to-the-genetic-risk-prediction-model-brcapro-article-a4822.
36. http://bayesmendel.dfci.harvard.edu/risk/.
37. https://www4.utsouthwestern.edu/breasthealth/cagene.
38. http://hughesriskapps.com/.
39. http://www.progenygenetics.com/clinical/risk.

40. Antoniou AC, Pharoah PP, Smith P, Easton DF. The BOADICEA model of genetic susceptibility to breast and ovarian cancer. Br J Cancer. 2004;91(8):1580–90.
41. Antoniou AC, Pharoah PD, McMullan G, et al. A comprehensive model for familial breast cancer incorporating BRCA1, BRCA2 and other genes. Br J Cancer. 2002;86(1):76–83.
42. Antoniou AC, Hardy R, Walker L, et al. Predicting the likelihood of carrying a BRCA1 or BRCA2 mutation: validation of BOADICEA, BRCAPRO, IBIS, Myriad and the Manchester scoring system using data from UK genetics clinics. J Med Genet. 2008;45(7):425–31.
43. Lee AJ, Cunningham AP, Kuchenbaecker KB, et al, Consortium of Investigators of Modifiers of BRCA1/2, Breast Cancer Association Consortium. BOADICEA breast cancer risk prediction model: updates to cancer incidences, tumour pathology and web interface. Br J Cancer. 2014;110(2):535–45.
44. Tyrer J, Duffy SW, Cuzick J. A breast cancer prediction model incorporating familial and personal risk factors. Stat Med. 2004;23(7):1111–30.
45. Warwick J, Birke H, Stone J, et al. Mammographic breast density refines Tyrer-Cuzick estimates of breast cancer risk in high-risk women: findings from the placebo arm of the International Breast Cancer Intervention Study I. Breast Cancer Res. 2014;16(5):451.
46. Subramanian J, Karmegam A, Papageorgiou E, et al. An integrated breast cancer risk assessment and management model based on fuzzy cognitive maps. Comput Methods Programs Biomed. 2015;118(3):280–97.
47. Brentnall AR, Harkness EF, Astley SM, et al. Mammographic density adds accuracy to both the Tyrer-Cuzick and Gail breast cancer risk models in a prospective UK screening cohort. Breast Cancer Res. 2015;17(1):147.
48. Panchal SM, Ennis M, Canon S, Bordeleau LJ. Selecting a BRCA risk assessment model for use in a familial cancer clinic. BMC Med Genet. 2008;9:116.
49. Barcenas CH, Monawar GM, Arun B, et al. Assessing BRCA carrier probabilities in extended families. J Clin Oncol. 2006;24(3):354–60.
50. Schneegans SM, Rosenberger A, Engel U, et al. Validation of three BRCA1/2 mutation-carrier probability models Myriad, BRCAPRO and BOADICEA in a population-based series of 183 German families. Fam Cancer. 2012;11(2):181–8.
51. Judkins T, Rosenthal E, Arnell C, et al. Clinical significance of large rearrangements in BRCA1 and BRCA2. Cancer. 2012;118(21):5210–6.
52. Campeau PM, Foulkes WD, Tischkowitz MD. Hereditary breast cancer: new genetic developments, new therapeutic avenues. Hum Genet. 2008;124(1):31–42.
53. Li J, Meeks H, Feng BJ, et al. Targeted massively parallel sequencing of a panel of putative breast cancer susceptibility genes in a large cohort of multiple-case breast and ovarian cancer families. J Med Genet. 2016;53(1):34–42.
54. Mazzola E, Chipman J, Cheng SC, Parmigiani G. Recent BRCAPRO upgrades significantly improve calibration. Cancer Epidemiol Biomark Prev. 2014;23(8):1689–95.
55. Antoniou AC, Kuchenbaecker KB, Soucy P, et al, CIMBA, SWE-BRCA, HEBON, EMBRACE, GEMO Collaborators Study, kConFab Investigators. Common variants at 12p11, 12q24, 9p21, 9q31.2 and in ZNF365 are associated with breast cancer risk for BRCA1 and/or BRCA2 mutation carriers. Breast Cancer Res. 2012;14(1):R33.
56. Antoniou AC, Kartsonaki C, Sinilnikova OM, et al. Common alleles at 6q25.1 and 1p11.2 are associated with breast cancer risk for BRCA1 and BRCA2 mutation carriers. Hum Mol Genet. 2011;20(16):3304–21.
57. Gaudet MM, Kirchhoff T, Green T, et al. Common genetic variants and modification of penetrance of BRCA2-associated breast cancer. PLoS Genet. 2010;6(10):e1001183.
58. Cox DG, Simard J, Sinnett D, et al, Consortium of Investigators of Modifiers of BRCA1/2. Common variants of the BRCA1 wild-type allele modify the risk of breast cancer in BRCA1 mutation carriers. Hum Mol Genet. 2011;20(23):4732–47.
59. Prosperi MC, Ingham SL, Howell A, et al. Can multiple SNP testing in BRCA2 and BRCA1 female carriers be used to improve risk prediction models in conjunction with clinical assessment? BMC Med Inform Decis Mak. 2014;14:87.

60. Kotsopoulos J, Lubinski J, Lynch HT, et al. Age at first birth and the risk of breast cancer in BRCA1 and BRCA2 mutation carriers. Breast Cancer Res Treat. 2007;105(2):221–8.
61. Kotsopoulos J, Lubinski J, Lynch HT, et al. Age at menarche and the risk of breast cancer in BRCA1 and BRCA2 mutation carriers. Cancer Causes Control. 2005;16(6):667–74.
62. Cullinane CA, Lubinski J, Neuhausen SL, et al. Effect of pregnancy as a risk factor for breast cancer in BRCA1/BRCA2 mutation carriers. Int J Cancer. 2005;117(6):988–91.
63. Jernstrom H, Lubinski J, Lynch HT, et al. Breast-feeding and the risk of breast cancer in BRCA1 and BRCA2 mutation carriers. J Natl Cancer Inst. 2004;96(14):1094–8.
64. Kotsopoulos J, Lubinski J, Salmena L, et al, Hereditary Breast Cancer Clinical Study Group. Breastfeeding and the risk of breast cancer in BRCA1 and BRCA2 mutation carriers. Breast Cancer Res. 2012;14(2):R42.
65. Kotsopoulos J, Lubinski J, Moller P, et al, Hereditary Breast Cancer Clinical Study Group. Timing of oral contraceptive use and the risk of breast cancer in BRCA1 mutation carriers. Breast Cancer Res Treat. 2014;143(3):579–86.
66. Kotsopoulos J, Huzarski T, Gronwald J, et al, The Hereditary Breast Cancer Clinical Study Group. Bilateral oophorectomy and breast cancer risk in BRCA1 and BRCA2 mutation carriers. J Natl Cancer Inst. 2017;109(1):djw177.
67. Kim SJ, Zuchniak A, Sohn KJ, et al. Plasma folate, vitamin B-6, and vitamin B-12 and breast cancer risk in BRCA1- and BRCA2-mutation carriers: a prospective study. Am J Clin Nutr. 2016;104(3):671–7.
68. Cybulski C, Lubinski J, Huzarski T, et al. Prospective evaluation of alcohol consumption and the risk of breast cancer in BRCA1 and BRCA2 mutation carriers. Breast Cancer Res Treat. 2015;151(2):435–41.
69. Nikitina D, Chen Z, Vallis K, et al. Relationship between caffeine and levels of DNA repair and oxidative stress in women with and without a BRCA1 mutation. J Nutrigenet Nutrigenomics. 2015;8(4–6):174–84.
70. Dennis J, Ghadirian P, Little J, et al. Alcohol consumption and the risk of breast cancer among BRCA1 and BRCA2 mutation carriers. Breast. 2010;19(6):479–83.
71. Nkondjock A, Ghadirian P, Kotsopoulos J, et al. Coffee consumption and breast cancer risk among BRCA1 and BRCA2 mutation carriers. Int J Cancer. 2006;118(1):103–7.
72. Kotsopoulos J, Olopado OI, Ghadirian P, et al. Changes in body weight and the risk of breast cancer in BRCA1 and BRCA2 mutation carriers. Breast Cancer Res. 2005;7(5):R833–43.
73. Welch BM, Kawamoto K. Clinical decision support for genetically guided personalized medicine: a systematic review. J Am Med Inform Assoc. 2013;20(2):388–400.
74. Welch BM, Kawamoto K. The need for clinical decision support integrated with the electronic health record for the clinical application of whole genome sequencing information. J Pers Med. 2013;3(4):306–25.

Genetic Evaluation for Women at Increased Risk

2

Karina L. Brierley, Claire L. Healy and Erin W. Hofstatter

Hereditary Cancer Risk Assessment

The US Preventative Services Task Force recommends providers perform a risk assessment in order to identify those who have a personal and/or family history of cancer that may be associated with a hereditary cancer predisposition. The collection of an accurate, cancer-focused family history is the foundation of this risk assessment [1–3].

What to Collect

Several physician organizations, including the American Society of Clinical Oncology (ASCO), have developed standards for the minimum collection requirements for an adequate family history. At a minimum, these include the following [4]:

K.L. Brierley · C.L. Healy
Smilow Cancer Genetics and Prevention Program, Yale New Haven Hospital,
330 Orchard Street, New Haven, CT 06511, USA
e-mail: karina.brierley@ynhh.org

C.L. Healy
e-mail: claire.healy@ynhh.org

E.W. Hofstatter (✉)
Department of Medical Oncology, Smilow Cancer Genetics
and Prevention Program, Yale New Haven Hospital/Yale School of Medicine,
333 Cedar Street, New Haven, CT 06520, USA
e-mail: erin.hofstatter@yale.edu

A.B. Chagpar (ed.), *Managing BRCA Mutation Carriers*,
DOI 10.1007/978-3-319-59198-8_2

1. Personal cancer history;
2. Cancer history of first degree relatives, i.e., siblings, parents, and children;
3. Cancer history of second degree relatives, i.e., grandparents, aunts, uncles, grandchildren, nieces, nephews, and half-siblings;
4. Inclusion of relatives on both the maternal and paternal side; and
5. Inclusion of information on the patient's ethnicity.

For each individual in the family where a cancer diagnosis is reported by the patient, it is important to collect the following information [4]:

1. Type of primary cancer;
2. Age at diagnosis of each primary cancer; and
3. Lineage (i.e., is the family member a maternal or paternal relative).

While not included in the minimum collection requirements, providers should consider asking whether there is a known hereditary cancer predisposition syndrome in the family and whether any family members have had genetic testing [4].

An accurate risk assessment is also dependent on family size. It is often just as important to know how many unaffected relatives there are in the family as there are individuals with cancer. For example, the risk for a hereditary cancer syndrome in an individual with a mother and aunt with breast cancer is substantially lower if the mother also had five sisters who have not developed cancer [1].

When to Collect

Cancer family history is typically collected at the initial visit and/or at the time of a patient's cancer diagnosis. However, after a patient has been diagnosed with cancer and has shared the diagnosis with other family members, conversations regarding the family history of cancer may occur which may alter the patient's initial report. Relatives may be more likely to share stories of other family members who have had cancer after learning of a recent diagnosis. For this reason, providers should consider reassessing the family history after the initial stress of a cancer diagnosis has abated [1].

It is also important to periodically reassess the family history, as cancer history can change significantly over time. Reassessment should include elicitation of any new family history information, as well as a determination of whether advances in genetic testing technology or the discovery of additional genes linked to hereditary cancer predispositions have occurred since the last evaluation, which may require re-referral and/or updated testing [4].

How to Collect

With limited time available, providers may struggle with how to collect adequate family history information. Some providers prefer to utilize a patient-centered collection tool prior to the visit. Patients can complete these questionnaires in the waiting room or at home prior to seeing their provider, and the collected information can then be reviewed and expanded upon in the visit. There are a number of organizations that have developed patient-friendly questionnaires for this purpose, including the following:

1. Cancer.Net Cancer Family History Questionnaire which is based on ASCO's recommendations for a minimum adequate family history [5]: http://www.cancer.net/sites/cancer.net/files/cancer_family_history_questionnaire.pdf
2. The US Surgeon General's My Family Health Portrait tool which can be accessed online or printed out and given to patients [6]: http://www.hhs.gov/programs/prevention-and-wellness/family-health-history/family-health-portrait-tool/index.html#

Regardless of whether the provider is reviewing information, the patient has provided through a questionnaire, or whether they are starting the family history collection from scratch, setting expectations with the patient prior to collecting or reviewing the information can be helpful in streamlining the intake of information [1]. Patients are more likely to provide concise responses if they know what information the provider is looking for ahead of time.

Limitations

There can be a number of barriers to obtaining an accurate and useful hereditary cancer risk assessment.

1. ***Patient Barriers***

Patients may have limited or no knowledge of their family history due to:

(a) Adoption or conception through donor eggs/sperm;
(b) Family estrangement; and
(c) Cultural barriers that prevent the discussion of cancer diagnoses.

In these situations, providers should focus on the information that is available while making note of the barriers that limit the risk assessment [1].

2. ***Provider Barriers***

Obtaining an accurate family history and determining who would benefit from additional risk assessment and genetic testing can be time consuming. In a world where providers are asked to do more with less time, it may be challenging to collect the necessary information. Utilizing patient questionnaires and tools like the ones mentioned above may help to reduce the amount of time providers spend on this task [4].

Guidelines for Further Risk Assessment and Genetic Testing

Many professional organizations have created guidelines outlining when patients should be referred to a provider with expertise in hereditary cancer genetics for further risk assessment, as well as when genetic testing should be performed. It is important to note that the criteria for further risk assessment are not identical to the criteria for genetic testing. Some patients who are referred for further risk assessment may not meet guidelines for genetic testing but may still be candidates for increased screening, behavior modifications, or medical interventions due to their personal and/or family history [2].

In some instances, the a priori risk for a mutation will be high enough based on a patient's personal history such that no further family history is needed to warrant further risk evaluation. This can include the following [7]:

1. Any individual with or without a cancer diagnosis who has a known mutation in a cancer susceptibility gene within their family;
2. Any man with a diagnosis of breast cancer;
3. Any woman with ovarian cancer;
4. Any woman with breast cancer diagnosed ≤ 45 years old; and/or
5. Any woman with a triple negative breast cancer diagnosed ≤ 60 years old.

Individuals of Ashkenazi Jewish descent also have a higher a priori risk due to the increased frequency of founder mutations in the *BRCA1* and *BRCA2* genes among this population. Approximately, 1 in 40 (2.5%) of individuals of Ashkenazi Jewish descent will carry a mutation in *BRCA1* or *BRCA2* versus the approximate carrier frequency of 1 in 400 (0.25%) in the Western European population [3]. Due to this increased frequency of mutations within this population, it is recommended that any individual of Ashkenazi Jewish descent with a diagnosis of breast, ovarian, or pancreatic cancer, regardless of age or family history, be referred for further risk assessment [7].

Oftentimes, the decision to refer a patient for further risk evaluation is based on a combination of personal and/or family history information. However, it is difficult to convey the myriad possible combinations of personal and family history of cancer that should prompt a referral to a specialist in cancer genetics for further risk evaluation. Table 2.1 lists common scenarios for which providers should pursue referral for risk assessment [7]. Providers may also wish to familiarize themselves with and utilize the 2015 practice guidelines published by the American College of

Table 2.1 Common scenarios for which providers should pursue referral for risk assessment [7]

Individuals with a personal history of breast cancer at any age plus any of the following in a first, second, or third degree relative:	Individuals without a personal diagnosis of cancer who have a family history of the following in a first, second, or third degree relative:
At least one relative with breast cancer diagnosed ≤ 50 years old	$\geq$ Two breast cancers in a single relative
At least one relative with invasive ovarian cancer at any age	$\geq$ Two relatives with breast cancer at least one of whom was diagnosed ≤ 50 years old
$\geq$ Two relatives with breast cancer at any age	A relative with ovarian cancer
$\geq$ Two relatives with pancreatic cancer at any age	A relative with male breast cancer

Medical Genetics & Genomics and the National Society of Genetic Counselors [8]. This document was created in an easy-to-read table and is designed to allow providers to cross-reference a specific type of cancer against the family history necessary to warrant referral.

Provision of Cancer Genetic Counseling and Testing

Over the past 20 years, the field of cancer genetic counseling and testing has grown exponentially and changed rapidly. Throughout its growth and evolution, there has been debate about which health care providers should provide these services, namely, only genetics specialists versus all health care providers regardless of specialty [9, 10]. Recent decisions by several large insurers to require genetic counseling by a certified genetics provider prior to cancer genetic testing, as well as the increasing complexity of the available cancer genetic testing options, have sparked renewed interest in this debate.

Some argue that all health care providers should provide genetic counseling and testing services based on the potential benefits of increased access to genetic services, cost efficiency, a more holistic approach, and better knowledge of patients' overall health due to existing long-term relationships [10]. On the other hand, there is much literature and expert opinion to support the belief that cancer genetic counseling and testing should ideally be provided by genetics specialists [9]. Specifically, numerous studies have demonstrated that many providers lack the training in and knowledge of genetics to adequately provide cancer genetic counseling and testing services to their patients [9, 11–13]. This includes data even on those providers who arguably have the most current genetics education and training, such as medical residents, and includes key concepts such as associated cancer risks and inheritance patterns [12]. Many providers also self-report lack of adequate time as a barrier to providing cancer genetic counseling and testing services [9, 11]. In addition, existing data suggest that many providers are not sufficiently familiar with the complex ethical and psychosocial issues that often accompany genetic counseling and testing, such as genetic discrimination concerns and the existing laws, concerns and policies regarding testing minors for adult-onset conditions [9, 14–16].

Although the availability of cancer genetics professionals is increasing and access to cancer genetics professionals is readily available in many areas, there may still be locations where these services are not as readily available. In locations where geography presents concerns regarding adequate access to genetics professionals, telemedicine genetic counseling services with board certified professionals are now available and covered by several major insurers. Recent studies suggest that telemedicine genetic counseling is cost-effective, associated with high patient satisfaction, and is equally effective as in-person genetic counseling [17, 18]. Ultimately, multidisciplinary teams, increased genetics education for all providers, self-awareness, close collaborations, and open lines of communication and referral will likely best serve patients and providers alike. Table 2.2 lists information about locating in-person- and/or telemedicine-based cancer genetic counseling professional services [19–22].

Cancer Genetic Counseling and Testing Process

Cancer genetic counseling has been described as a multistep communication process between a clinician and a patient/family of which the actual genetic testing is only one component [23]. For some individuals, this process occurs even without actual

Table 2.2 Resources for locating a genetic counselor [19–22]

Resource
National Society of Genetic Counselors (NSGC) [19] http://nsgc.org/page/find-a-gc-search Database of genetic counselors who are members of the NSGC that is searchable by geographic location and specialty
National Cancer Institute (NCI) Cancer Genetics Services Directory [20] https://www.cancer.gov/about-cancer/causes-prevention/genetics/directory Directory of professionals who provide services related to cancer genetics (including genetic counselors)
Informed DNA [21] http://www.informeddna.com/ Nationwide network of genetic counselors that provides telephone and Web-based genetic counseling services to patients and providers that are covered by some major insurers
GeneTests [22] https://www.genetests.org/ Gene Tests has an international directory of genetics professionals searchable by location, role, and specialty as well as an international directory of genetics clinics searchable by location and keywords

genetic testing as it may be determined that testing is not warranted, the patient is not the best candidate for testing in their family, and/or the patient is not interested in pursuing testing. Some of the essential elements in the cancer genetic counseling and testing process include the following: intake, risk assessment, pretest counseling and informed consent, and result disclosure and interpretation [24].

The intake/history component of the genetic counseling process should include collection of a detailed personal medical history and a 3–4 generation family history in order to provide accurate risk assessment, differential diagnosis, and development of a personalized management plan. The intake should also include an assessment of the patient's concerns, motivations, needs, values, and knowledge or understanding of the pertinent information related to cancer genetics [23, 24]. The personal history should include any diagnoses of cancer, benign tumors, or unusual findings that may be relevant to risk assessment (e.g., multiple colon polyps, skin findings), frequency of cancer surveillance, surgical interventions, environmental exposures (e.g., tobacco use, occupational exposures), and reproductive information (e.g., oral contraceptive use, tubal ligation) [24]. An accurate family history is an essential tool in the hereditary cancer risk assessment and result interpretation. The 3–4 generation family history should include information on both affected and unaffected individuals, their relationship to the patient, current age or age at death, the site and age at diagnosis for any cancer diagnoses, ancestry/ethnicity, consanguinity, surgical interventions (which may reduce the cancer incidence), any findings that may be relevant to differential diagnoses under consideration (e.g., multiple polyps, unusual skin findings, autism spectrum disorders, benign tumors), and the results of any prior genetic testing on family members [24]. Information about family history may be inaccurate and thus, efforts should be made to confirm family history information with medical records or death certificates when possible to improve accuracy of risk assessment [24]. Interestingly, cancer type, gender of historian, education level, family size, and degree of relatedness to affected relative have all been shown to impact the accuracy of reporting of cancer diagnoses among relatives [25]. Reporting of results of prior genetic testing on relatives is also often inaccurate or incomplete and thus should also be confirmed with records.

Based on the personal and family history information collected, a risk assessment and differential diagnosis should be generated. In general, the risk assessment should distinguish between individuals at: high risk (personal and/or family history consistent with a highly penetrant hereditary cancer syndrome), moderately increased risk (history consistent with either a multifactorial cause or a low- to moderate-penetrance mutation), and average risk [23, 24]. The risk assessment, differential diagnosis, ideal testing strategy, and available testing options should be discussed with the patient.

Whom to Test

In order to obtain the most accurate interpretation of genetic test results, it is preferable to start genetic testing with an individual in the family who is most likely to carry a mutation. This may not always be the individual who presents for the initial risk assessment. Beginning genetic testing in an individual who has had a cancer diagnosis most closely related to the hereditary cancer syndrome in question (i.e., breast or ovarian cancer in *BRCA1* and *BRCA2*) is likely to yield the most informative results for the family. If there are multiple relatives in a family with an associated cancer diagnosis who are available for testing, priority could be given to those with bilateral disease, multiple primary cancers, or the youngest age at diagnosis [25].

In some instances, there will not be an affected individual available for testing, or the results of an affected relatives genetic testing may be inaccessible, due to death, estrangement, or a refusal to pursue testing. In these situations, testing an individual without a cancer diagnosis may be appropriate but the limitations of a negative genetic test result should be clearly reviewed.

When there is significant suspicion for a hereditary cancer predisposition in a family, a negative genetic test results in an unaffected individual could be explained in two ways:

1. There is a mutation in a hereditary cancer gene in the family which the patient did not inherit. In this instance, the patient's risk to develop cancer would be the same as an individual in the general population; or,
2. There is no currently identifiable mutation in a hereditary cancer gene in the family. In this instance, the patient's risk to develop cancer would still be considered elevated above the general population risk, and screening and prevention decisions would be based on the family history.

Given that it is not possible to distinguish between these two explanations when the only genetic testing that has been completed in a family was in an individual without a cancer diagnosis, providers should err on the side of caution and follow their patients based on family history despite their negative genetic test results.

In general, a detailed informed consent process should accompany any genetic testing and in some states informed consent is required by law [24]. The informed consent process should include a discussion of the genes being testing, the possible test results [positive, negative, variant of uncertain significance (VUS)], how results may impact the individual's cancer risks and medical management options, how results may impact family members' risks, ethical/legal/psychosocial aspects (e.g., discrimination issues and protections, family issues), economic considerations (e.g., potential costs and coverage), and a review of the benefits, risks, limitations, and alternatives to genetic testing [24]. For patients who choose not to proceed with testing, recommendations for cancer screening, and prevention based on personal and family history alone should be reviewed as well as recommendations for genetic counseling and testing for other relatives, if applicable [24].

Disclosure of the results of any genetic testing, regardless of the test result (positive, negative, or VUS), should be accompanied by a thorough discussion including the following: a personalized interpretation of the results in the context of the individual's personal and family history, revised cancer risk assessment, medical management guidelines/recommendations, identification of at-risk relatives and/or other relatives who may benefit from genetic counseling and testing, and tools to assist the patient in informing family members (e.g., family letter, online resources, referrals to genetics providers) [24].

Hereditary Breast and Ovarian Cancer Testing Options

The intersection of the introduction of mainstream use of next-generation sequencing technology and the Supreme Court ruling overturning gene patenting in 2013 has led to exponential growth of the available testing options and laboratory choices for hereditary breast cancer testing [26, 27]. In addition, these developments have also led to increased availability of low-cost testing options, with out-of-pocket costs ranging in several hundred dollars, rather than several thousand dollars. There are now more than 10 laboratories offering *BRCA1* and *BRCA2* testing as a targeted test or as part of one of the dozens of multi-gene panel test options, ranging from 6 to 100+ genes. These developments mean that broader genetic testing options for *BRCA1* and *BRCA2* and other hereditary cancer genes are available to clinicians and patients, and that testing is likely to be accessible and affordable for more patients even if they lack insurance coverage for testing. However, it also means that navigating the available choices can be complicated, particularly in the context of aggressive marketing efforts by the commercial testing laboratories who are trying to secure business in a competitive marketplace. In the face of this multitude of laboratory and testing choices, there are a number of aspects to consider when choosing a laboratory and test. These include quality, methods, data sharing, cost, insurance verification process, genes included in available panels, variant classification, variant analysis and reporting process, and family studies programs for variants [28, 29].

Multi-gene panel tests offer the advantage of cost- and time-efficient testing for multiple genes. Several studies have now demonstrated that multi-gene panel testing does provide additional diagnostic yield compared to a syndrome-specific gene testing approach, with an absolute additional yield of identification of a deleterious mutation in ~4–16% of individuals, meaning that this approach to testing may identify the causative mutation in additional individuals/families [26, 30, 31]. At least one of these studies also demonstrated that this additional yield of mutations changed the management for the patient and/or close relatives in many cases [31]. However, the benefits of multi-gene panel testing over syndrome-specific gene testing must be balanced by the limitations, particularly several layers of complexity and uncertainty that can come with this testing [26, 29]. One important layer of complexity is that multi-gene panel testing is currently associated with a high rate of identification of variants of uncertain significance, ranging from 28 to 40% [27, 30, 31]. These variants often can be misinterpreted by providers and patients as clinically relevant (i.e., potentially associated with high cancer risks), when typically these variants are later reclassified as normal, benign variants [29, 32]. Thus, this can be an important source of unnecessary worry, anxiety, and, most critically, unnecessary medical interventions, including invasive "prophylactic" surgeries [29, 32].

Another layer of uncertainty that arises frequently with the advent of multi-gene panel testing is the identification of deleterious mutations in genes where the clinical implications are less clear [29]. Many gene panels include more newly described, "moderate-penetrance" genes for which data are often more limited regarding the exact cancer risks, range of associated cancers, and appropriate management recommendations [29]. Thus, determining how to use this information can be challenging for patients and providers alike. Identification of a mutation in a more newly described or lower penetrance gene may also pose other result interpretation and/or medical management challenges for providers, patients, and family members as it is not always clear if the identified mutation completely explains the personal and/or family history that prompted testing. This leads to difficulties in making decisions about whether or not to test other relatives, interpreting "true negative" test results, and determining residual risks and appropriate management [26].

An additional challenge posed by multi-gene panel testing is the possibility of an unexpected mutation in high-penetrance gene that is not consistent with the known history that prompted testing [26]. These unexpected results again can be challenging to interpret in terms of advising patients and their family members regarding expected cancer risks and appropriate management, as there is very little data at this time regarding whether the presentation, severity, and risks will be different in families where mutation is an "incidental finding" and thus, whether management should be based on genotype alone, phenotype alone, or some combination of the two.

A sometimes less recognized or appreciated but important challenge of current cancer genetic testing choices is that the classification of a given variant can differ from one laboratory to the next, with different laboratories classifying the same genetic change or variant as a variant of uncertain significance, a likely pathogenic variant, or a pathogenic variant (mutation) [27]. These discrepancies can occur based on conflicting interpretations of available data.

For all of these reasons, testing should be ideally offered in the context of care by professionals with genetics expertise [26, 27, 29, 33]. In addition, determination of the most appropriate testing options should be made by the clinician based on the patient's clinical and family history. When a choice between more limited syndrome-specific testing and broader multi-gene panel testing is reasonable, the clinician should help the patient make an informed choice based on a discussion of the benefits and limitations of the available options and the patient's values and preferences [29].

Hereditary Breast and/or Ovarian Cancer Genes Frequently Included in Multi-gene Panels

The focus of this volume is the management of individuals with *BRCA* mutations, as *BRCA1* and *BRCA2* are the most common genes associated with hereditary breast and/or ovarian cancer. However, any current discussion of hereditary breast and ovarian cancer would not be complete without mention of other rare high-penetrance genes and moderate-penetrance genes that are now included in many routine clinical genetic testing options for hereditary breast and ovarian cancer. In addition to *BRCA1* and *BRCA2*, there are several rare hereditary cancer syndromes that place individuals at high risk of developing breast cancer including the following: Li-Fraumeni Syndrome, Cowden Syndrome, Peutz-Jeghers Syndrome, and hereditary diffuse gastric cancer.

Li-Fraumeni Syndrome (LFS) is caused by mutations in the *TP53* gene and is associated with a diverse range of cancers [34]. The lifetime risk of developing cancer with a *TP53* mutation is ~90% and individuals are at high risk to develop multiple primary cancers [34, 35]. The core cancers associated with LFS are soft tissue sarcomas, osteosarcomas, brain tumors, very early-onset breast cancer, and adrenal cortical carcinoma. However, individuals with LFS can develop a wide range of cancers. LFS is a rare hereditary cancer syndrome with an estimated prevalence of ~1/5000–1/20,000 and accounts for ~1% or less of breast cancer cases [34, 35]. However, breast cancer is the most frequent cancer among female carriers of *TP53* mutations and in many cases breast cancer occurs before age 30 [34].

Cowden Syndrome is a rare hereditary cancer syndrome (prevalence of ~1/200,000–1/250,000) caused by mutations in the *PTEN* gene [33, 36]. *PTEN* hamartoma tumor syndrome refers to a broader range of syndromes, including Cowden Syndrome and Bananayan-Riley-Ruvalcaba syndrome, that can be associated with *PTEN* mutations. Cowden Syndrome/*PTEN* hamartoma tumor syndrome is associated with multiple hamartomas and a high risk of benign and cancerous tumors in a variety of tissues including the breast, thyroid, and endometrium [35, 36]. The lifetime risk of breast cancer associated with *PTEN* mutations is ~25–50% by most estimates, although a few studies report higher risks, as high as ~75–85% [36, 37]. *PTEN* hamartoma tumor syndrome is also associated with a wide range of features including unusual mucocutaneous features (oral papillomas, trichilemmomas, penile freckling), macrocephaly, developmental delay, autism spectrum disorders, multiple gastrointestinal polyps (including hamartomas and ganglioneuromas), and vascular malformations [36, 38].

Peutz-Jeghers Syndrome (PJS) is caused by mutations in the *STK11* (or *LKB1*) gene and is a rare autosomal dominant hereditary cancer syndrome (prevalence estimates of ~1/25,000–1/280,000) [39]. PJS is associated with mucocutaneous hyperpigmentation (melanocytic macules on buccal mucosa, lips, nostrils, fingers) and multiple hamartomatous gastrointestinal polyps (especially in the small intestine) often resulting in symptoms (e.g., intussusception, obstruction, gastrointestinal bleeding) [39]. The lifetime cancer risk associated with PJS is ~50–85%, with the highest risks being for colorectal and breast cancers, but risks for stomach, small intestine, pancreatic, gynecologic, testicular, and lung cancers are also increased [39, 40]. Women with PJS have a lifetime risk of breast cancer of ~32–54% and are at increased risk for ovarian sex cord tumors with annular tubules (SCTATs) and adenoma malignum of the cervix [33, 39].

Hereditary diffuse gastric cancer is associated with mutations in the E-cadherin (*CDH1*) gene. Individuals with a germline *CDH1* mutation have a ~65–85% lifetime risk of developing diffuse gastric cancer with an average age of diagnosis of 40 years old [35, 41, 42]. Women who carry a *CDH1* mutation have a ~40–54% lifetime risk of developing breast cancer, primarily of the lobular subtype [35, 41, 42]. Mutations in the *CDH1* gene are thought to be rare with prevalence of <0.1/100,000.

Over recent years, other genes associated with hereditary breast and/or ovarian cancer, most of which are currently considered "moderate-penetrance" genes, have been identified and are now included on many multi-gene hereditary cancer testing panels. The distinction between "moderate" and "high" risk is somewhat arbitrary as the lifetime risk ranges for several "moderate" risk genes overlap with "high" risk genes [43]. However, current data suggest that the risks associated with some of these genes may vary significantly based on specific mutation and/or family history [43]. These "moderate-penetrance" genes have mainly been identified by searching for mutations in genes that share some functionality with *BRCA1* and *BRCA2* either by directly interacting with the *BRCA* proteins and/or being involved in the Fanconi anemia pathway which is involved in double strand DNA break repair and homologous recombination. Their association with hereditary breast and/or ovarian cancers has been strengthened by the identification of mutations in these genes in individuals and/or families whose history was suspicious for hereditary breast and/or ovarian cancer and had negative *BRCA1/2* testing. These genes and current information about the associated cancer risks are listed in Table 2.3 [27, 34–50].

Table 2.3 Associated cancer risks of high- and moderate-penetrance genes [27, 34–50]

Gene(s)	Breast cancer lifetime risk	Ovarian cancer lifetime risk	Other associated cancers/features
High-penetrance genes			
BRCA1	55–87%	15–60%	Prostate
			Male breast
			Pancreas
BRCA2	45–82%	15–40%	Prostate
			Male breast
			Pancreas
			Melanoma

(continued)

Table 2.3 (continued)

Gene(s)	Breast cancer lifetime risk	Ovarian cancer lifetime risk	Other associated cancers/features
CDH1	39–53% (particularly lobular)	No known increase	Diffuse gastric (55–85%)
			Possibly colon
PTEN	~25–50% by most estimates (some higher estimates)	No known increase	Thyroid, endometrial, renal, and possibly colorectal cancer
			Benign breast and thyroid disease
			Uterine fibroid tumors
			Skin findings (oral papillomas, facial trichilemmomas)
			Macrocephaly
			Developmental delay
			Autism spectrum disorders
			Multiple GI polyps (including hamartomas and ganglioneuromas)
			Vascular malformations
STK11	32–54%	18–21% (mainly sex cord stromal)	Colorectal, gastric, pancreatic, uterine, small intestine, testicular, and lung cancers
			Multiple hamartomatous GI polyps mucocutaneous hyperpigmentation
TP53	Significantly increased, may be as high as 79%	Unknown/not well defined	Sarcoma, brain, adrenal cortical carcinoma, leukemia, lung, and other cancers
			Childhood onset cancers
			Multiple primary cancers
Moderate-penetrance genes			
ATM	17–52%	No known increase	Possibly pancreas and prostate, but limited data
BARD1	Increased	Unknown/ insufficient data	None known
BRIP1	Possibly increased/ insufficient and conflicting data	Increased (up to ~10–13%)	None known
CHEK2	18–40%	No known increase	Possibly colon, melanoma, male breast, and others (prostate, kidney, thyroid)
MRE11A	Possibly increased/ insufficient data	Unknown/ insufficient data	None known
NBN	Increased (may be as high as ~30%)	No known increase	None known

(continued)

Table 2.3 (continued)

Gene(s)	Breast cancer lifetime risk	Ovarian cancer lifetime risk	Other associated cancers/features
NF1	Increased/not well defined (may be as high as ~26%)	No known increase	Neurofibromas
			Multiple café-au-lait spots
			Axillary and inguinal freckling
			Optic nerve and CNS gliomas
			Malignant peripheral nerve sheath tumors
			GIST
PALB2	30–58%	Unknown/insufficient data	Possibly pancreas and male breast
RAD50	Possibly increased/ insufficient data	Unknown/insufficient data	None known
RAD51C	Unknown/insufficient data	Increased/not well defined (may be ~6–7%)	None known
RAD51D	Unknown/insufficient data	Increased/not well defined (may be ~7–14%)	None known

References

1. Schneider K. Counseling about cancer: strategies for genetic counseling. 3rd ed. Hoboken: Wiley-Blackwell; 2012.
2. The American College of Obstetricians and Gynecologists (ACOG) Committee on Genetics. Hereditary cancer syndromes and risk assessment. Obstetr Gynecol. 2015;125:1538–43.
3. U.S. Preventative Services Task Force. Risk assessment, genetic counseling, and genetic testing for BRCA-related cancer in women: recommendation statement. Am Fam Physician. 2015;91:118A–E.
4. Lu KH, Wood ME, Daniels M, et al. American Society of Clinical Oncology Expert Statement: collection and use of a cancer family history for oncology providers. J Clin Oncol. 2014;32:833–40.
5. Cancer.Net Family History Questionnaire, last accessed 10/2016: http://www.cancer.net/sites/cancer.net/files/cancer_family_history_questionnaire.pdf.
6. US Surgeon General My Family Health Portrait Tool, last accessed 10/2016: http://www.hhs.gov/programs/prevention-and-wellness/family-health-history/family-health-portrait-tool/index.html#.
7. National Comprehensive Cancer Network. Genetic/Familial High-Risk Assessment: Breast and Ovarian. Version 1.2017. https://www.nccn.org/professionals/physician_gls/pdf/genetics_screening.pdf. Accessed October 7, 2016.
8. Hampel H, Bennett R, Buchanan A, et al. A practice guideline from the American College of Medical Genetics and Genomics and the National Society of Genetic Counselors: referral indications for cancer predisposition assessment. Genet Med. 2014;17:70–87.
9. Greendale K, Pyeritz RE. Empowering primary care health professionals in medical genetics: How soon? How fast? How far? Am J Med Genet. 2001;106:223–32.
10. Guttmacher AE, Jenkins J, Uhlmann WR. Genomic medicine: Who will practice it? A call to open arms. Am J Med Genet. 2001;106:216–22.

11. Suther S, Goodson P. Barriers to the provision of genetic services by primary care physicians: a systematic review of the literature. Genet Med. 2003;5(2):70–6.
12. Ready KJ, Daniels MS, Sun CC, et al. Obstetrics/Gynecology residents' knowledge of hereditary breast and ovarian cancer and Lynch syndrome. J Cancer Educ. 2010;25:401–4.
13. Wideroff L, Vadaparampil ST, Greene MH, et al. Hereditary breast/ovarian and colorectal cancer genetics knowledge in a national sample of US physicians. J Med Genet. 2005;42:749–55.
14. Lowstuter KJ, Sand S, Blazer KR, et al. Influence of genetic discrimination perceptions and knowledge on cancer genetics referral practice among clinicians. Genet Med. 2008;10 (9):691–8.
15. O'Neill SC, Peshkin BN, Luta G, et al. Primary care providers' willingness to recommend BRCA1/2 testing to adolescents. Fam Cancer. 2010;9:43–50.
16. Laedtke AL, O'Neill SM, Rubinstein WS, Vogel KJ. Family physicians' awareness and knowledge of the Genetic Information Non-Discrimination Act (GINA). J Genet Counsel. 2012;21(2):345–52.
17. Platten U, Rantala J, Lindblom A, et al. The use of telephone in genetic counseling versus in-person counseling: a randomized study on counselees' outcome. Fam Cancer. 2012;11 (3):371.
18. Buchanan AH, Datta SK, Skinner CS, et al. Randomized trial of telegenetics vs. in-person cancer genetic counseling: cost, patient satisfaction and attendance. J Genet Counsel. 2015;24:961–970.
19. National Society of Genetic Counselors website, last accessed 10/2016: http://nsgc.org.
20. National Cancer Institute website, last accessed 10/2016: https://www.cancer.gov.
21. Informed DNA website, last accessed 10/2016: http://www.informeddna.com.
22. GeneTests website, last accessed 10/2016: https://www.genetests.org.
23. Christinat A, Pagani O. Practical aspects of genetic counseling in breast cancer: lights and shadows. Breast. 2013;22:375–82.
24. Riley BD, Culver JO, Skrzynia C, et al. Essential elements of genetic cancer risk assessment, counseling, and testing: updated recommendations of the National Society of Genetic Counselors. J Genet Counsel. 2012;21:151–61.
25. Schneider KA, DiGianni LM, Patenaude AF, et al. Accuracy of cancer family histories: comparison of two breast cancer syndromes. Genet Test. 2004;8(3):222–8.
26. Kurian AW, Kingham KE, Ford JM. Next-generation sequencing for hereditary breast and gynecologic cancer risk assessment. Curr Opin Obstet Gynecol. 2015;27(1):23–33.
27. Hall MJ, Obeid EI, Schwartz SC, et al. Genetic testing for hereditary cancer predisposition: BRCA1/2, Lynch syndrome, and beyond. Gynecol Oncol. 2016;140:565–74.
28. Raineville IR, Rana HQ. Next-generation sequencing for inherited breast cancer risk: Counseling through the complexity. Curr Oncol Rep. 2014;16:371–82.
29. Hall MJ, Forman AD, Pilarski R, et al. Gene panel testing for inherited cancer risk. J Natl Compr Canc Netw. 2014;12(9):1339–46.
30. Yurgelun MB, Allen B, Kaldate RR, et al. Identification of a variety of mutations in cancer predisposition genes in patients with suspected Lynch Syndrome. Gastroenterology. 2015;149:604–13.
31. Desmond A, Kurian AW, Gabree M, et al. Clinical actionability of multigene panel testing for hereditary breast and ovarian cancer risk assessment. JAMA Oncol. 2015;1(7):943–51.
32. Norquist BM. More genes, more problems? Benefits and risks of multiplex genetic testing. Gynecol Oncol. 2013;139:209–10.
33. Economopoulou P, Dimitriadis G, Psyrri A. Beyond BRCA: new hereditary breast cancer susceptibility genes. Cancer Treat Rev. 2015;41:1–8.
34. Schneider K, Zelley K, Nichols KE, et al. Li-Fraumeni syndrome. 1999 Jan 19 [Updated 2013 Apr 11]. In: Pagon RA, Adam MP, Ardinger HH, et al, editors. GeneReviews® [Internet]. Seattle (WA): University of Washington, Seattle; 1993–2016. Available from: https://www.ncbi.nlm.nih.gov/books/NBK1311/.

35. Apostolou P, Fostira F. Hereditary breast cancer: the era of new susceptibility genes. Biomed Res Int. 2013;2013:747318. doi:10.1155/2013/747318 Epub 2013 Mar 21.
36. Pilarski R. Cowden syndrome: a critical review of the clinical literature. J Genet Counsel. 2009;18:13–27.
37. Tan MH, Mester JL, Ngeow J, et al. Lifetime cancer risks in individuals with germline PTEN mutations. Clin Cancer Res. 2012;18(2):400–7.
38. Eng C. PTEN hamartoma tumor syndrome. 2001 Nov 29 [Updated 2016 Jun 2]. In: Pagon RA, Adam MP, Ardinger HH, et al, editors. GeneReviews® [Internet]. Seattle (WA): University of Washington, Seattle; 1993–2016. Available from: https://www.ncbi.nlm.nih.gov/books/NBK1488/.
39. McGarrity TJ, Amos CI, Baker MJ. Peutz-Jeghers syndrome. 2001 Feb 23 [Updated 2016 Jul 14]. In: Pagon RA, Adam MP, Ardinger HH, et al, editors. GeneReviews® [Internet]. Seattle (WA): University of Washington, Seattle; 1993–2016. Available from: https://www.ncbi.nlm.nih.gov/books/NBK1266/.
40. Hearle N, Schumacher V, Menko FH, et al. Frequency and spectrum of cancers in the Peutz-Jeghers syndrome. Clin Cancer Res. 2006;12(10):3209–15.
41. Pharoah PD, Guilford P, Caldas C. International Gastric Cancer Linkage Consortium. Incidence of gastric cancer and breast cancer in CDH1 (E-cadherin) mutation carriers from hereditary diffuse gastric cancer families. Gastroenterology. 2001;121(6):1348–53.
42. Kaurah P, Huntsman DG. Hereditary diffuse gastric cancer. 2002 Nov 4 [Updated 2014 Jul 31]. In: Pagon RA, Adam MP, Ardinger HH, et al, editors. GeneReviews® [Internet]. Seattle (WA): University of Washington, Seattle; 1993–2016. Available from: https://www.ncbi.nlm.nih.gov/books/NBK1139/.
43. Tung N, Domchek SM, Stadler Z, et al. Counselling framework for moderate-penetrance cancer-susceptibility mutations. Nat Rev Clin Oncol. 2016;13:581–8.
44. Petrucelli N, Daly MB, Feldman GL. BRCA1 and BRCA2 hereditary breast and ovarian cancer. 1998 Sep 4 [Updated 2013 Sep 26]. In: Pagon RA, Adam MP, Ardinger HH, et al, editors. GeneReviews® [Internet]. Seattle (WA): University of Washington, Seattle; 1993–2016. Available from: https://www.ncbi.nlm.nih.gov/books/NBK1247/.
45. King MC, Marks JH, Mandell JB, New York Breast Cancer Study Group. Breast and ovarian cancer risks due to inherited mutations in BRCA1 and BRCA2. Science 2003;302 (5645):643–6.
46. Ford D, Easton DF, Bishop DT, Narod SA, Goldgar DE. Risks of cancer in BRCA1-mutation carriers. Breast Cancer Linkage Consortium. Lancet. 1994 19;343(8899):692–5.
47. Mavaddat N, Peock S, Frost D, et al. Cancer risks for BRCA1 and BRCA2 mutation carriers: results from prospective analysis of EMBRACE. J Natl Cancer Inst. 2013;105(11):812–22.
48. Chen S, Parmigiani G. Meta-analysis of BRCA1 and BRCA2 penetrance. J Clin Oncol. 2007;25(11):1329–33.
49. Rich TA, Woodson AH, Litton J, Arun B. Hereditary breast cancer syndromes and genetic testing. J Surg Oncol. 2015;111(1):66–80.
50. Easton DF, Pharoah PDP, Antoniou AC, et al. Gene-panel sequencing and the prediction of breast-cancer risk. N Engl J Med. 2015;372(23):2243–57.

Imaging Screening and Surveillance

3

Ellen Warner

Introduction

Risk-reducing mastectomy is highly effective for preventing breast cancer (BC) and can now be accompanied by a number of cosmetically excellent reconstruction procedures. Nevertheless, many *BRCA* mutation carriers decline this option even after a BC diagnosis. Other women choose to postpone preventive surgery until a specific milestone is achieved, such as finding a partner or completing childbearing and breastfeeding. Another group of women who require highly sensitive breast screening are the first-degree relatives of known *BRCA* mutation carriers who decline genetic testing. For all these groups, an effective breast screening regimen is essential. Although breast *screening* technically refers to women without any history of breast problems (unaffected women) and breast *surveillance* to women with a previous premalignant breast lesion or malignancy, the term *surveillance* has also often been used for unaffected women known to be at higher-than-average risk. For simplicity, the term *screening* will be used in this chapter to refer to the early detection of cancer in both unaffected and previously affected *BRCA* mutation carriers.

While multiple studies have looked at breast screening for *BRCA* mutation carriers, there is a paucity of data on ovarian screening. The limited data we have on this subject will be summarized at the end of the chapter.

E. Warner (✉)
Medical Oncologist, Sunnybrook Health Sciences Centre,
Department of Medicine, Division of Medical Oncology,
University of Toronto, 2075 Bayview Avenue,
Toronto, ON M4N 3M5, Canada
e-mail: ellen.warner@sunnybrook.ca

A.B. Chagpar (ed.), *Managing BRCA Mutation Carriers*,
DOI 10.1007/978-3-319-59198-8_3

Table 3.1 Recent breast screening guidelines for *BRCA* mutation carriers [3, 14, 27, 30]

Group	Year	Country	Annual mammography	Annual MRI	CBE	BSE
NCCN [3]	2016	US	Age 30–75	Age 25–75	Q6-12 months age 25+	–
NICE [27]	2013	UK	Age 40–69[a] Consider for age 30–39	Age 30–49 Age 50–69 if dense breasts	–	–
ESMO [30]	2011	Europe	Age 25–30+	Age 25–30+	Q6 months age 25–30+	Age 25–30+
OBSP high risk [14]	2011	Canada	Age 30–74	Age 30–69	–	–

CBE clinical breast examination; *BSE* monthly breast self-examination
OBSP Ontario breast screening program
[a]Age 70+ revert to population screening guidelines

Breast Screening

Starting Age

The great majority of guidelines currently recommend that screening mammography for the *general population* begin sometime between ages 45 and 50. These women have an annual incidence of breast cancer between 0.1 and 0.2% [1]. It would, therefore, be reasonable to begin screening *BRCA* mutation carriers when their annual risk is similar. That age is somewhere between 25 and 30 [2], which is when most guidelines recommend that screening begin (see Table 3.1). Some groups have advocated 'breast awareness' as early as age 18–20 [3]. While this may be reassuring to anxious physicians and parents, there is no available evidence that it is effective.

What Screening Modalities Should Be Used?

Screening a large, relatively low-risk population for disease can only be cost-effective if the screening regimen has high specificity, since the overwhelming majority of people screened will be disease-free, and even small drops in specificity will translate into a very large number of additional false positives. However, when screening a small, high-risk population such as *BRCA* mutation carriers, the most important performance characteristic is sensitivity. One would certainly not expect mammography alone to have adequate sensitivity to screen mutation carriers under age 50, given its fairly low sensitivity in women aged 40–49 in the general population (and likely even poorer performance in women under age 40) [4]. Indeed, in case series of screening mammography with or without clinical breast examination for *BRCA* mutation carriers, the interval cancer rate ranged from 35 to 50%, few cases of DCIS were detected, 40–78% of the invasive cancers were greater than 1 cm in size, and 20–56% had lymph node involvement [5–8].

Unlike mammography, which relies on anatomic density, distortions, and secondary byproducts of malignancy (calcifications), contrast-enhanced magnetic resonance imaging (MRI) provides insight into tissue functionality by creating a 'blood flow map' detecting tumor neovascularity and peritumoral inflammation, rendering its sensitivity relatively independent of breast density and higher than that of any other breast imaging modality [9]. Approximately 50% of invasive cancers demonstrate a classic MRI pattern of early contrast enhancement and early washout, because the contrast agent accumulates faster and washes out faster from the more vascular tumor than from the normal or benign tissues. Certain morphologic features are also typical of malignancy including spiculated or irregular lesion margins and increased enhancement of the lesion's periphery. Absence of enhancement correlates well with absence of invasive breast cancer with a negative predictive value of over 95% [10].

Accordingly, multiple non-randomized observational studies were started in the mid- to late 1990s in which *BRCA* mutation carriers were screened annually with both MRI and mammography concurrently and the performance of each imaging modality evaluated. A meta-analysis of the results of these studies [11] found that the sensitivity of mammography ranged from 25 to 59% with a pooled sensitivity of 39% (95% CI 37–41), while MRI had a sensitivity of 68–100% with a pooled sensitivity of 77% (95% CI 70–84). The two modalities were complementary as the sensitivity of the combination was 94% (95% CI 90–97). Most importantly, the majority of cancers detected in these studies were either non-invasive or very early invasive cancers with a node-positive rate of 12–26%. Based on these results, annual MRI has now been incorporated into all recent screening guidelines for mutation carriers (Table 3.1). A dedicated breast coil, capacity for MRI-directed biopsies, and radiologists experienced in reading breast MRI must all be available for a center to be able to offer reliable screening MRI. It has been suggested that high-risk screening centers should offer at least 150 screening breast MRI examinations per year and perform at least 10 MRI-guided breast biopsies [12].

The biggest drawback of MRI, besides its high cost, is its low specificity. The overall reported false-positive recall rate for MRI in the above meta-analysis averaged over all rounds of screening was 13.7% (95% CI 8.3–19.1%), compared to 5.3% (95% CI 3.5–7.0%) for mammography. However, the false-positive biopsy rate of MRI was only 3.9% (95% CI 2.6–5.2) compared to 1.5% (95% CI 0.8–2.2) for mammography [11]. Recall and biopsy rates were generally substantially higher in the first year of screening in the absence of a previous MRI study for comparison, dropping by approximately one-third on subsequent rounds of screening [13]. These rates also tended to be higher in North American centers than in Europe [11].

Screening MRI does have additional drawbacks. The need to perform the test during the second week of the menstrual cycle to optimize sensitivity and specificity makes it logistically difficult to schedule the test too far in advance for the great number of women who have irregular cycles. This may be particularly annoying for women who have to travel a considerable distance to get to a screening center. The need for an intravenous line to inject the gadolinium-based contrast agent is another disadvantage. Approximately 5–10% of women get so claustrophobic lying prone in the magnet that they require mild sedation. Usually 1 to 2 mg of lorazepam before the procedure suffices. There are also contraindications to MRI such as indwelling metal devices and renal failure.

One question that has arisen from the formal observational studies of MRI screening, that were generally done at large academic centers, is whether the sensitivity and specificity of MRI would be comparable when performed in the community setting. The Ontario High-Risk Breast Screening Program is a population-based screening program for high-risk women (known mutation carriers or calculated lifetime breast cancer risk of 25% or higher) that was established in 2011 and is currently available at 28 centers across Canada [14]. The program offers annual MRI and mammography from ages 30 to 69. In the first round of screening, 35 cancers were detected in 2150 women with 71% of the cancers diagnosed in known mutation carriers. No cancer was detected by mammography alone, but 66% were detected by MRI alone. Specificity of MRI was 82%, which compares favorably to reports in the North American observational studies [13, 15].

Several screening studies of annual MRI plus mammography for very high-risk women also included screening ultrasound [13, 15–19] and/or clinical breast examination [13, 15, 18–20]. Neither modality was found to add significantly to the cancer detection rate of MRI plus mammography, but each additional modality did increase the number of false positives [21]. Nonetheless, because screening ultrasound can detect a significant proportion of cancers missed by mammography, it should be performed annually in place of MRI for women who are unable to access or tolerate MRI (even with sedation) or for whom MRI is contraindicated [22]. In one study of women at moderately increased risk, 18.5% (95% CI 16.4–20.8%) were unable to tolerate MRI because of claustrophobia, a metallic implant (e.g., pacemaker, aneurysm clip), impaired renal function, gadolinium intolerance, or other reasons [23]. On the other hand, in the Ontario High-Risk Breast Screening Program, 94.2% of 2359 eligible women had at least one screening MRI [14]. It is likely that higher-risk women are motivated to put up with the discomfort and/or fear of MRI. Among the 50 women who underwent ultrasound instead of MRI, the false-positive ultrasound recall rate was 6% and one invasive mammographically occult cancer was diagnosed by ultrasound alone.

Although ultrasound has a very limited role as a screening modality if MRI is being done, it is the most important modality for the investigation of mass lesions detected by screening MRI. This topic has been extensively reviewed by Leung [24]. Because MRI is performed with the patient prone and the breast hanging, while ultrasound is performed with the patient in the supine, oblique, or lateral decubitus positions, localizing the MRI lesion may be challenging. Lesion size may also vary by up to 20%. Approximately two-thirds of MRI-detected lesions will be visible on ultrasound, which may be able to give a definitive diagnosis of a benign lesion such as a cyst or fibroadenoma without the need for biopsy. While larger lesions are more likely to have an ultrasound correlate, the single most important predictor of identification of an ultrasound correlate is lesion type, with an ultrasound correlate significantly more likely for masses, including small foci, than for non-mass enhancement. Biopsies, when necessary, are also most easily done by ultrasound if the lesion is visible sonographically. Most investigators have reported a higher rate of malignancy for lesions with a non-benign sonographic correlate than for those that are sonographically occult, but suspicious lesions visualized only on MRI must be biopsied under MRI guidance. Centers that lack the capability of performing MRI-guided biopsies should not offer breast screening MRI as the major benefit of MRI is its ability to detect non-calcified DCIS and tiny invasive lesions not visible with other modalities.

There is no role for screening with any imaging modality in asymptomatic women who have undergone risk-reducing mastectomy.

At What Age Can MRI Be Stopped?

One might think that MRI would be necessary for women only up until age 50, the age from which mammography performs quite well in the general population. However, in formal observational studies, the incremental benefit of MRI over mammography in terms of disease detection was at least as great for mutation carriers over age 50 as it was for the younger women [25]. The finding that the benefit of MRI is not confined to younger women with generally greater breast density is supported by our finding in the Toronto study that, although the performance of screening mammography was better in *BRCA* mutation carriers with fatty breasts than in those with dense breasts, mammography still missed 50% of the cancers in women with breasts not considered to be dense [26]. This makes it difficult to understand the rationale of the NICE guidelines [27] for not continuing MRI after age 50 for women who do not have dense breasts (Table 3.1).

As none of the screening studies that included women over age 50 had a significant number of participants above age 60, it is impossible to know precisely when MRI can be safely discontinued beyond that age. Since the sensitivity of mammography improves with age [28], while the growth rate of *BRCA*-related cancers slows with age [29], and breast screening has not demonstrated a mortality benefit in women over age 69 in the general population, it might be reasonable to discontinue MRI after age 70. This is the current practice in the Ontario High-Risk Breast Screening Program [14], while the NCCN continues MRI until age 75 [3] and ESMO does not specify a cutoff age [30].

Screening Interval

In the observational screening studies, all imaging was performed annually with each modality done either the same day or within a very short time frame. Since the combination of concurrent annual MRI and mammography fails to reach a sensitivity of 100%, it has been suggested that the interval cancer rate might be reduced if screening was performed in a staggered manner, i.e., with mammography and MRI alternating every six months. Although this is the practice in many centers [10] and may be reassuring to patients and physicians, there is no evidence to date that this approach is more effective than concurrent imaging.

Because in many studies the interval cancer rate was highest in *BRCA1* mutation carriers under age 50 [13, 31–33], the suggestion has been made that MRI should be done every 6 months in this population [32]. This approach has not been officially recommended by any group, and its superiority has yet to be documented.

A corollary of the above is that, with the known age-related slowing in the growth rate of *BRCA*-related breast cancers with age [29], it might be safe to perform MRI less frequently in women over age 60, particularly for *BRCA2* mutation carriers whose cancers tend to remain in situ for a longer period of time. This approach has yet to be tested, but would not be unreasonable in a setting with limited resources. What does seem clear, however, is that in the absence of MRI, screening mammography must be continued on an annual basis. In a case–control study, *BRCA* mutation carriers aged 60 and older who underwent biennial screening mammography in the Netherlands, as per their national guidelines, had twice the interval cancer rate and were 2.5 times as likely to have unfavorable breast cancer histology as mutation carriers who received annual mammography. Too few women had annual MRI for the authors to analyze that group separately [34].

Role of Mammography

Given the very high sensitivity of MRI and the limited sensitivity of mammography, particularly for young *BRCA1* mutation carriers, the role of screening mammography has been questioned. This is compounded by worry about the cumulative effect of annual breast irradiation started at a very young age, particularly for *BRCA* mutation carriers who may be even more susceptible to the carcinogenic effects of radiation because of reduced ability to repair DNA. In the GENE-RAD-RISK study which evaluated 1993 female *BRCA* mutation carriers, exposure to diagnostic radiation before the age of 30 was associated with an increased risk of breast cancer with a dose–response relationship (HR 1.9, 95% CI 1.2–3.0). Exposure to mammography before the age of 30 also increased the risk of breast cancer independent of family history [35].

The counter-arguments from radiologists for continuing to screen young women with mammography have been that:

1. Mammography is still the only breast screening modality that has been shown to reduce breast cancer mortality in randomized controlled trials. There has never been a randomized trial of MRI screening with mortality as an endpoint, and such a trial would no longer be considered ethical.
2. Mammography done concurrently with MRI helps with interpretation of the MR images.
3. Mammography may detect lesions such as low-grade DCIS that are not seen on MRI.

In a retrospective study of all *BRCA1* mutation carriers who developed breast cancer while on surveillance with annual MRI and mammography at one of three centers in the Netherlands, subsequent to the introduction of digital mammography at those centers, 82 invasive cancers and 12 cases of DCIS were found. MRI detected 88 of the 94 cancers (sensitivity 95.7%), but mammography detected only 48 cancers (sensitivity 51%). Mammography alone detected 2 of the cases of DCIS, both of which were in patients 50 years of age or older. Based on these results, the investigators concluded that annual MRI should be the only screening modality for *BRCA1* mutation carriers between the ages of 25 and 39 [36]. Similarly, in the Italian HIBCRIT-1 screening study, mammography did not significantly improve the receiver operating characteristics (ROC) analysis when added to MRI in either *BRCA1* or *BRCA2* mutation carriers [16]. Accordingly, the authors recommended that MRI should be the only annual imaging screening test for high-risk women outside a clinical trial [37]. However, this recommendation has not been accepted in North America to date. A randomized trial is currently underway in Italy comparing MRI alone versus MRI plus ultrasound up to age 35 and MRI plus ultrasound and mammography after age 35 [37].

Long-Term Outcome of Screen-Detected Cancers

As MRI screening trials for *BRCA* mutation carriers only began in the late 1990s, to date none of these trials has reported long-term survival data. In interpreting the data, it is important to distinguish 'prevalent cancers,' which are those diagnosed on the first round of screening and are more likely to be more advanced and at higher risk for recurrence, from 'incident cancers,' which are cancers diagnosed on subsequent screens.

Reported survival data are summarized in Table 3.2. In the MRISC study [38] distant recurrences occurred in 10% of the *BRCA* mutation carriers at a median follow-up of 9 years, with no difference between *BRCA1* and *BRCA2* mutation carriers. No distinction was made by the authors between prevalent and incident cancers. In the Toronto study [39], at a median follow-up of over 8 years, only one of 41 mutation carriers had developed metastatic disease from a prevalent cancer. In the combined UK MARIBS and NICE studies [40] in which 45 breast cancers were detected, only 2 deaths (both in *BRCA1* mutation carriers with tumors 3 cm or greater in size) were observed. The percentage of women who had developed metastatic disease, and whether the deaths were related to prevalent or incident cancers, was not stated. In contrast to these encouraging results, the Norwegian group reported 10 deaths at a median follow-up of 4.2 years in 68 *BRCA1* mutation carriers for an estimated 5-year breast cancer-specific survival of only 75% (95% CI 56–86%) and 10-year survival of 69% (95% CI 48–83%). The 10-year survival was 67% for women with prevalent cancers versus 76% for women with incident cancers and 62% for women diagnosed between ages 35 and 49 versus 81% for women diagnosed at age 50 or older. Of the women who died, two had prevalent cancers, six had incident cancers, and two had interval cancers [41]. There are several possible explanations for these disappointing results. The sensitivity of MRI may have been lower than in other studies as the interval cancer rate was higher at 12%. Also, facilities were not in place to perform MRI-guided biopsies. Although the authors claimed that all suspicious lesions were biopsied, scans with areas of non-mass enhancement may have been ignored due to the low likelihood of being able to image such lesions with other modalities. This may explain the low rate of DCIS detection (7% of all cancers).

Given the frequency of metastatic breast cancer recurrences 15 or more years after diagnosis, particularly in the case of estrogen receptor-positive tumors, as well as the relatively long natural history of metastatic breast cancer, these results likely underestimate the true recurrence and mortality rates, particularly for the *BRCA2* mutation carriers. On the other hand, it is over 15 years since most of these studies began accrual. In that time, there has been significant improvement in the technical quality of MRI examinations, in the experience of radiologists with interpreting MRI, and in breast cancer treatment. Consequently, results of these older studies could overestimate the risk of cancer recurrence for women being screened today. For example, in the Toronto study [39], the sensitivity of MRI increased from 74 to 94% ($p < 0.0001$) between the years 1997–2002 and 2003–2009 with a concomitant increase in the proportion of women diagnosed with in situ disease.

Knowledge of the long-term outcome of breast screening is essential for two distinct groups: health-care policy makers and *BRCA* mutation carriers. For the former group, the critical question is whether the high cost of screening MRI translates into a significant survival benefit over screening with mammography alone or simply provides lead time. In the absence of any randomized controlled trials of screening with or without MRI, the most accurate answer to this question can be obtained by comparing matched cohorts of *BRCA* mutation carriers screened with mammography, with or without MRI over the same time period. Our Toronto group compared the breast cancer stage at diagnosis of *BRCA* mutation carriers screened with MRI plus mammography in our screening trial, to the stage at diagnosis of age- and gene-matched mutation carriers, enrolled in Dr. Steven Narod's database, who were screened with mammography alone [42]. Patients screened with MRI were significantly less likely to have large tumors and/or node-positive disease at diagnosis (1.9% vs. 6.6%, $p = 0.02$). We assume that this will ultimately translate into a significant distant recurrence and survival difference,

Table 3.2 Long-term outcomes of MRI-based breast screening [38–41]

Study	No. of cancers	Mean/median age at diagnosis	*BRCA1*/ *BRCA2*	Median f/u (years)	% distant recurrence	% died breast cancer
MRISC [38]	51[a]	44 (26–67)	33/18	9 (0–14)	10	ns
Toronto [39]	41[a]	46 (32–68)	21/20	8 (2–13)	2	2
MARIBS + NICE [40]	45[a]	ns	24/21	12 (0.3–19)	ns	4
Norway [41]	68[a]	<50	68/0	4	ns	15

[a]Interval cancers are included

but follow-up of these 2 cohorts is still too short with too few events to detect meaningful differences. In the UK studies [40], *BRCA* mutation carriers who were screened by MRI had an overall 10-year survival of 95.3% compared to 87.7% for a matched cohort screened with mammography alone (HR 0.21, $p = 0.03$).

Probably the most important reason patients choose risk-reducing mastectomy over breast screening is the concern that screening will fail to detect the cancer at a curable stage. Thus, from the patients' point of view, it is essential to know how the long-term survival of patients getting breast screening is compared to that of patients undergoing risk-reducing mastectomy. The latter group of patients is still at some risk for breast cancer and at risk for other cancers. As the ideal randomized trial will, for obvious reasons, never be done, indirect comparisons are necessary. A Monte Carlo computer simulation model estimating survival among *BRCA1* and *BRCA2* mutation carriers undergoing different risk-reducing strategies found that the combination of risk-reducing mastectomy and risk-reducing salpingo-oophorectomy at age 30 provided the greatest gains in life expectancy, but that substituting intensive breast screening for mastectomy would only reduce life expectancy by a maximum of 1.5 years for *BRCA1* mutation carriers and 0.7 years for *BRCA2* mutation carriers [43].

Cost-Effectiveness

High-risk screening is expensive. The major cost is the screening MRI examinations themselves, but MRI recalls and MRI-guided biopsies of mammographically and sonographically occult lesions add significantly to the cost. Some groups have attempted to estimate the cost-effectiveness of various screening strategies. Such estimates are clearly limited by the lack of data on the long-term survival of women with screen-detected cancers and the lack of a concurrent control group screened with digital mammography alone.

A cost-effectiveness analysis of the UK MARIBS screening study estimated an additional cost of £11,731 and £15,302 per cancer detected for *BRCA1* and *BRCA2* mutation carriers, respectively, with the addition of MRI to film-screen mammography [44]. Using a decision analytic model based on US costs, Taneja et al. [45] estimated that the cost per quality-adjusted life year gained by adding MRI to standard film-screen mammography for a single round of screening was a modest $25,277, which is well within the $50,000–$100,000 considered to be acceptable. Plevritis et al. [46] only a few years earlier, using a different model but with similar costs, estimated that the cost per QALY gained with annual MRI screening from the ages of 25 to 69 was $88,651 for *BRCA1* mutation carriers, but $188,034 for *BRCA2* mutation carriers.

Chubiz et al. [47] in another simulation model, compared the costs and benefits of digital mammography (DM) and MRI alternating at 6-month intervals beginning at age 25 (Alt 25) versus annual MRI beginning at age 25 alternating with DM added at age 30 (MRI25/Alt 30) versus DM and MRI alternating at 6-month intervals beginning at age 30 (Alt 30) and found the latter strategy had the lowest cost per QALY for *BRCA1* mutation carriers ($74,200), but a higher cost per QALY for *BRCA2* mutation carriers ($215,700).

Pyschosocial Effects of Screening

There is no doubt that patients undergo some stress around the time of any screening examination, lasting at least until they receive a report that no evidence of malignancy was found. In the case of women with *BRCA* mutations, who are aware of their very high risk of being diagnosed with cancer, one might expect the level of stress to be even greater. This is compounded by the claustrophobic nature of the MRI scanning procedure and by the very high incidence of false-positive tests over many years of screening.

As part of the UK MARIBS study, the 600+ participants were assessed psychologically with standardized questionnaires at baseline (4 weeks before screening), immediately before, immediately after, and 6 weeks after the scans. High levels of satisfaction were reported for both MRI (96.3%) and mammography (97.7%). Low levels of self-reported distress were reported for both procedures, though MRI was distressing to more women than mammography (7.8% vs. 3.5%, $p = 0.005$). Higher anticipatory anxiety was reported before MRI than before mammography ($p = 0.0003$), and MRI-related distress was more likely to persist at 6 weeks after the scans in the form of intrusive MRI-related thoughts ($p = 0.006$) and total MRI-related distress ($p = 0.014$). More women stated that they intended to return in one year for mammography (96.3%) than for MRI (88%). These effects were more marked in the first year of screening, but were also statistically significant in the subsequent years [48].

In the Dutch MRISC study, 334 of 519 high-risk women undergoing screening were assessed with questionnaires at baseline, on the day of screening, and 4 weeks after screening. Scores of the study population showed significantly better generic health-related quality of life (QOL) than age-/sex-matched reference scores from the general population. Thirty percent of women described mammography as 'quite' or 'very' painful compared to 0.9% of MRI, but the proportion of participants experiencing screening-related anxiety ('quite' or 'very') was higher for MRI (10.2%) than for mammography (5.2%). Neither QOL nor distress scores changed significantly over time. Specifically, women who were recalled for additional diagnostic evaluations did not have a significant change in their scores [49].

Of the 236 women in the Toronto study, 55 completed questionnaires related to global and breast cancer-specific anxiety and quality of life 1–2 weeks before, 4–6 weeks after, and 6 months after screening. The 18 (32%) women recalled at some point in time for further imaging had a significant increase in global anxiety 4–6 weeks after the initial screening, but this returned to baseline by 6 months. No change over time was observed for any of the other psychological measures [50].

Although the probability of women who are on long-term MRI screening being recalled for a false-positive result is high, false positives have not been shown to increase the probability that a woman will undergo risk-reducing mastectomy [51, 52].

Rapid MRI

The high direct and indirect costs of MRI limit clinical access to screening MRI. One of the factors accounting for this high cost is the long time it takes to acquire MR images (20–40 min) and to read the hundreds of images that are generated. Recently, several groups have reported results of abbreviated MRI protocols limited to the early post-contrast period, followed by standard image reconstruction tools to allow a rapid overview of the imaging volume. In a recent study of 443 women at mildly to moderately increased breast cancer risk, the abbreviated protocol (AP) consisted of a stack of 27–33 axial T1-weighted gradient echo images acquired once before and repeated immediately after contrast injection. These two image stacks were subtracted to yield 27–33 individual first post-contrast subtracted (FAST) images. These fast images were then fused into a single summation image, the maximum-intensity projection (MIP). Expert radiologists read the MIP first to search for significant enhancement and then reviewed the AP (consisting of the MIP and FAST images) and only afterward read the full diagnostic protocol (FDP). Acquisition time for the AP was 3 min, compared to 17 min for the FDP. Average reading time was 2.8 s for the MIP and only 28 s for the AP. MIP readings were positive for 10 of the 11 cancers, and all 11 were detected on the AP. Specificity and positive predictive value of the AP and FDP protocols were equivalent [53]. Before this rapid protocol can be adopted for *BRCA* mutation carriers, however, it needs to be validated in that population as well as in a multicenter trial [54].

Ovarian Screening

Risk-reducing salpingo-oophorectomy is currently recommended for *BRCA* mutation carriers between ages 35 and 40. This enables most women to complete their families prior to undergoing the surgery, and for those who have not, egg or embryo freezing is an increasingly effective option. However, for those women who wish to delay the surgery or decline it altogether, an effective way to screen for ovarian, fallopian tube, or primary peritoneal cancer (which for simplicity will simply be referred to as 'ovarian cancer') would be extremely welcome. Unfortunately, there is much less data on screening for ovarian cancer than there is for breast cancer, and the available data are far from encouraging.

In a prospective study of 981 *BRCA* mutation carriers at five European centers screened with annual serum CA125 and transvaginal ultrasound, 49 epithelial ovarian cancers were diagnosed, of which 35 (71%) were stage 3 or 4. There was no difference in stage distribution between cancers diagnosed at the prevalent round of screening and those diagnosed at an incident round. Five-year survival was 59% (95% CI 51–66%) and 10-year survival 36% (95% CI 27–45%), similar to what one would expect in the absence of screening [55]. Similarly, in a study of 538 *BRCA* mutation carriers who underwent annual screening with CA125 and transvaginal ultrasound at 37 regional centers in the UK, only 2 of the 13 incident cancers were stage 1 [56]. Between 2007 and 2009, the protocol of the latter study was changed so that screening was conducted every 4 months. These results are yet to be reported.

The NCCN guidelines state 'For those patients who have not elected risk-reducing salpingo-oophorectomy, while there may be circumstances where clinicians find screening helpful, data do not support a positive recommendation, but screening with transvaginal ultrasound may be considered at the clinician's discretion starting at age 30–35. Serum CA-125 is an additional ovarian screening test with caveats similar to transvaginal ultrasound [3].' Interestingly, in a recent international survey of 22 centers, all but seven centers (from England, Germany, the Netherlands, Canada, Boston, and the two Australian centers) offered semiannual or annual gynecological examination, transvaginal ultrasound, and CA125 measurements for screening *BRCA* mutation carriers who had not undergone risk-reducing surgery, generally starting at age 30 or 35. After risk-reducing salpingo-oophorectomy, only 4 centers offered specific gynecologic surveillance to detect primary peritoneal cancer [57].

A recently reported randomized study of annual ovarian screening in the general population, which excluded women with increased hereditary risk, showed some evidence of a mortality reduction in the group receiving multimodality screening who had incident cancers [58]. Although the biology of *BRCA*-related ovarian cancer likely renders annual screening with any protocol inadequate, this trial may at least represent 'proof of principle' that an effective ovarian cancer screening regimen may be achievable for all risk groups in the not too distant future. Until then, *BRCA* mutation carriers need to understand that no screening regimen can be assumed to have any efficacy.

Conclusions

Annual breast screening is a very reasonable option for female *BRCA* mutation carriers who wish to avoid or delay risk-reducing mastectomy. While experts all agree on the need for annual MRI from ages 30 to 50, the most cost-effective screening regimen for specific subgroups based on age, breast density, and mutation type is yet to be determined. No ovarian cancer screening regimen has demonstrated efficacy in this population to date.

References

1. SEER Cancer Statistics 2009–2013. http://seer.cancer.gov/csr/1975_2013/results_merged/sect_04_breast.pdf.
2. Chen S, Parmigiani G. Meta-analysis of *BRCA1* and *BRCA2* penetrance. J Clin Oncol. 2007;11:1329–33.
3. *BRCA*-related Breast and/or Ovarian Cancer Syndrome. NCCN guidelines for detection, prevention and risk reduction; version 2. 2016. www.nccn.org/professionals.
4. Corsetti V, Houssami N, Ghirardi M, et al. Evidence of the effect of adjuvant ultrasound screening in women with mammography-negative dense breasts: interval breast cancers at 1 year follow-up. Eur J Cancer. 2011;47:1021–6.
5. Brekelmans CTM, Seynaeve C, Bartels CCM, et al. Surveillance in *BRCA1/2* gene mutation carriers and women with high familial risk. J Clin Oncol. 2001;19:924–30.
6. Scheuer L, Kauff N, Robson M, et al. Outcome of preventive surgery and screening for breast and ovarian cancer in *BRCA* mutation carriers. J Clin Oncol. 2002;20:1260–8.
7. Vasen HFA, Tesfay E, Bonnstra H, et al. Early detection of breast and ovarian cancer in families with *BRCA* mutations. Eur J Cancer. 2004;41:549–54.
8. Komenaka IK, Ditkoff B, Joseph K, et al. The development of interval breast malignancies in patients with *BRCA* Mutations. Cancer. 2004;100:2079–83.
9. Morris EA. Diagnostic breast MRI imaging. Current status and future directions. Magn Reson Imaging Clin N Am. 2010;18:57–74.
10. Heywang-Kobrunner SH, Hacker A, Sedlacek S. Magnetic resonance imaging: the evolution of breast imaging. The Breast. 2014;3:S77–82.
11. Warner E, Messersmith H, Causer P, Eisen A, Shumak R, Plewes D. Systematic review: using magnetic resonance imaging to screen women at high risk for breast cancer. Ann Int Med. 2008;148:671–9.
12. Sardanelli F, Boetes C, Borisch B, Decker T, Federico M, Gilbert FJ, et al. Magnetic resonance imaging of the breast: recommendations from the EUSOMA working group. Eur J Cancer. 2010;46:1296–315.
13. Warner E, Plewes DB, Hill KA. Surveillance of *BRCA1* and *BRCA2* mutation carriers with magnetic resonance imaging, ultrasound, mammography, and clinical breast examination. JAMA. 2004;292:1317–25.
14. Chiarelli AM, Prummesl MV, Muradali D, et al. Effectiveness of screening with annual magnetic resonance imaging and mammography: Results of the initial screen from the Ontario high risk breast screening program. J Clin Oncol. 2014;32:2224–30.
15. Lehman CD, Isaacs C, Schnall MD, et al. Cancer yield of mammography, MR, and US in high-risk women: prospective mutli-institution breast cancer screening study. Radiology. 2007;244:381–8.
16. Sardanelli F, Podo F, Santoro F, et al. Multicenter surveillance of women at high genetic breast cancer risk using mammography, ultrasonography, and contrast-enhanced magnetic resonance imaging (the high breast cancer risk Italian 1 study): final results. Invest Radiol. 2010;46:94–105.
17. Riedl CC, Luft N, Bernhart C, Weber M, Bernathova M, Muy-Khang MT, et al. Triple-modality screening trial for familial breast cancer underlines the importance of magnetic resonance imaging and questions the role of mammography and ultrasound regardless of patient mutation status, age, and breast density. J Clin Oncol. 2015;33:1128–35.
18. Kuhl CK, Schrading S, Leutner CC, et al. Mammography, breast ultrasound, and magnetic resonance imaging for surveillance of woman at high familial risk for breast cancer. J Clin Oncol. 2005;23:8469–76.
19. Trecate G, Vergnanghi D, Manoukian S, et al. MRI in the early detection of breast cancer in women with high genetic risk. Tumori. 2006;92:517–23.

20. Kriege M, Brekelmans CT, Boetes C, et al. Efficacy of MRI and mammography for breast cancer screening in women with a familial or genetic predisposition. N Engl J Med. 2004;351:427–37.
21. Berg WA. How well does supplemental screening magnetic resonance imaging work in high-risk women? J Clin Oncol. 2014;21:2193–6.
22. Lee CH, Dershaw DD, Kopans D, et al. Breast cancer screening with imaging: recommendations from the society of breast imaging and the ACR on the use of mammography, breast MRI, breast ultrasound, and other technologies for the detection of clinically occult breast cancer. J Am Coll Radiol. 2010;7:18–27.
23. Berg WA, Blume JD, Adams AM, et al. Reasons women at elevated risk of breast cancer refuse breast MR imaging screening; ACRIN 6666. Radiology. 2010;254:79–87.
24. Leung JWT. Utility of second-look ultrasound in the evaluation of MRI-detected breast lesions. Semin Roentgenol. 2011;260–274.
25. Phi X-A, Saadatmand S, De Bock GH, et al. Contribution of mammography to MRI screening in *BRCA* mutation carriers by *BRCA* status and age: individual patient data meta-analysis. Br J Cancer. 2016;114:631–7.
26. Bigenwald RZ, Warner E, Gunesakara A, et al. Is mammography adequate for screening women with inherited *BRCA* mutations and low breast density? Cancer Epidemiol Biomarkers Prev. 2008;17:706–11.
27. Clinical guideline: Familial Cancer: Classification and care of people at risk of familial breast cancer and management of breast cancer and related risks in people with a family history of breast cancer (last update 2013). www.nice.org.uk/guidance.
28. Nelson HD, O'Meara ES, Kerlikowske K, et al. Factors associated with rates of false-positive and false-negative results from digital mammography screening: an analysis of registry data. Ann Intern Med. 2016;164:226–35.
29. Tilanus-Linthorst MMA, Obdeijn I-M, Hop WCJ, et al. *BRCA1* mutation and young age predict faster cancer growth in the Dutch, United Kingdom and Canadian magnetic resonance imaging screening trials. Clin Cancer Res. 2007;13:7357–62.
30. Balmaña J, Díez O, Rubio IT, Cardoso F; ESMO Guidelines Working Group. *BRCA* in breast cancer: ESMO clinical practice guideline. Ann Oncol. 2011;22(suppl 6):vi31–4.
31. Rijnsburger AJ, Obdeijn I-M, Kaas R, et al. *BRCA1*-associated breast cancers present differently from *BRCA2*-associated and familial cases: long-term follow-up of the Dutch MRISC screening study. J Clin Oncol. 2010;28:5265–73.
32. Chereau E, Uzan C, Balleyguider C, et al. Characteristics, treatment, and outcome of breast cancers diagnosed in *BRCA1* and *BRCA2* gene mutation carriers in intensive screening programs including magnetic resonance imaging. Clin Breast Cancer. 2010;10:113–8.
33. Shah P, Rosen M, Sopfer J, et al. Prospective study of breast MRI in *BRCA1* and *BRCA2* mutation carriers: effect of mutation status on cancer incidence. Breast Cancer Res Treat. 2009;118:539–46.
34. Saadatmand S, Vos JR, Hooning MJ, et al. Relevance and efficacy of breast cancer screening in *BRCA1* and *BRCA2* mutation carriers above 60 years: a national cohort study. Int J Cancer. 2014;135:2940–9.
35. Pijpe A, Andrieu N, Easton DF, et al. Exposure to diagnostic radiation and risk of breast cancer among carriers of *BRCA1/2* mutations: retrospective cohort study (GENE-RAD-RISK). Br Med J. 2012;345:e5660.
36. Obdeijn I-M, Winter-Warnars GAO, Mann RM, Hooning MJ, Hunink MGM, Tilanus-Linthorst MMA. Breast Cancer Res Treat. 2014;144:577–82.
37. Sardanelli F, Santoro F, Podo F. Screening high risk women with MRI alone? Eur J Radiol. 2012;81(Suppl1):S137–8.
38. Saadatmand S, Obdeijn I-M, Rutgers EJ, et al. Survival benefit in women with *BRCA1* mutation or familial risk in the MRI screening study (MRISC). Int J Cancer. 2015;137:1729–38.
39. Passaperuma K, Warner E, Causer PA, et al. Long-term results of screening with magnetic resonance imaging in women with *BRCA* mutations. Br J Cancer. 2012;107:24–30.

40. Evans DG, Kesavan N, Yit L, et al. MRI breast screening in high-risk women: cancer detection and survival analysis. Breast Cancer Res Treat. 2014;145:663–72.
41. Moller P, Stormorken A, Jonstrud C, et al. Survival of patients with *BRCA1*-associated breast cancer diagnosed in an MRI-based surveillance program. Breast Cancer Res Treat. 2013;139:155–61.
42. Warner E, Hill K, Causer P, et al. Prospective study of breast cancer incidence in women with a *BRCA1* or *BRCA2* mutation under surveillance with and without magnetic resonance imaging. J Clin Oncol. 2011;29:1664–9.
43. Sigal BM, Munoz DF, Kurian AW, Plevritis SK. A simulation model to predict the impact of prophylactic surgery and screening on the life expectancy of *BRCA1* and *BRCA2* mutation carriers. Cancer Epidemiol Biomarkers Prev. 2012;21:1066–77.
44. Griebsch I, Brown J, Dixon A, et al. Cost-effectiveness of screening with contrast-enhanced magnetic resonance imaging versus Xray mammography of women at a high familial risk of breast cancer. Br J Cancer. 2006;95:801–10.
45. Taneja C, Edelsberg J, Weycker D, et al. Cost effectiveness of breast cancer screening with contrast-enhanced MRI in high-risk women. J Am Coll Radiol. 2009;6:171–9.
46. Plevritis SK, Kurian AW, Sigal BM, et al. Cost-effectiveness of screening *BRCA1/2* mutation carriers with breast magnetic resonance imaging. JAMA. 2006;295:2374–84.
47. Chubiz JEC, Lee JM, Gilmore ME, et al. Cost-effectiveness of alternating magnetic resonance imaging and digital mammography screening in *BRCA1* and *BRCA2* gene mutation carriers. Cancer. 2013;119:1266–76.
48. Hutton J, Walker LG, Gilber FJ, et al. Psychological impact and acceptability of magnetic resonance imaging and xray mammography: the MARIBS study. Br J Cancer. 2011;104:578–86.
49. Rijnsburger AJ, Essink-Bot ML, van Dooren S, et al. Impact of screening for breast cancer in high-risk women on health-related quality of life. Br J Cancer. 2004;91:69–76.
50. Spiegel TN, Esplen MJ, Hill KA, Wong J, Causer PA, Warner E. Psychological impact of recall on women with *BRCA* mutations undergoing MRI surveillance. Breast. 2011;20:424–30.
51. Hoogerbrugge N, Kamm YJL, Bult P, et al. The impact of a false-positive MRI on the choice for mastectomy in *BRCA* mutation carriers is limited. Ann Oncol. 2008;19:655–9.
52. Haroun I, Graham T, Poll A, et al. Reasons for risk-reducing mastectomy versus MRI-screening in a cohort of women at high hereditary risk of breast cancer. Breast. 2011;20:254–8.
53. Kuhl CK, Schrading S, Strobel K, Schild HH, Hilger R-D, Bieling HB. Abbreviated breast magnetic resonance imaging (MRI): first postcontrast subtracted images and maximum-intensity projection—a novel approach to breast screening with MRI. J Clin Oncol. 2014;32:2304–10.
54. Morris EA. Rethinking breast cancer screening: ultra FAST breast magnetic resonance imaging. J Clin Oncol. 2014;32:2281–3.
55. Evans DG, Gaarenstroom KN, Stirling D, et al. Screening for familial ovarian cancer: poor survival of *BRCA1/2* related cancers. J Med Genet. 2009;46:593–7.
56. Rosenthal AN, Fraser L, Manchanda R, et al. Results of annual screening in phase I of the United Kingdom familial ovarian cancer screening study highlight the need for strict adherence to screening schedule. J Clin Oncol. 2012;31:49–57.
57. Madorsky-Feldman D, Sklair-Levy M, Perri T, et al. An international survey of surveillance schemes for unaffected *BRCA1* and *BRCA2* mutation carriers. Breast Cancer Res Treat. 2016;157:319–27.
58. Jacobs IJ, Menon U, Ryan A, et al. Ovarian cancer screening and mortality in the UK Collaborative Trial of Ovarian Cancer Screening (UKCTOCS): a randomized controlled trial. Lancet. 2016;387:945–56.

Risk-Reducing Surgery for *BRCA1/2* Genetic Mutation Carriers

4

Cristina O'Donoghue, Sonia Orcutt, Tuya Pal and Christine Laronga

Introduction

Over the last few years, tremendous technological advances have led to plummeting costs and widespread availability of gene sequencing, including multigene panel testing. These tests include genes of variable cancer penetrance and wide cancer spectrum. Consequently, interpretation of results and formulation of an appropriate management plan has become a complex yet critical component of care among these patients, many of whom have not yet been diagnosed with cancer. Simultaneously, increasing social media awareness about genetic testing has led to growth of genetic risk assessment as part of standard care, resulting in higher numbers of patients identified with mutations in inherited cancer genes. Still today, the most common genetic mutations detected arise in the *BRCA1* and *BRCA2* genes which account for the largest proportion of inherited breast cancer, and yield the highest risk for development both breast and ovarian cancer. The focus of this chapter will

C. O'Donoghue · S. Orcutt · C. Laronga (✉)
Department of Breast Oncology, H. Lee Moffitt Cancer Center & Research Institute, 10290 North McKinley Drive, Tampa, FL 33612, USA
e-mail: christine.laronga@moffitt.org

C. O'Donoghue
e-mail: cristina.odonoghue@moffitt.org

S. Orcutt
e-mail: sonia.orcutt@moffitt.org

T. Pal
Department of Cancer Epidemiology, H. Lee Moffitt Cancer Center & Research Institute, 12902 Magnolia Drive, MRC-CANCONT, Tampa, FL 33612, USA
e-mail: tuya.pal@moffitt.org

A.B. Chagpar (ed.), *Managing BRCA Mutation Carriers*,
DOI 10.1007/978-3-319-59198-8_4

be limited to discussion of risk-reducing prophylactic surgery for breast cancer prevention in *BRCA* mutation carriers.

Discussion of Risk-Reducing Options

In the context of identifying a *BRCA* mutation, women and their treating healthcare provider make many decisions. Even before testing, women must discuss the pros and cons of testing with their genetic counselor or other healthcare provider to have a genetic risk assessment through collection of a comprehensive family history and make an informed decision to choose whether they would like to proceed with testing. Among women who are tested, those identified to have a *BRCA* mutation still have many more decisions to contemplate. Equally important (although not the focus of this chapter) are those who test negative yet have a striking family history thus remain at high risk for inherited cancer (often referred to as "uninformative negative" patients) who must consider many of the same decisions as those identified with *BRCA* mutations.

First, women must decide if they even want risk-reducing surgery. Carrying a genetic mutation is not a guarantee of primary breast cancer development in that woman's future. Rather, it is a prediction of risk during one's lifetime in the range of 60–70% [1–4]. Risk-reducing mastectomy had been used for decades before the *BRCA* gene mutations were identified but the impetus to determine the efficacy of prophylactic mastectomies has increased. Initial practice guidelines for *BRCA* carriers did not make recommendations for or against prophylactic surgeries because of insufficient evidence [5]. In the following decade, research data accumulated for risk-reducing mastectomy in unaffected women with moderate-to-high risk for breast cancer including those with *BRCA* mutations [6]. One such retrospective study found that risk-reducing mastectomy decreased the risk of developing breast cancer by over 90% for high-risk and known *BRCA* mutation carriers with a median follow-up of over 13 years [7]. One study found that, for every six patients treated with a risk-reducing mastectomy, one case of breast cancer could be averted [8]. Smaller prospective studies of unaffected *BRCA* mutation carriers showed risk-reducing mastectomy reduced the incidence of breast cancer by over 90% at a mean follow-up of 3–6 years [9, 10]. The PROSE study group found that risk-reducing mastectomy reduced the risk of breast cancer in *BRCA* carriers by 95% in those that had an oophorectomy and 90% in those that had intact ovaries [10]. A meta-analysis of four prospective studies with a total of 2635 patients solidified the evidence that *BRCA* mutation carriers who undergo risk-reducing mastectomy have a significant reduction in the incidence of breast cancer (HR 0.07; 95% CI 0.01–0.44; p = 0.004), and this risk reduction remains significant even for those without previous risk-reducing salpingo-oophorectomy (RRSO) [11]. Currently, the National Comprehensive Cancer Network (NCCN) guidelines recommend discussing the option of risk-reducing mastectomy including the degree of protection, reconstruction options, and risks of surgery with *BRCA* mutation-positive women [12]. The NCCN guidelines now also specifically mention discussing the risks and benefits of a nipple-sparing mastectomy which until recently was considered investigational.

In the context of a breast cancer diagnosis, women with a *BRCA* mutation must discuss surgical options for both the appropriate treatment of their breast cancer and the consideration of a contralateral prophylactic mastectomy (see Chap. 7) [13].

Psychosocial Impact

The impact of social factors in the decision to undergo risk-reducing mastectomy cannot be underestimated. A mastectomy is a body image/quality of life altering event and regardless of how perfect the reconstruction is (if reconstruction is performed), the reconstructed breast will not feel like the woman's natural breast to herself. Studies have shown that women who elect to have a risk-reducing mastectomy do not regret their decision from a fear-of-cancer development standpoint [14–17]. Yet, they also acknowledge the impact that surgical choice has had on their body image, intimacy, and quality of life.

Some studies have demonstrated that being a *BRCA* mutation carrier can be associated with increased cancer-related stress after surgery but lower general distress 6–18 months after prophylactic surgery [14–17]. One systematic review of the literature on patient-reported outcomes after bilateral prophylactic mastectomy found that the majority of women (70%) were satisfied with the outcomes, reported high psychosocial well-being and positive body image [18] (Fig. 4.1). This review of 22 studies found that sexual well-being and somatosensory function were

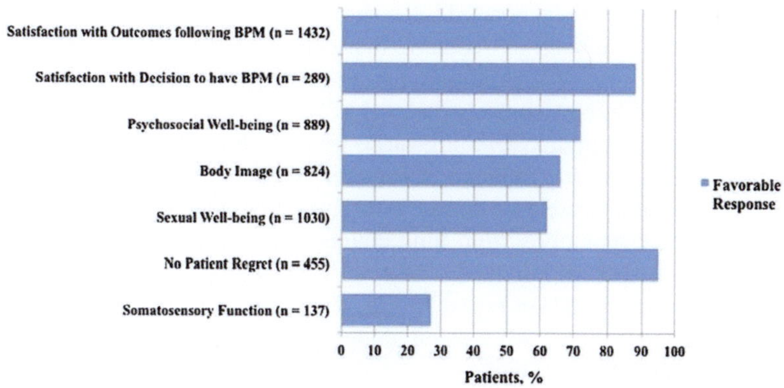

Fig. 4.1 Percent of patients who have had bilateral prophylactic mastectomy reporting favorable results for each quality of life domain [18]. BPM, bilateral prophylactic mastectomy. Reprinted from Quality of Life Research, Quality of life among patients after bilateral prophylactic mastectomy: a systematic review of patient-reported outcomes; Vol. 25/No. 6, © 2016, pp. 1409–1421, Razdan SN, Patel V, Jewell S, McCarthy CM, with permission of Springer

negatively affected, and high preoperative cancer distress and vulnerability on body image scale were significant negative predictors of quality of life after surgery. Following risk-reducing mastectomy, 95% of patients did not report any regret and between 86 and 100% would make the same choice and/or would recommend risk-reducing mastectomy to another woman at high-risk for developing breast cancer.

Decision Making

The decision to have a prophylactic mastectomy is influenced by many factors, and many women may have already decided their preferences for management prior to meeting her healthcare team. One study showed that 90% of *BRCA* mutation carriers indicated a preference for risk management at 1 week after DNA testing; prior to obtaining results, most women had stable preferences over time [19]. Decision aids prior to disclosure of *BRCA* status had some influence on preference to receiving risk-reducing mastectomy with 47% of women receiving a brochure and video prior to results opting for risk-reducing mastectomy compared to 35% of women who received the materials after results ($X^2 = 4.83$; $p = 0.028$). Women choosing a risk-reducing mastectomy were more likely to have young children (<13 years), rate their hypothetical breast cancer risk high, and have a high amount of anticipated regret if they did not undergo surgery [19]. *BRCA* mutation carrier women participating in shared decision-making interventions have better general health, are less depressed, held stronger treatment preferences, and agreed that they weighed pros and cons to surgery [20]. Decision aids have little impact on women who have already decided management but for undecided women, decision aids assist in reaching a management decision, and can lead to decreased decisional conflict and increased satisfaction [21]. Psychological consultation and talking with other women who have had risk-reducing mastectomyis of benefit to some women in their decision making [22]. Additionally, a majority of women feel postsurgical psychological consultation would be helpful [22].

Uptake in risk-reducing mastectomy varies by race/ethnicity and nationality. It is important to understand the interaction between patient preferences, provider differences, and access to care that drives such variations. One study surveyed 2677 women with *BRCA* mutations from nine countries and found that only 18.0% of eligible women had a prophylactic bilateral mastectomy; roughly half of the patients in this study relied on breast cancer screening alone. Women in the USA had the highest percentage of women undergoing a risk-reducing mastectomy (36%); interestingly, Norway had a lower rate of risk-reducing mastectomy (4.5%) but a higher uptake of bilateral prophylactic oophorectomy (73.5%) compared to the US oophorectomy percentage (71.1%) [23]. The PROSE consortium study estimated approximately 46% of women worldwide have a risk-reducing mastectomy by age 70 [24] (Fig. 4.2).

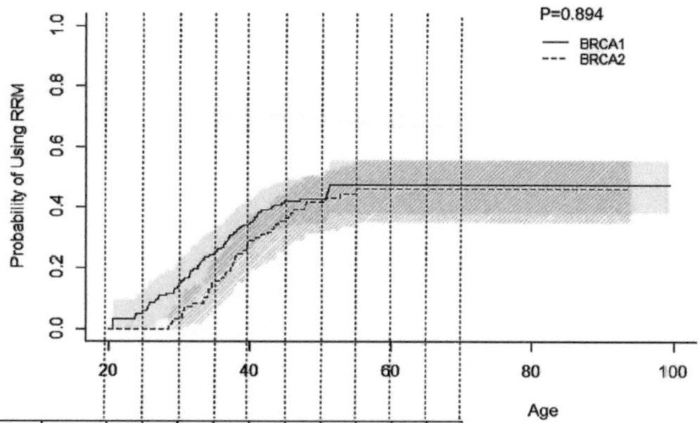

	Age	20-25	25-30	30-35	35-40	40-45	45-50	50-55	55-60	60-65	65-70
Probability of RRM Occurring in 5-Year Window	BRCA1	0.054	0.084	0.104	0.101	0.063	0.022	0.036	0	0	0
	BRCA2	0	0.042	0.105	0.133	0.085	0.059	0.031	0.010	0	0

Cumulative Probability of RRM	BRCA1	0.054	0.138	0.242	0.343	0.406	0.428	0.464	0.464	0.464	0.464
	BRCA2	0	0.042	0.147	0.280	0.365	0.424	0.455	0.465	0.465	0.465

Fig. 4.2 Kaplan–Meier estimates for the cumulative probability of risk-reducing mastectomy by age and *BRCA* gene mutation [24]. Reprinted from Breast Cancer Research and Treatment, use of risk-reducing surgeries in a prospective cohort of 1499 *BRCA1* and *BRCA2* mutation carriers, Vol. 148/No. 2, © 2014, pp. 397–406, Chai X, Friebel TM, Singer CF, Evans DG, Lynch HT, Isaacs C, et al., with permission of Springer

Several studies in the USA have looked at risk reduction in different racial and ethnic groups. Minority women may be less informed about risk-reduction options because of decreased rates of referrals to genetic counseling or genetic testing which impacts their knowledge regarding their risk and ultimately their decision making regarding whether to undergo risk-reducing mastectomy [25, 26]. Additionally, more research needs to be done to determine if young African American women with triple negative breast cancers are at higher risk of being *BRCA* mutation carriers as has been found in the Florida Cancer Registry [27]. One study conducted telephone interviews of both high- and normal-risk women from four racial/ethnic groups to assess their ability to recognize modalities to reduce breast cancer risk. The study concluded that race/ethnicity, health insurance coverage (public versus private), and interview language correlated with the ability to identify risk-reducing measures. Asian–American women were the least likely to have heard of prophylactic surgery 50.8% (45.5–56.2, 95% CI) as compared to African American 67.8% (62.4–72.8, 95% CI), Latina 56.4% (50.8–61.8, 95% CI), or White 85.8 (82.9–88.3, 95% CI) women [28]. One study of African American women from a single large kindred found that 7/7 (100%) *BRCA1* mutation carriers opted for surveillance over prophylactic surgery or chemoprevention [29]. The study concluded that patient–provider communication about genetic test results was suboptimal. Another study that included patients with breast cancer, found that 41.7% of African American women who were affected by breast cancer underwent a risk-reduction mastectomy [30]. Women with breast or ovarian cancer and a *BRCA* mutation were significantly more likely to have a risk-reducing mastectomy than women with breast or ovarian cancer without a *BRCA* mutation (41.7% vs. 9.9%, $p < 0.01$). African American women who choose to undergo bilateral mastectomy often have a higher number of relatives with breast and ovarian cancer, ($p = 0.024$) in one study [25]. This same study of African American women noted those that had bilateral mastectomy had a higher household income ($p = 0.009$) and concluded that clinical factors such as family history and financial means may influence surgery recommendations and decisions at both the patient and provider levels. A population-based sample of *BRCA* mutation carriers found that black women had lower rates of risk-reduction mastectomy (67%) compared to Hispanics (83%) and non-Hispanic whites (94%) [31]. Little research has been done to evaluate risk-reducing mastectomy uptake in the Latino community, and it is an important area to understand attitudes and beliefs and improve awareness in a culturally appropriate manner [32]. Important variables that influence differences in uptake of risk-reducing mastectomy include patient–physician communication, perceived risk of cancer, cultural beliefs, provider knowledge, and access to surgery with insurance coverage.

The Patient Protection and Affordable Care Act mandates coverage of genetic testing for women deemed to be at increased risk; however, it does not address insurance coverage for preventative interventions [33]. Though federal law does not mandate insurance coverage for risk-reducing mastectomy, most insurance companies cover the procedure [34] and some states do mandate coverage. Prior to proceeding with risk-reducing surgery, patients and providers should take steps to determine whether prophylactic surgery is covered by the patient's insurance or pursue other institutional or charitable resources so as to not cause undue financial burden or stress for the patient. Reconstruction coverage is mandated by the federal Women's Health and Cancer Rights Act (WHCRA), enacted in 1999, which requires breast reconstruction coverage after mastectomy [35].

Timing of Surgery

Consideration for the age of the woman when the carrier status is identified, whether or not she currently has breast or ovarian cancer, [36, 37] incidence of breast cancer within the family [38], and the woman's general health all play into her decision-making process for surgery [39]. Timing of breast surgery is the next hurdle. If she currently has breast cancer, then the "when" is relative to the sequencing of her cancer treatment. However. if she is a previvor (an unaffected genetic mutation carrier), then timing will be relative to her current age, age of the youngest family member affected with breast cancer, previous personal history of benign breast disease and biopsies, and social factors (stable relationship, completed child-bearing, financial/insurance concerns, desire for and availability of reconstructive options). The timing of a bilateral risk-reducing salpingo-oophorectomy also should be discussed with the patient's multidisciplinary team.

Surgical Options

Once the woman has decided on risk-reducing mastectomy, she should meet with a (breast) surgeon and a plastic surgeon (if immediate reconstruction desired). As per NCCN guidelines, women should have bilateral mammogram within 6 months of surgery [12] and many would recommend an MRI within 1 year of planned surgery because of the risk of an occult malignancy.

Simple Versus Subcutaneous Mastectomy

Given that there are many options for breast reconstruction, the patient together with her surgical team will need to decide on the type of mastectomy and the type of reconstruction desired/recommended. The type of mastectomy centers on what is removed and what remains in situ. A "subcutaneous" mastectomy removes most of the breast tissue but leaves a rim of tissue attached to the under surface of the native breast skin. This operation was originally championed in 1962 by Freeman before genetic testing was known and thus available [40]. Considering this operation was performed primarily for benign disease, the reasons for leaving tissue behind was to minimize the risk of skin necrosis from vascular compromise, leave the nipple/areolar disc in situ, and improve acceptance of the reconstruction from a cosmetic standpoint. At that time breast reconstruction was in its infancy and options were limited. Depending on the amount of residual breast tissue, women still needed to be screened by mammography and a small portion later developed breast cancer. As surgical techniques advanced both for the mastectomy itself and for reconstruction, the mastectomy performed evolved into a "simple" or "total" mastectomy (the same operation performed for breast cancer). A total mastectomy attempts to remove all breast tissues, but studies have shown that a small amount of scattered cells will still remain behind on the undersurface of the skin and the anterior surface of the pectoralis muscle [41–44]. The amount is far less than with a subcutaneous mastectomy and thus translates into a predicted lower rate of primary breast cancer development in the woman's future. Currently, most risk-reducing mastectomies in the USA are performed as a total or simple mastectomy.

Skin, Areolar, Nipple-Sparing Mastectomy

The next decision after choosing between a subcutaneous or total mastectomy is to decide the amount of residual native breast skin. If no reconstruction is planned, then the goal is to remove the nipple, areola, and the necessary amount of skin to allow for the skin to lie flat against the chest wall without being too tight (limiting range of motion) or too loose. However, if immediate reconstruction is planned, then providing the plastic surgeon with a skin envelope would allow for a more natural and esthetically pleasant appearance. In 1984, Toth and Lappert championed the skin-sparing mastectomy through which the nipple and areolar disc are removed but takes less than 20% of the native breast skin [45]. They also recommended avoiding placement of scars in the upper poles of the breast that would otherwise detract from the cosmetic appearance of the breast reconstruction. Initial concerns centered on the oncologic safety of preserving additional native breast skin and the potential for increasing the overall volume of residual isolated breast cells. However, studies failed to demonstrate an increase in local recurrence or new primary cancer development in skin-sparing mastectomies versus non-skin-sparing techniques; therefore, the technique is deemed oncologically safe [46–49]. Over the past 2 decades, skin-sparing mastectomy with immediate reconstruction has become the standard practice for many breast cancer patients and previvors alike (Fig. 4.3).

During the same time frame as the *BRCA* genes were being discovered in the mid-1990s, reconstructive techniques were advancing at a rapid pace and consideration turned to the possibility of preserving the nipple areolar complex. Since the majority of mastectomies performed are for the treatment of breast cancer, the primary concern was the potential increased local recurrence rate incurred with preserving the nipple areolar complex. Investigation determined the areolar disc to

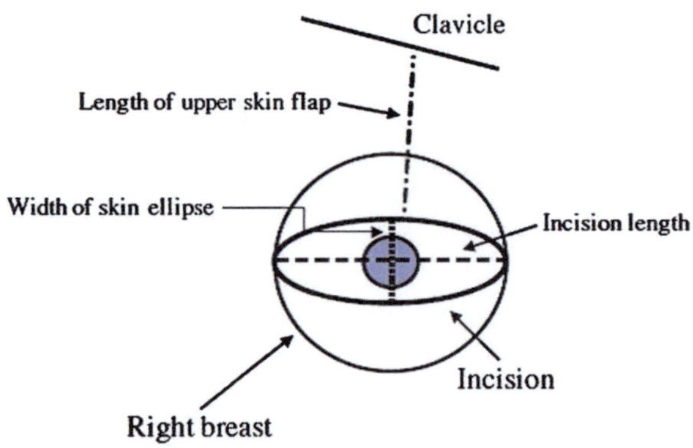

Fig. 4.3 Designing skin-sparing mastectomy. Diagram of pre-incision measurements [50]. Reprinted from Annals of Surgical Oncology, Skin flap necrosis after mastectomy with reconstruction: a prospective study, Vol. 23/No. 1, © 2016, pp. 257–264, Matsen CB, Mehrara B, Eaton A, Capko D, Berg A, Stempel M, et al., with permission of Springer

be a skin appendage and not breast tissue. Therefore, although a primary breast cancer can extend into the areolar disc, it cannot of itself make a de novo breast cancer [49, 51]. This opened the door in the 1990s for surgeons to conceptualize designing incisions that would leave the areolar disc in situ but allow for removal of the entire breast and nipple. The plastic surgeon would then be able to use the areolar disc to recreate the nipple, and the areolar disc could later be tattooed or a skin graft used to simulate an areola. Over the subsequent years, areolar-sparing mastectomies have been shown to be low risk with regard to the incidence of local recurrence or new primary cancer development [51].

The next step in the evolution of preserving the breast skin envelope was the nipple-sparing mastectomy. With nipple-sparing mastectomy, one preserves all of the native breast skin and the nipple areolar complex. From a technical standpoint, this is a much more complex procedure with a significant learning curve. Early pioneers grappled with eligibility criteria, patient selection, location and length of the incision, technical constraints of available instrumentation, type of reconstruction, and outcomes from both a cosmetic and an oncologic perspective. Studies from the 1970s and 1980s demonstrating an 8–50% risk of occult cancer found beneath the nipple of mastectomy specimens questioned the risk of nipple preservation [52, 53]. These studies used varying definitions of the distance from the nipple base to the "occult" cancer or extension of the primary cancer toward the nipple to label the pathology results as nipple involvement. In fact, mastectomy flap thickness was much larger back then (upwards of 1 cm), and as such anything within 10–20 mm of the nipple base was considered in these studies occult nipple involvement. Mastectomy flap thickness by today's standards would be closer to 3–5 mm; thus, the definition of occult cancer of the nipple needs to redefined [54]. One study, in women with *BRCA* mutations, measured the amount of breast tissue that remains when the nipple is spared with a standard retroareolar margin of 5 mm and found that this only encompasses 1.3% less of the total at-risk breast tissue [55]. The nipple proper has 8–10 milk ducts traversing it, and given that the most common histology of breast cancer is of ductal origin, initial reports of the nipple-sparing mastectomy technique discussed "coring" the nipple proper to minimize the residual ductal channels. This, however, came at the cost of increased nipple necrosis (partial and complete) and loss of nipple projection, pigmentation, and sensation [56]. A landmark study by Stolier and colleagues identified that terminal duct lobular unit (the progenitor of most breast cancers) is present in only 25% of nipples and more importantly, demonstrated that when present, the terminal duct lobular unit is always located at the base of the nipple not within the nipple proper [57]. Armed with this information, nipple "coring" is no longer required unless atypical cells are found intraoperatively on frozen section of the base of the nipple. Some centers have deferred intraoperative to permanent pathology to minimize the need for nipple "coring."

Nipple "coring" is not the only factor affecting the viability of the nipple. Placement of the skin incision along the lateral inframammary fold with lateral extension toward the axilla [58], large pendulous breasts, autologous reconstruction, direct-to-implant reconstruction [59], and a patient history of smoking all attenuates the vascular supply to the nipple areola disc [60]. The blood supply to the nipple is variable with some breasts having the vascular supply originating solely from the surrounding skin, while others have it arising solely from the native breast tissue. Most breasts have a combination of sources from breast tissue and skin. Patients whose breasts rely on breast tissue vascular supply would need to have a subcutaneous mastectomy if they wish to retain viability of their native nipple. Otherwise, total mastectomy with reconstruction of a nipple remains an alternative option. In this context, some institutions have adopted a procedure whereby they take the woman to the operating room for a sentinel lymph node biopsy (if applicable) and dissection of the nipple areolar complex off the breast mound with sampling of the sub-areolar breast tissue for pathologic evaluation. This "delay" procedure allows for oncologic evaluation before proceeding with a nipple-sparing mastectomy with immediate reconstruction and an interval development of collateral blood supply to the nipple areolar disc via the skin [61] (Fig. 4.4). Others use indocyanine green (ICG) dye and a specialized infrared camera-computer system (SPY *Elite*™) to direct placement of mastectomy incisions to minimize ischemic complications involving the nipple-areolar complex [60]. In several series of women with breast cancer, preservation of the skin and

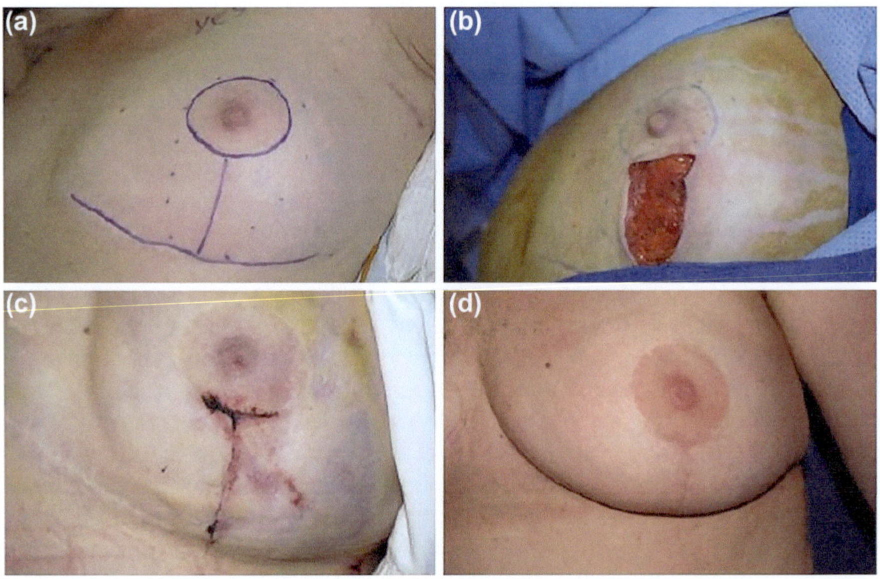

Fig. 4.4 Planned nipple-sparing mastectomy after previous mastopexy with circumareolar inverted T scar (**a**). Surgical delay procedure to ensure perfusion of the nipple-areolar complex performed 1 week prior to mastectomy as outpatient. (**b**, **c**). Final reconstruction: immediate, postoperative follow-up (**d**). *Note* this is a representative photograph and not from the author's institution [61]. Reprinted from Annals of Surgical Oncology, Nipple-sparing mastectomy in 99 patients with a mean follow-up of 5 years, Vol. 18/No. 6, © 2011, pp. 1665–1670, Jensen JA, Orringer JS, Giuliano AE, with permission of Springer

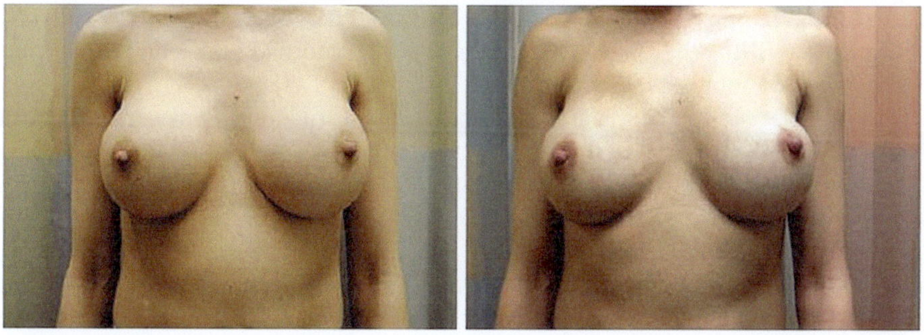

Fig. 4.5 Patient preoperative (*left*) and postoperative (*right*) results with nipple-sparing mastectomy after 8 months. *Note* this is a representative photograph and not from the author's institution [73]. Reprinted from Annals of Surgical Oncology, Total skin-sparing mastectomy in *BRCA* mutation carriers, Vol. 21/No. 1, © 2014, pp. 37–41, Peled AW, Irwin CS, Hwang ES, Ewing CA, Alvarado M, Esserman LJ, with permission of Springer

nipple-areolar complex did not result in an increased local recurrence rate or a change in survival [49, 62–69]. Nipple-sparing mastectomy has also been shown to be safe for risk reduction including in *BRCA*-positive patients [70–74] (Fig. 4.5).

Although a previvor does not have cancer at the time of planned risk-reducing mastectomy, , her primary goal is risk reduction and thus oncologic safety is paramount. Her secondary goal is aesthetic outcome and minimizing impact on body image, quality of life, and intimacy. In counseling a woman about her decision for surgery, it is important that several factors about mastectomy and reconstruction are well explained. First, the native breast skin, and thus the nipple if preserved, will most likely be insensate. Over time, some sensation may return but it will most likely be altered. One study evaluating nipple sensation found that some sensation was preserved in those who underwent nipple-sparing mastectomy in both nipples for 26% of patients and in one of the two nipples in 68% of patients [75]. Second, the reconstructed breast will not feel like a natural breast to the woman herself. There have been mixed reports of women's overall quality of life with nipple-sparing compared to skin sparing mastectomies. One single-center study of women who completed the BREAST-Q reconstruction module [76, 77] found that nipple-sparing compared to skin-sparing mastectomy patients reported significantly higher scores for psychosocial ($p = 0.01$) and sexual well-being ($p = 0.02$), but no difference in physical well-being, satisfaction with the breast, satisfaction in outcomes. Finding slightly different results, another study of 53 women prospectively used the Breast Evaluation Questionnaire and the Body Image after Breast Cancer Questionnaire [78, 79] and found that quality of life and satisfaction did not differ between women who underwent nipple-sparing mastectomy versus skin-sparing mastectomy [75].

Many factors impact a woman and her surgeon's decision on the type of mastectomy chosen for risk-reduction. In one study of women eligible to receive both skin-sparing mastectomy and nipple-sparing mastectomy, patients choosing skin-sparing mastectomy were younger (43 ± 10 vs. 49 ± 10 years, $p = 0.05$), had a higher BMI (26 ± 4 vs. 23 ± 3, $p = 0.02$), and had larger breasts on final pathology (564 g/breast vs. 366 g/breast, $p < 0.001$) [75].

Another decision to be made is the type of reconstruction (if any) that would be performed concomitantly with the risk reducing mastectomy. Women will have different goals for the size and shape of their reconstructed breast which will inform which surgery will serve them best. Factors that impact the type of reconstruction offered by plastic surgeons include body type, medical history, previous radiation, available soft tissue donor sites, skin deficit and quality including the laxity and thickness of chest wall skin. Women and their surgeons will also consider the overall recovery time and various risks to determine which surgery to have. The three most common options include the following: (1) two-stage tissue expander and then implant or direct-to-implant, (2) Latissimus dorsiflap with or without expander, and an (3) autologous alone flap such as a transverse rectus abdominis muscle (TRAM), deep inferior epigastric perforator (DIEP) or superior epigastric artery perforator (SGAP) flap. The most common reconstruction for women undergoing risk-reducing mastectomyis an implant reconstruction [80]. For nipple-sparing mastectomy, some institutions have found that they have improved outcomes and less nipple necrosis with a two-stage tissue expander then delayed implant reconstruction as compared to direct-to-implant or autologous reconstructions [59].

As mentioned above, women who are found to have invasive cancer or ductal carcinoma in situ at the time of risk-reducing mastectomy should be treated accordingly for breast cancer. In a recent publication from Sloan Kettering, ductal carcinoma in situ or invasive cancer was found in 4.8 and 1.7%, respectively, in 459 prophylactic nipple-sparing mastectomy specimens [81].

Sentinel Lymph Node Biopsy

Sentinel lymph node biopsy in the setting of risk-reducing mastectomy has limited indications. As mentioned prior to risk-reduction surgery, screening imaging should be within 6 months of surgery to detect any malignancy. In the setting of a discovered malignancy in the workup for risk-reducing mastectomy, sentinel lymph node biopsy would be indicated for the treatment and staging of malignancy as per NCCN guidelines for breast cancer [82]. Women with abnormal findings on imaging who do not have a preoperative biopsy or those without a previous MRI are at risk for an occult primary tumor, and a sentinel lymph node biopsy is reasonable to stage the axilla with a risk-reducing mastectomy. There is no evidence to suggest an increased risk of lymphedema in patients undergoing sentinel node biopsy in this setting [83].

Surveillance

After risk-reducing mastectomy, women should have an annual clinical examination of the chest/reconstructed breast because there is still a small risk of the future development of breast cancer emanating from the residual ductal cells on the undersurface of the mastectomy flap. Mammography, however, is not required after mastectomy, whether with reconstruction or not. Women should continue to follow with her gynecologist for ovarian cancer screening unless she has also had a risk-reducing salpingo-oophorectomy.

References

1. Rebbeck TR, Mitra N, Wan F, Sinilnikova OM, Healey S, McGuffog L, et al. Association of type and location of *BRCA1* and *BRCA2* mutations with risk of breast and ovarian cancer. JAMA. 2015;313(13):1347–61.
2. King M-C, Marks JH, Mandell JB. Breast and ovarian cancer risks due to inherited mutations in *BRCA1* and *BRCA2*. Science. 2003;302(5645):643–6.
3. Struewing JP, Hartge P, Wacholder S, Baker SM, Berlin M, McAdams M, et al. The risk of cancer associated with specific mutations of *BRCA1* and *BRCA2* among Ashkenazi Jews. N Engl J Med. 1997;336(20):1401–8.
4. Chen S, Parmigiani G. Meta-analysis of *BRCA1* and *BRCA2* penetrance. J Clin Oncol. 2007;25(11):1329–33.
5. Burke W, Daly M, Garber J, Botkin J, Kahn MJE, Lynch P, et al. Recommendations for follow-up care of individuals with an inherited predisposition to cancer: II. *BRCA1* and *BRCA2*. JAMA. 1997;277(12):997–1003.
6. Hartmann LC, Schaid DJ, Woods JE, Crotty TP, Myers JL, Arnold PG, et al. Efficacy of bilateral prophylactic mastectomy in women with a family history of breast cancer. N Engl J Med. 1999;340(2):77–84.
7. Hartmann LC, Sellers TA, Schaid DJ, Frank TS, Soderberg CL, Sitta DL, et al. Efficacy of Bilateral Prophylactic Mastectomy in *BRCA1* and *BRCA2* gene mutation carriers. J Nat Cancer Inst. 2001;93(21):1633–7.
8. Hamm RM, Lawler F, Scheid D. Prophylactic mastectomy in women with a high risk of breast cancer. N Engl J Med. 1999;340(23):1837–8; author reply 9.
9. Meijers-Heijboer H, van Geel B, van Putten WL, Henzen-Logmans SC, Seynaeve C, Menke-Pluymers MB, et al. Breast cancer after prophylactic bilateral mastectomy in women with a *BRCA1* or *BRCA2* mutation. N Engl J Med. 2001;345(3):159–64.
10. Rebbeck TR, Friebel T, Lynch HT, Neuhausen SL, van 't Veer L, Garber JE, et al. Bilateral prophylactic mastectomy reduces breast cancer risk in *BRCA1* and *BRCA2* mutation carriers: the PROSE Study Group. J Clin Oncol. 2004;22(6):1055–62.
11. De Felice F, Marchetti C, Musella A, Palaia I, Perniola G, Musio D, et al. Bilateral risk-reduction mastectomy in *BRCA1* and *BRCA2* mutation carriers: a meta-analysis. Ann Surg Oncol. 2015;22(9):2876–80.
12. Network NCC. NCCN Clinical Practice Guidelines in Oncology: Breast Cancer Risk Reduction (Version I. 2016) 2016 [National Comprehensive Cancer Network]. Available from: https://www.nccn.org/professionals/physician_gls/pdf/breast_risk.pdf.
13. Pal T, Vadaparampil ST. Genetic risk assessments in individuals at high risk for inherited breast cancer in the breast oncology care setting. Cancer Control. 2012;19(4):255.
14. Bresser PJ, Van Gool AR, Seynaeve C, Duivenvoorden HJ, Niermeijer MF, van Geel AN, et al. Who is prone to high levels of distress after prophylactic mastectomy and/or salpingo-ovariectomy? Ann Oncol. 2007;18(10):1641–5.

15. Hatcher MB, Fallowfield L, A'Hern R. The psychosocial impact of bilateral prophylactic mastectomy: prospective study using questionnaires and semistructured interviews. BMJ. 2001;322(7278):76.
16. Gopie JP, Mureau MA, Seynaeve C, ter Kuile MM, Menke-Pluymers MB, Timman R, et al. Body image issues after bilateral prophylactic mastectomy with breast reconstruction in healthy women at risk for hereditary breast cancer. Fam Cancer. 2013;12(3):479–87.
17. Brandberg Y, Sandelin K, Erikson S, Jurell G, Liljegren A, Lindblom A, et al. Psychological reactions, quality of life, and body image after bilateral prophylactic mastectomy in women at high risk for breast cancer: a prospective 1-year follow-up study. J Clin Oncol. 2008; 26(24):3943–9.
18. Razdan SN, Patel V, Jewell S, McCarthy CM. Quality of life among patients after bilateral prophylactic mastectomy: a systematic review of patient-reported outcomes. Quality Life Res. 2016;25(6):1409–21.
19. van Dijk S, van Roosmalen MS, Otten W, Stalmeier PF. Decision making regarding prophylactic mastectomy: stability of preferences and the impact of anticipated feelings of regret. J Clin Oncol. 2008;26(14):2358–63.
20. Van Roosmalen M, Stalmeier P, Verhoef L, Hoekstra-Weebers J, Oosterwijk J, Hoogerbrugge N, et al. Randomized trial of a shared decision-making intervention consisting of trade-offs and individualized treatment information for *BRCA1/2* mutation carriers. J Clin Oncol. 2004;22(16):3293–301.
21. Schwartz MD, Valdimarsdottir HB, DeMarco TA, Peshkin BN, Lawrence W, Rispoli J, et al. Randomized trial of a decision aid for *BRCA1/BRCA2* mutation carriers: impact on measures of decision making and satisfaction. Health Psychol. 2009;28(1):11.
22. Patenaude AF, Orozco S, Li X, Kaelin CM, Gadd M, Matory Y, et al. Support needs and acceptability of psychological and peer consultation: attitudes of 108 women who had undergone or were considering prophylactic mastectomy. Psychooncology. 2008;17(8): 831–43.
23. Metcalfe KA, Birenbaum-Carmeli D, Lubinski J, Gronwald J, Lynch H, Moller P, et al. International variation in rates of uptake of preventive options in *BRCA1* and *BRCA2* mutation carriers. Int J Cancer. 2008;122(9):2017–22.
24. Chai X, Friebel TM, Singer CF, Evans DG, Lynch HT, Isaacs C, et al. Use of risk-reducing surgeries in a prospective cohort of 1499 *BRCA1* and *BRCA2* mutation carriers. Breast Cancer Res Treat. 2014;148(2):397–406.
25. Pal T, Bonner D, Kim J, Monteiro AN, Kessler L, Royer R, et al. Early onset breast cancer in a registry-based sample of African-American women: *BRCA* mutation prevalence, and other personal and system-level clinical characteristics. Breast J. 2013;19(2):189–92.
26. Hall MJ, Olopade OI. Disparities in genetic testing: thinking outside the *BRCA* box. J Clin Oncol. 2006;24(14):2197–203.
27. Pal T, Bonner D, Cragun D, Monteiro AN, Phelan C, Servais L, et al. A high frequency of *BRCA* mutations in young black women with breast cancer residing in Florida. Cancer. 2015;121(23):4173–80.
28. Kaplan CP, Haas JS, Pérez-Stable EJ, Gregorich SE, Somkin C, Des Jarlais G, et al. Breast cancer risk reduction options: awareness, discussion, and use among women from four ethnic groups. Cancer Epidemiol Biomarkers Prev. 2006;15(1):162–6.
29. Kinney AY, Simonsen SE, Baty BJ, Mandal D, Neuhausen SL, Seggar K, et al. Risk reduction behaviors and provider communication following genetic counseling and *BRCA1* mutation testing in an African American kindred. J Genet Couns. 2006;15(4):293–305.
30. Lynce F, Smith KL, Stein J, DeMarco T, Wang Y, Wang H, et al. Deleterious *BRCA1/2* mutations in an urban population of black women. Breast Cancer Res Treat. 2015; 153(1):201–9.
31. Pal T, Cragun D, Lewis C, Bonner D, Camperlengo LP, Vadaparampil ST, editors. Disparities in cancer risk management among *BRCA* carriers across a diverse sample of young black,

Hispanic, and non-Hispanic white breast cancer survivors. ASCO Annual Meeting Proceedings; 2016.

32. Kinney AY, Gammon A, Coxworth J, Simonsen SE, Arce-Laretta M. Exploring attitudes, beliefs, and communication preferences of Latino community members regarding *BRCA1/2* mutation testing and preventive strategies. Genet Med. 2010;12(2):105–15.
33. Prince AE. Prevention for those who can pay: insurance reimbursement of genetic-based preventive interventions in the liminal state between health and disease. J Law Biosci. 2015; 2(2):365–95.
34. Malanga S, Sprissler R, Robertson C, Joiner K. A problem not yet manifest: gaps in insurance coverage of medical interventions after genetic testing. J Law Biosci. 2015:lsv043.
35. Labor USDo. Your Rights After a Mastectomy. Available from: https://www.dol.gov/ebsa/publications/whcra.html.
36. Heemskerk-Gerritsen BA, Rookus MA, Aalfs CM, Ausems MG, Collée JM, Jansen L, et al. Improved overall survival after contralateral risk-reducing mastectomy in *BRCA1/2* mutation carriers with a history of unilateral breast cancer: a prospective analysis. Int J Cancer. 2015;136(3):668–77.
37. Sigal BM, Munoz DF, Kurian AW, Plevritis SK. A simulation model to predict the impact of prophylactic surgery and screening on the life expectancy of *BRCA1* and *BRCA2* mutation carriers. Cancer Epidemiol Biomarkers Prev. 2012;21(7):1066–77.
38. Metcalfe K, Gershman S, Lynch H, Ghadirian P, Tung N, Kim-Sing C, et al. Predictors of contralateral breast cancer in *BRCA1* and *BRCA2* mutation carriers. Br J Cancer. 2011; 104(9):1384–92.
39. Kurian AW, Munoz DF, Rust P, Schackmann EA, Smith M, Clarke L, et al. Online tool to guide decisions for *BRCA1/2* mutation carriers. J Clin Oncol. 2012:JCO. 2011.38. 6060.
40. Freeman BS. Subcutaneous mastectomy for benign breast lesions with immediate or delayed prosthetic replacement. Plast Reconstr Surg Transplant Bull. 1962;30(6):676–82.
41. Cheng JC-H, Chen C-M, Liu M-C, Tsou M-H, Yang P-S, Jian JJ-M, et al. Locoregional failure of postmastectomy patients with 1–3 positive axillary lymph nodes without adjuvant radiotherapy. Int J Radiat Oncol Biol Phys. 2002;52(4):980–8.
42. Fisher B, Redmond C, Fisher ER, Bauer M, Wolmark N, Wickerham DL, et al. Ten-year results of a randomized clinical trial comparing radical mastectomy and total mastectomy with or without radiation. N Engl J Med. 1985;312(11):674–81.
43. Fisher B, Jeong J-H, Anderson S, Bryant J, Fisher ER, Wolmark N. Twenty-five-year follow-up of a randomized trial comparing radical mastectomy, total mastectomy, and total mastectomy followed by irradiation. N Engl J Med. 2002;347(8):567–75.
44. Barton FE Jr, English JM, Kingsley WB, Fietz M. Glandular excision in total glandular mastectomy and modified radical mastectomy: a comparison. Plast Reconstr Surg. 1991; 88(3):389–92.
45. Toth BA, Lappert P. Modified skin incisions for mastectomy: the need for plastic surgical input in preoperative planning. Plast Reconstr Surg. 1991;87(6):1048–53.
46. Kroll SS, Schusterman MA, Tadjalli HE, Singletary SE, Ames FC. Risk of recurrence after treatment of early breast cancer with skin-sparing mastectomy. Ann Surg Oncol. 1997; 4(3):193–7.
47. Carlson GW, Styblo TM, Lyles RH, Jones G, Murray DR, Staley CA, et al. The use of skin sparing mastectomy in the treatment of breast cancer: the Emory experience. Surg Oncol. 2003;12(4):265–9.
48. Kroll SS, Khoo A, Singletary ES, Ames FC, Wang B-G, Reece GP, et al. Local recurrence risk after skin-sparing and conventional mastectomy: a 6-year follow-up. Plast Reconstr Surg. 1999;104(2):421–5.
49. Simmons RM, Fish SK, Gayle L, Trenta GSL, Swistel A, Christos P, et al. Local and distant recurrence rates in skin-sparing mastectomies compared with non-skin-sparing mastectomies. Ann Surg Oncol. 1999;6(7):676–81.

50. Matsen CB, Mehrara B, Eaton A, Capko D, Berg A, Stempel M, et al. Skin flap necrosis after mastectomy with reconstruction: a prospective study. Ann Surg Oncol. 2016;23(1):257–64.
51. Simmons RM, Brennan M, Christos P, King V, Osborne M. Analysis of nipple/areolar involvement with mastectomy: can the areola be preserved? Ann Surg Oncol. 2002;9(2): 165–8.
52. Cense H, Rutgers ET, Cardozo ML, Van Lanschot J. Nipple-sparing mastectomy in breast cancer: a viable option? Eur J Surg Oncol (EJSO). 2001;27(6):521–6.
53. Laronga C, Robb G, Singletary S. Feasibility of skin-sparing mastectomy with preservation of the nipple-areola complex. Breast Dis Yearb Q. 1998;19(2):125–7.
54. Laronga C, Kemp B, Johnston D, Robb GL, Singletary SE. The incidence of occult nipple-areola complex involvement in breast cancer patients receiving a skin-sparing mastectomy. Ann Surg Oncol. 1999;6(6):609–13.
55. Zavos A, Valachis A. Risk of chemotherapy-induced amenorrhea in patients with breast cancer: a systematic review and meta-analysis. Acta Oncol. 2016;55(6):664–70.
56. Crowe JP, Patrick RJ, Yetman RJ, Djohan R. Nipple-sparing mastectomy update: one hundred forty-nine procedures and clinical outcomes. Arch Surg. 2008;143(11):1106–10.
57. Stolier AJ, Wang J. Terminal duct lobular units are scarce in the nipple: Implications for prophylactic nipple-sparing mastectomy. Ann Surg Oncol. 2008;15(2):438–42.
58. Garcia-Etienne CA, Cody HS III, Disa JJ, Cordeiro P, Sacchini V. Nipple-sparing mastectomy: initial experience at the memorial sloan-kettering cancer center and a comprehensive review of literature. Breast J. 2009;15(4):440–9.
59. Wang F, Peled AW, Garwood E, Fiscalini AS, Sbitany H, Foster RD, et al. Total skin-sparing mastectomy and immediate breast reconstruction: an evolution of technique and assessment of outcomes. Ann Surg Oncol. 2014;21(10):3223–30.
60. Wapnir I, Dua M, Kieryn A, Paro J, Morrison D, Kahn D, et al. Intraoperative imaging of nipple perfusion patterns and ischemic complications in nipple-sparing mastectomies. Ann Surg Oncol. 2014;21(1):100–6.
61. Jensen JA, Orringer JS, Giuliano AE. Nipple-sparing mastectomy in 99 patients with a mean follow-up of 5 years. Ann Surg Oncol. 2011;18(6):1665–70.
62. Laronga C, Kemp B, Johnston D, Robb GL, Singletary SE. The incidence of occult nipple-areola complex involvement in breast cancer patients receiving a skin-sparing mastectomy. Ann Surg Oncol. 1999;6(6):609–13.
63. Garcia-Etienne CA, Borgen PI. Update on the indications for nipple-sparing mastectomy. J Support Oncol. 2006;4(5):225–30.
64. Petit JY, Veronesi U, Orecchia R, Luini A, Rey P, Intra M, et al. Nipple-sparing mastectomy in association with intra operative radiotherapy (ELIOT): A new type of mastectomy for breast cancer treatment. Breast Cancer Res Treat. 2006;96(1):47–51.
65. Margulies AG, Hochberg J, Kepple J, Henry-Tillman RS, Westbrook K, Klimberg VS. Total skin-sparing mastectomy without preservation of the nipple-areola complex. Am J Surg. 2005;190(6):907–12.
66. Gerber B, Krause A, Reimer T, Muller H, Kuchenmeister I, Makovitzky J, et al. Skin-sparing mastectomy with conservation of the nipple-areola complex and autologous reconstruction is an oncologically safe procedure. Ann Surg. 2003;238(1):120–7.
67. Crowe JP Jr, Kim JA, Yetman R, Banbury J, Patrick RJ, Baynes D. Nipple-sparing mastectomy: technique and results of 54 procedures. Arch Surg. 2004;139(2):148–50.
68. Sacchini V, Pinotti JA, Barros AC, Luini A, Pluchinotta A, Pinotti M, et al. Nipple-sparing mastectomy for breast cancer and risk reduction: oncologic or technical problem? J Am Coll Surg. 2006;203(5):704–14.
69. Peled AW, Foster RD, Stover AC, Itakura K, Ewing CA, Alvarado M, et al. Outcomes after total skin-sparing mastectomy and immediate reconstruction in 657 breasts. Ann Surg Oncol. 2012;19(11):3402–9.
70. Rusby JE, Smith BL, Gui GP. Nipple-sparing mastectomy. Br J Surg. 2010;97(3):305–16.

71. Murthy V, Chamberlain RS. Defining a place for nipple sparing mastectomy in modern breast care: an evidence based review. Breast J. 2013;19(6):571–81.
72. Manning A, Wood C, Eaton A, Stempel M, Capko D, Pusic A, et al. Nipple-sparing mastectomy in patients with *BRCA1/2* mutations and variants of uncertain significance. Br J Surg. 2015;102(11):1354–9.
73. Peled AW, Irwin CS, Hwang ES, Ewing CA, Alvarado M, Esserman LJ. Total skin-sparing mastectomy in *BRCA* mutation carriers. Ann Surg Oncol. 2014;21(1):37–41.
74. Yao K, Liederbach E, Tang R, Lei L, Czechura T, Sisco M, et al. Nipple-sparing mastectomy in *BRCA1/2* mutation carriers: an interim analysis and review of the literature. Ann Surg Oncol. 2015;22(2):370–6.
75. Dossett LA, Lowe J, Sun W, Lee M, Smith PD, Jacobsen PB, et al. Prospective evaluation of skin and nipple-areola sensation and patient satisfaction after nipple-sparing mastectomy. J Surg Oncol. 2016.
76. Cano SJ, Klassen AF, Scott AM, Cordeiro PG, Pusic AL. The BREAST-Q: further validation in independent clinical samples. Plast Reconstr Surg. 2012;129(2):293–302.
77. Pusic AL, Klassen AF, Scott AM, Klok JA, Cordeiro PG, Cano SJ. Development of a new patient-reported outcome measure for breast surgery: the BREAST-Q. Plast Reconstr Surg. 2009;124(2):345–53.
78. Anderson RC, Cunningham B, Tafesse E, Lenderking WR. Validation of the breast evaluation questionnaire for use with breast surgery patients. Plast Reconstr Surg. 2006;118 (3):597–602.
79. Baxter NN, Goodwin PJ, McLeod RS, Dion R, Devins G, Bombardier C. Reliability and validity of the body image after breast cancer questionnaire. Breast J. 2006;12(3):221–32.
80. Cemal Y, Albornoz CR, Disa JJ, McCarthy CM, Mehrara BJ, Pusic AL, et al. A paradigm shift in US breast reconstruction: Part 2. The influence of changing mastectomy patterns on reconstructive rate and method. Plast Reconstr Surg. 2013;131(3):320e–6e.
81. Manning AT, Sacchini VS. Conservative mastectomies for breast cancer and risk-reducing surgery: the memorial sloan kettering cancer center experience. Gland Surg. 2016;5(1):55.
82. Network NCC. NCCN Clinical Practice Guidelines in Oncology: Breast Cancer 2016 [2.2016]. Available from: https://www.nccn.org/professionals/physician_gls/pdf/breast.pdf.
83. Miller CL, Specht MC, Skolny MN, Jammallo LS, Horick N, O'Toole J, et al. Sentinel lymph node biopsy at the time of mastectomy does not increase the risk of lymphedema: implications for prophylactic surgery. Breast Cancer Res Treat. 2012;135(3):781–9.

Prophylactic Oophorectomy for Patients with Germline *BRCA* Mutations

5

Dario R. Roque and Don S. Dizon

Introduction

Ovarian cancer is the second most common gynecologic malignancy and the most common cause of gynecologic cancer death in the USA [1]. During 2016, an estimated 22,280 women will be diagnosed with ovarian cancer in the USA, and approximately 14,240 patients will succumb to their disease [1]. Based on data from the Surveillance, Epidemiology, and End Results (SEER) Program, women in the USA have a 1.4% lifetime risk of developing ovarian cancer; the average age of diagnosis is 63 years old. Over 95% of ovarian malignancies have an epithelial origin, and serous carcinoma is the most common type of epithelial ovarian cancer (EOC) [2]. Clinically and histologically, serous carcinomas of the ovary exhibit very similar behavior to fallopian tube and primary peritoneal serous carcinomas. Therefore, for simplicity, throughout this chapter the term "ovarian cancer" will be used to refer to carcinomas of epithelial origin arising in the ovaries, fallopian tubes, and peritoneal lining.

In addition to older age, there are a number of risk factors that have been associated with the development of sporadic ovarian cancer, including the reproductive and environmental factors listed in Table 5.1. However, while the majority of ovarian cancer cases are sporadic, germline mutations in several genes, including

D.R. Roque (✉)
Department of Obstetrics and Gynecology, Program in Women's Oncology,
Women and Infants Hospital/The Warren Alpert Medical School at Brown University,
101 Dudley Street, Providence, RI 02906, USA
e-mail: drroque@wihri.org

D.S. Dizon
Department of Medicine, Massachusetts General Hospital Cancer Center, 55 Fruit Street,
Yawkey 9E, Boston, MA 02114, USA
e-mail: ddizon@partners.org

A.B. Chagpar (ed.), *Managing BRCA Mutation Carriers*,
DOI 10.1007/978-3-319-59198-8_5

BRCA1, *BRCA2*, *BRIP1*, *RAD51C*, and *RAD51D*, as well as in mismatch repair genes, are associated with increased susceptibility and *account for* approximately 25% of ovarian carcinomas [3–5]. Of the currently known ovarian cancer susceptibility genes, germline mutations in *BRCA1* and *BRCA2* account for a significant number of hereditary ovarian cancers and have been found in approximately 15% of women with this malignancy [3, 6, 7].

Prevention and early detection are paramount in reducing the high mortality rate associated with ovarian cancer [8]. This high mortality rate is in part due to the fact that over 70% of women with this malignancy have advanced-stage disease at the time of diagnosis. To date, there are no reliable screening tests for ovarian cancer, and prior to the discovery of the *BRCA* genes, there were no primary prevention methods for this malignancy. Therefore, recognizing that mutated *BRCA1* and *BRCA2* predispose to hereditary ovarian cancer prompted genetic assessment of patients diagnosed with ovarian cancer and subsequent genetic screening of family members for those patients who were found to carry a deleterious mutation. In doing so, more and more women have been identified as *BRCA* carriers and offered risk-reducing surgery to minimize their risk of developing ovarian cancer. Because of this intervention, we should see a small decrease in the incidence of ovarian cancer as more women who are genetic carriers undergo risk-reducing surgery before the potential onset of this aggressive disease.

However, despite the tremendous benefits derived from risk-reducing bilateral salpingo-oophorectomy (rrBSO) in this patient population, the decision to undergo surgery and timing of such procedure involves complex, emotionally charged, and often life-changing decision making [9]. Furthermore, the resulting surgical menopause can have significant psychological and physical health consequences that should be taken into account during preoperative counseling and postoperative follow-up. Therefore, our goal for this chapter will be to discuss the relationship

Table 5.1 Risk factors associated with ovarian cancer [98–105]

Risk factors	Relative risks
Infertility	2.67 [98]
Polycystic ovarian syndrome	2.52 [99]
Endometriosis[a]	2.04–3.05 [100]
Cigarette smoking[b]	2.1 [101]
Protective factors	
History of contraceptive use	0.73 [102]
Breast feeding > 12 months	0.72 [103]
Tubal ligation	0.69 [104]
Pregnancy	0.71[c] [105]

[a]Increase in risk of clear cell, endometrioid, and low-grade serous carcinomas

[b]Increase in risk of mucinous carcinoma

[c]Parous versus nulliparous women. Risk appears to decrease further with increasing parity

between *BRCA* mutations and ovarian cancer as well as to review screening and management recommendations for these patients, with special emphasis on the role, benefits, and risks associated with risk-reducing bilateral salpingo-oophorectomy.

Germline Mutations in *BRCA1* & *BRCA2*

The gene encoding for *BRCA1* was first identified in 1994 and mapped to the long arm of chromosome 17. A year later, *BRCA2* was identified on the long arm of chromosome 13 [10]. The proteins encoded by both *BRCA* genes have been implicated in DNA repair processes [11]. *BRCA1* functions in the signaling of DNA damage and its repair by homologous recombination, nucleotide excision repair, and non-homologous end-joining. Meanwhile, *BRCA2* has a more specific role in DNA repair, regulating the activity of *RAD51*, a protein required for homologous recombination [8, 11]. However, the *BRCA* genes are also involved in cell cycle checkpoint regulation, chromosomal segregation, and estrogen metabolism [12, 13]. Therefore, a deleterious mutation in either of these genes can lead to disruption of the genomic integrity and subsequent oncogenesis.

Mutations in the *BRCA* genes are highly penetrant, and all germline mutations identified to date have been inherited, suggesting the possibility of a large "founder" effect in which a certain mutation is common to a small number of individuals who established a new population. *BRCA* mutation prevalence in the general population ranges between 1/300 and 1/800 [14]. However, the frequency and specific type of mutation occurs much more commonly in certain populations because of the previously mentioned founder effect. For example, the frequency of a *BRCA1* or *BRCA2* mutation is 1 in 40 individuals in Ashkenazi Jews, with 1 of 3 founder mutations most commonly identified: *BRCA1* 187delAG, *BRCA1* 5385insC, and *BRCA2* 6174delT [14–16]. Other groups with increased mutational frequency include women of French Canadian, Polish, Icelandic, and Hispanic descent [14]. However, the likelihood that ovarian cancer will develop in a mutation carrier varies even among families with the same gene. This suggests that certain factors may influence whether cancer develops in *BRCA* carriers and there are a number of risk modifiers that are the subject of ongoing research [9, 17]. Nevertheless, having a family history of ovarian cancer is a strong predictor of future risk in mutation carriers [9].

BRCA Function and Risk Associated with Carrier Status

Both *BRCA1* and *BRCA2* encode proteins involved in tumor suppression. Mutations in these genes are inherited in an autosomal dominant fashion. The susceptibility to cancer results from the inheritance of one mutant allele of either *BRCA1* or *BRCA2* followed by loss of heterozygosity in breast or ovarian epithelial cells, ultimately leading to complete inactivation of the gene [18]. Female mutation carriers have increased susceptibility to breast, ovarian, and pancreatic carcinoma [10]. However, in this chapter, we will focus on the ovarian cancer susceptibility associated with *BRCA* mutations.

Histologically, both mutations are associated with the development of predominantly serous ovarian carcinoma [19]. A number of studies have reported on the lifetime risk of developing ovarian cancer. A meta-analysis of 10 studies combining for a total of 1641 carriers from multiple countries reported a mean cumulative risks for ovarian cancer of 40% for *BRCA1* and 18% for *BRCA2* [20]. Meanwhile, a prospective trial in the UK following 1887 carriers reported the lifetime risk of ovarian cancer to be 59% for patients with *BRCA1* and 16.5% among *BRCA2* carriers [21]. In contrast, the average risk of developing ovarian cancer by age 70 in the general population is less than 1%. Overall, the risk of developing ovarian cancer in *BRCA* carriers before the age of 40 is small. However, after the fourth decade of life, the risk increases significantly [20]. Furthermore, the risk of ovarian cancer is not only higher in *BRCA1* mutations, but these women also tend to develop an ovarian malignancy at an earlier age than *BRCA2* carriers (Fig. 5.1). Rebbeck et al. found that the mean age of diagnosis of ovarian cancer was 50.8 years for *BRCA* carriers [22].

Since the risk of developing ovarian cancer varies by age, a woman's current age should be taken into consideration when counseling her about her lifetime risk. For example, using the estimates from Chen et al. [20], an unaffected 30-year-old *BRCA2* carrier has a 16% risk of developing ovarian cancer by age 70. Meanwhile, a 60-year-old unaffected *BRCA2* carrier's risk of developing ovarian cancer by age 70 should be quoted at approximately 9% [20]. In addition to family history of ovarian cancer, other factors may increase the individual lifetime risk of genetic carriers. Recently, two genetic modifier mutations were found (*BRCA1*-specific SNP rs4691139 and SNP rs17631303), which confer an even greater lifetime risk of ovarian cancer [23]. Identifying these types of mutation modifiers will further enhance our ability to counsel patients appropriately about their lifetime risks.

Fig. 5.1 Ovarian cancer risk by mutated gene and decade of life

Guidelines for Genetic Testing

The National Comprehensive Cancer Network (NCCN) Guidelines recommend risk assessment, genetic counseling, and *BRCA* testing for any woman with a family member who is carrier of a deleterious *BRCA* mutation as well as all women with a personal history of ovarian cancer. In addition, women with a personal history of breast cancer or pancreatic cancer should also undergo *BRCA* mutation testing if they meet any of the criteria listed in Table 5.2 [24].

Table 5.2 Criteria for offering *BRCA* mutation genetic testing per NCCN guidelines [24]

Personal/family history	*BRCA* testing criteria
Family with a known deleterious gene mutation	No further criteria needed. *Test all women.*
Personal history of ovarian carcinoma[c]	No further criteria needed. *Test all women.*
Personal history of breast cancer, including ductal carcinoma in situ (DCIS)	Test all women who meet at least *ONE* of the following four criteria: (1) Diagnosed at age ≤ 45 (2) Diagnosed at age ≤ 50 with: • An additional breast cancer primary • ≥ 1 close blood relative[a] with breast cancer at any age • ≥ 1 close relative with pancreatic cancer • ≥ 1 relative with prostate cancer (Gleason score ≥ 7) • An unknown or limited family history (3) Diagnosed at age ≤ 60 with: • Triple negative breast cancer (4) Diagnosed at any age with: • ≥ 1 close blood relative[a] with breast cancer diagnosed at age ≤ 50 • ≥ 2 close blood relatives[a] with breast cancer at any age • ≥ 1 close blood relative[a] with ovarian carcinoma^ • ≥ 2 close blood relatives[a] with pancreatic cancer and/or prostate cancer (Gleason score ≥ 7) at any age • A close male blood relative[a] with breast cancer • For an individual of ethnicity associated with higher mutation frequency (e.g., Ashkenazi Jewish) no additional family history may be required[b]

(continued)

Table 5.2 (continued)

Personal/family history	*BRCA* testing criteria
Personal history of pancreatic cancer	*Test all women who meet at least ONE of the following four criteria:* (1) Ashkenazi Jewish ancestry (2) ≥ 1 close blood relative[a] with ovarian carcinoma[c] at any age (3) ≥ 1 close blood relative[a] with breast cancer diagnosed at age ≤ 50 (4) ≥ 2 relatives with breast, pancreatic cancer or prostate cancer (Gleason score ≥ 7) at any age
Family history only	*Test all women who meet at least ONE of the following two criteria:* (1) First- or second-degree blood relative meeting any of the above criteria (2) Third-degree blood relative who has breast cancer and/or ovarian carcinoma[c] and who has ≥ 2 close blood relatives with breast cancer (at least one with breast cancer diagnosed at age ≤ 50) and/or ovarian carcinoma[c]

[a]Close relative is defined as a first-degree (parent, sibling, offspring), second-degree (grandparent, grandchild, uncle, aunt, nephew, niece, half-sibling), or third-degree (first cousin, great-grandparent or great-grandchild) relative
[b]Testing for Ashkenazi Jewish founder-specific mutation(s) should be performed first. Comprehensive genetic testing may be considered if ancestry also includes non-Ashkenazi Jewish relatives or if other *BRCA*-related criteria are met. Founder mutations exist in other populations [24]
[c]Includes fallopian tube and primary peritoneal cancers. *BRCA*-related ovarian cancers are associated with epithelial non-mucinous histology

Once a patient undergoes *BRCA* testing, if a mutation is found, current guidelines from the NCCN and the Society of Gynecologic Oncology (SGO) recommend for that patient to undergo risk-reducing bilateral salpingo-oophorectomy (rrBSO) once she is between 35 and 40 years of age or sooner if she has completed childbearing [24, 25]. In addition, family members should be offered testing for the specific familial mutation. This algorithm is outlined in Fig. 5.2. The recommended timing of rrBSO is aimed at maximizing the reduction in ovarian cancer risk for *BRCA* carriers. As previously discussed, the risk starts to increase significantly after age 40. More importantly and perhaps the biggest determinant of surgical timing recommendation is the fact that a number of patients are diagnosed with an occult malignancy at the time of their rrBSO. Given the aggressive nature of ovarian cancer, rrBSO should ideally be performed at a time prior to the development of the malignancy. In *BRCA1* mutation carriers, prevalence of ovarian cancer found during rrBSO was 1.5% for those younger than age 40 and 3.8% in those between the ages of 40 and 49. *BRCA1* mutation carriers between the ages of 50 and 59 years had the highest incidence rate at 6.7% [26]. Meanwhile, the highest incidence rate in *BRCA2* mutation carriers was observed between the ages of 60 and 69 years at 4.9%, and no occult malignancies were diagnosed in women who

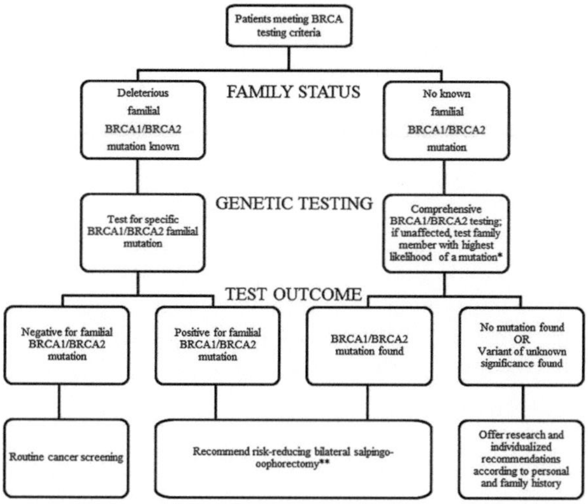

Fig. 5.2 Testing and management algorithm to decrease the risk of ovarian cancer in women who meet criteria for *BRCA* genetic evaluation by NCCN guidelines [24]. *Single asterisk* For both affected and unaffected individuals of Ashkenazi Jewish descent with no known familial mutation, first test for the three common mutations. Then, if negative for the three mutations and ancestry also includes non-Ashkenazi Jewish relatives or other *BRCA*-related criteria are met, consider comprehensive genetic testing. For both affected and unaffected individuals who are non-Ashkenazi Jewish and who have no known familial mutation, comprehensive genetic testing is the approach, if done. *Double asterisk* Risk-reducing salpingo-oophorectomy should be completed between ages 35–40, and upon completion of childbearing. Because ovarian cancer onset in patients with *BRCA2* mutations is an average of 8–10 years later than in patients with *BRCA1* mutations, it is reasonable to delay RRSO until age 40–45 in patients with *BRCA2* mutations who have already maximized their breast cancer prevention [24]

underwent rrBSO prior to age 60 years [26]. Therefore, the recommended age for rrBSO could be younger for women with a *BRCA1* mutation than for women with a *BRCA2* mutation [24, 26]. However, the age at which rrBSO is performed may also be individualized according to the earliest age of onset in the family [25].

Outcomes After Risk-Reducing Bilateral Salpingo-Oophorectomy

Risk-reducing bilateral salpingo-oophorectomy is the most effective intervention in reducing the risk of ovarian cancer in *BRCA* mutation carriers. The benefit from rrBSO is likely due to removal of the fallopian tubes, as data have now emerged that the fallopian tubes are the likely origin of precursor lesions that lead to the development of ovarian cancer [27]. In addition, rrBSO may have a role in further decreasing breast cancer risk in women with *BRCA* mutations. Lastly, given the associated reduction in the risk of developing ovarian cancer, and the fact that many patients are diagnosed with an early ovarian cancer at the time of their rrBSO, this procedure may also confer *BRCA* carriers a survival advantage compared to those women with *BRCA* mutations who do not undergo rrBSO.

Ovarian Cancer Precursor Lesions and *BRCA*

Historically, several theories have been presented regarding ovarian carcinogenesis. The majority attempted to describe how the ovarian mesothelium (also known as the ovarian surface epithelium) underwent metaplasia and dysplastic changes [28, 29]. None of them were able to identify a precursor lesion for high-grade serous carcinoma of the ovary (HGSC). In 2001, Piek et al. reported areas of cellular dysplasia and hyperplastic lesions in the fallopian tubal epithelium of women undergoing rrBSO for *BRCA* or a strong family history of ovarian cancer [30]. The lesions identified had a histologic resemblance to HGSC without the invasive component. In addition, further studies confirmed the presence of early fallopian tube malignancies in some patients with *BRCA* mutations undergoing rrBSO. The majority of these malignancies had a noninvasive, dysplastic component at the fimbriated end of the fallopian tube [31, 32]. Thus, it appeared that these patients had an increased risk for the development of a HGSC of the fallopian tube rather than the ovary [33]. The dysplastic regions within the fallopian tube became known as "serous tubal intraepithelial carcinoma" (STIC), and it has been hypothesized that it represents a precursor lesions for most HGSC that arise in the pelvis (i.e., ovary, fallopian tube, and peritoneal lining) [34–36].

Multiple findings have helped support this hypothesis. Among women undergoing rrBSO for increased risk of ovarian carcinoma, up to 17% are found to have a pelvic serous pre-invasive or invasive lesion and 80% of these involved the fallopian tubes [32, 33, 37, 38]. The incidence of STIC in non-*BRCA* mutation carriers is not yet well established; however, Kindelberger et al. examined the pathology of 55 women with advanced-stage serous ovarian, tubal, or primary peritoneal carcinoma and found that 75% of all cases of pelvic serous carcinomas contained areas of STIC, suggesting that STIC may represent a precursor lesion even in women at average risk of developing HGSC [35]. Lastly, at the molecular level, STIC and extrauterine pelvic serous carcinoma diagnosed in the same patient often share identical TP53 mutations, further suggesting a common origin [35, 39].

Ovarian Cancer Risk Reduction

In women who are *BRCA* carriers undergoing rrBSO, the resulting risk reduction for the development of ovarian cancer ranges between 72 and 96% as reported by multiple studies (Table 5.3). A meta-analysis of ten observational studies evaluating ovarian cancer outcomes in *BRCA* mutation carriers who had undergone rrBSO included three non-overlapping data sets [40–42] based on which the authors found a risk reduction for ovarian cancer of approximately 79% (HR = 0.21; 95% CI =

Table 5.3 Reported ovarian cancer risk reduction in *BRCA* carriers as a result of risk-reducing bilateral salpingo-oophorectomy [22, 26, 41–45, 47, 106]

Study	# of patients		# of ovarian cancers		Hazard ratio (95% CI)
	rrBSO	No rrBSO	rrBSO	No rrBSO	*BRCA1* & *BRCA2*
Rebbeck et al. [22]	259	292	2	58	0.04 (0.01–0.16)
Finch et al. [42]	1041	779	7	32	0.20 (0.07–0.58)
Kauff et al. [106]	98	72	1	5	0.15 (0.02–1.31)
Domchek et al. [47]	155	271	2	16	0.11 (0.03–0.47)
Kauff et al. [41]	509	283	3	12	0.12 (0.03–0.41)
Domchek et al. [45]	939	1678	6	63	0.28 (0.12–0.69)
Finch et al. [26]	3513	2270	32	108	0.20 (0.13–0.30)
Rebbeck et al. [43]	1555	1285	NR	NR	0.21 (0.12–0.39)
Marchetti et al. [44]	4961	4231	NR	NR	0.19 (0.13–0.27)

0.12 to 0.39) [43]. A more recent meta-analysis of three prospective studies found a similar reduction of 81% in ovarian cancer risk after rrBSO (HR 0.19, 95% CI 0.13–0.27) [44].

Whether the risk for ovarian cancer after rrBSO decreases equally for *BRCA1* and *BRCA2* mutation carriers is less clear because few studies have reported separate estimates. However, in a subgroup analysis of the study by Domchek et al., which included 939 *BRCA* carriers who underwent rrBSO, ovarian cancer occurred in 1–2% of *BRCA1* carriers who underwent rrBSO versus 7–8% in those who did not undergo prophylactic surgery. Meanwhile, there were no ovarian cancer cases in *BRCA2* carriers who underwent rrBSO compared with 3% of those who did not undergo rrBSO [45]. These findings suggest that *BRCA1* carriers may get a more significant reduction because their baseline risk is higher than women with *BRCA2* mutations. However, the higher risk appears to persist after rrBSO in patients with mutations in *BRCA1* compared to *BRCA2*. Further data are needed to evaluate whether this difference in risk truly persists following rrBSO.

Breast Cancer Risk Reduction After Prophylactic Bilateral Salpingo-Oophorectomy

Women with *BRCA1* and *BRCA2* mutations have a lifetime risk of approximately 57–60 and 49–55%, respectively, of developing breast cancer. In addition, *BRCA* carriers who are diagnosed and treated for breast cancer remain at risk of developing a second breast cancer [46]. Therefore, these women undergo frequent screening with annual mammograms and breast MRI, and many of them choose to undergo risk-reducing bilateral mastectomy. The reduction in breast cancer risk after undergoing prophylactic mastectomy is approximately 90% [9]. Patients who choose to either postpone or completely forego risk-reducing mastectomy should continue to follow the recommended breast screening guidelines, but there are data to suggest that in these patients, rrBSO may also offer some reduction in their risk of developing breast cancer.

Several observational studies have reported a decrease in breast cancer risk in *BRCA* carriers with no prior history of breast cancer who underwent rrBSO. The reported reduction in breast cancer risk has ranged between 37 and 62% in *BRCA1* mutation carriers, 21–64% for *BRCA2* carriers, and a reduction between 38 and 64% for all *BRCA* carriers [21, 22, 45, 47–49]. Furthermore, the majority of these trials showed a risk reduction of approximately 50% when the rrBSO was performed in pre-menopausal women [22, 45, 47–49]. Greater reductions in breast cancer risk were observed in women with a *BRCA1* mutation who had an rrBSO at age 40 years or younger (OR, 0.36; 95% CI, 0.20–0.64) relative to those who underwent the procedure at age 41–50 (OR, 0.50; 95% CI, 0.27–0.92) [50]. In addition, some studies suggest that rrBSO after age 50 is not associated with a substantial decrease in breast cancer risk [48]. These findings suggest that the decrease in breast cancer risk may be linked to decreased hormonal exposure following surgical removal of the ovaries.

A more recent study by Heemskerk-Gerritsen et al. also looked at the effect of rrBSO on later risk of breast cancer in *BRCA* carriers and failed to show any reduction in risk, HR 1.09 (95% CI 0.67–1.77) [51]. This study employed more stringent inclusion criteria to minimize bias and utilized person-time before rrBSO in order to calculate a more accurate measure of risk. However, a subsequent trial that adjusted for person-time bias as a result of the Heemskerk-Gerritsen analysis continued to find a protective effect of rrBSO on breast cancer incidence in *BRCA1/2* mutation carriers (HR, 0.59; 95% CI, 0.42–0.82) [52]. Therefore, while most published data have shown a risk reduction in breast cancer in *BRCA* carriers who undergo rrBSO, the findings from Heemskerk-Gerritsen et al. should be taken into consideration and mentioned when counseling patients about this possible benefit [9].

Effects of Risk-Reducing Salpingo-Oophorectomy on Mortality

A reduction in mortality is ultimately the outcome that we as providers are most interested in when offering rrBSO to our patients. Several studies have looked at the impact of rrBSO on mortality; the estimates are very reassuring and further solidify the role of rrBSO as a worthwhile intervention.

Among a cohort of 5783 women with a *BRCA* mutation who were observed prospectively for an average of 5.6 years, the risk of death for those who were unaffected with cancer at study entry fell by 77% (HR 0.23; 95% CI, 0.13–0.39) after rrBSO. This decrease in mortality was the result in large part to reduction in the incidence of ovarian cancer, but the authors also demonstrated reduced breast cancer incidence and mortality as a result of rrBSO [26]. Similarly, the Prevention and Observation of Surgical Endpoints (PROSE) multicenter prospective cohort study of 2482 *BRCA* carriers also demonstrated lower all-cause mortality (HR, 0.40; 95% CI, 0.26–0.61), breast cancer-specific mortality (HR, 0.44; 95% CI,

0.26–0.76), and ovarian cancer-specific mortality (HR, 0.21; 95% CI, 0.06–0.80) after 3.7 years in patients who underwent rrBSO [45]. Based on these results, it can be concluded that rrBSO is an effective preventive strategy, and in the absence of other interventions with similar risk reductions in ovarian cancer and in ovarian cancer-related mortality, it should continue to be offered as the standard of care for women with *BRCA* mutations.

Preoperative Counseling

Despite the benefits derived from rrBSO in *BRCA* carriers, the decision to undergo surgery and timing of such procedure involves complex, emotionally charged and often life-changing decision making [9]. As such, the discussion with the patient seeking consultation should address not only the benefits from undergoing the procedure but also the risks of the procedure as well as the implications of ovarian removal, particularly in younger women who will effectively go into a surgical menopause as a result. Given the complexity of the discussion as well as the possibility of an occult ovarian cancer diagnosis at the time of rrBSO, the NCCN recommends for patients to seek consultation with a gynecologic oncologist [24]. The counseling should address all of the topics discussed in this chapter including:

1. Baseline lifetime cancer risks associated with the specific *BRCA* mutation that the patient carries. The lifetime risks quoted should be adjusted to the patient's age.
2. The likely disease course of ovarian cancer as well as how rrBSO is the most effective intervention in *BRCA* carriers to prevent (and in some cases, detect) ovarian cancer
3. Reduction in ovarian cancer risk and breast cancer risk as a result of rrBSO.
4. Reduction in all-cause, ovarian cancer-related and breast cancer-related mortality as a result of rrBSO.
5. Prevalence of occult ovarian cancers found during rrBSO as it relates to the patient's specific *BRCA* mutation. Discussion of this issue should also highlight recommended timing for the surgery based on specific mutations and whether or not the patient would be willing to undergo an ovarian cancer surgical staging simultaneously or as an interval procedure if an occult malignancy was identified during rrBSO.

In addition, in pre-menopausal women, the discussion should address the reproductive and medical implications of surgical menopause. Current guidelines recommend that women undergo rrBSO between the ages of 35–40 and upon completion of childbearing. These recommendations can be especially distressing for women who have yet to start or have not completed childbearing by the time they learn they are a *BRCA* mutation carrier. These patients are faced with the prospect of altering their reproductive plans as they pursue risk-reducing surgery and most would prefer fertility preservation until they have completed childbearing

[53]. Therefore, for women with *BRCA* mutations who have not fulfilled their reproductive wishes by age 35, consultation with a reproductive endocrinologist should be offered.

Women with *BRCA* mutations also have to face the prospect of passing their mutation down to their children since it is inherited in an autosomal dominant fashion. In addition to the fear of passing the gene to their offspring, they may also worry about disclosing their carrier status to their current or future partners as it could affect their relationship. Therefore, women should be offered the support of specialists so that they are made aware of the different resources available to them including counseling through therapists [14].

Unlike decisions about fertility preservation, all women who undergo rrBSO by the recommended age go into an immediate surgical menopause and will deal with consequences of estrogen deficiency. Over 80% of women will experience vasomotor symptoms as a result of menopause [54, 55], including night sweats and hot flashes. Other symptoms include sleep disturbances, depression, and vaginal dryness [55]. Vaginal dryness can be a major source of dyspareunia and may subsequently lead to sexual dysfunction. The long-term side effects of premature menopause have been well characterized in the general population and include increased risks of osteoporosis and cardiovascular disease. Both of these side effects should be discussed with the patient as part of their preoperative counseling and consenting process.

Increased bone loss is a direct result of decreased circulating levels of estrogen in women who go through surgical menopause. Bone mineral density has been shown to decrease by as much as 6.7% at 12 months in women who undergo pre-menopausal oophorectomies [56]. Consequently, oophorectomy before the age of 45 years has been associated with an increased prevalence of osteoporosis within 6 years of surgery. The risk for fracture is also substantially increased in women who undergo oophorectomy prior to age 45 years (OR, 3.64; 95% CI: 1.01–13.04) compared with women who undergo surgery after age 45 years [57, 58]. Therefore, during their preoperative counseling, patients should be advised about this risk and proper follow-up should be arranged once patients undergo rrBSO.

Bilateral oophorectomy has also been associated with increased risk of cardiovascular disease in several observational studies. In the Nurses Health Study, after controlling for age and cigarette smoking, women who underwent bilateral oophorectomy had an increased risk of cardiovascular disease compared with women with intact ovaries (Rate Ratio, 2.2; 95% CI: 1.2–4.2) [59]. Similarly, the Mayo Clinic Cohort of Oophorectomy and Aging showed that women who underwent bilateral oophorectomy before the age of 45 years experienced increased mortality due to cardiovascular disease. (HR, 1.44; 95% CI: 1.01–2.05) [57, 60]. This increased risk is also secondary to estrogen deficiency, and the pathophysiology appears to be mediated in part by menopause-related changes in lipid profile, including higher levels of low-density lipoprotein (LDL) as well as a possible decline in the protective effect of high-density lipoprotein (HDL) [61, 62].

Surgical Planning

In patients with *BRCA* mutations, risk-reducing salpingo-oophorectomy involves complete removal of the fallopian tubes and ovaries. The surgery may be performed either via laparotomy or laparoscopy. However, a laparoscopic approach (with or without robotic assistance) is preferred given the benefits of laparoscopic surgery including reductions in recovery time and hospital stay [63–65]. As opposed to laparotomy, laparoscopic salpingo-oophorectomy is considered an outpatient procedure. In addition, compared to laparotomy, laparoscopic surgery is associated with less postoperative pain and postoperative complications [65].

Intraoperatively, once access has been gained into the abdomen, peritoneal washings should be obtained. Although the sensitivity, specificity, and prognostic values of positive lavage cytology have not been determined, washings are part of the current International Federation of Gynecology and Obstetrics (FIGO) staging system for ovarian cancer and should be collected in these patients to assist with staging if an occult ovarian cancer is identified. Multiple small series of patients undergoing rrBSO have reported finding positive peritoneal washings in several women diagnosed with an occult ovarian or fallopian tube malignancy at the time of their surgery [66–68]. Inspection of the surfaces of the diaphragm, upper abdomen, paracolic gutters, and then pelvis should then be performed, and any lesions that appear suspicious should be biopsied [14]. The pelvic retroperitoneum should then be accessed to visualize the ureter and isolate the infundibulopelvic ligament before transection. All ovarian tissue should be removed along with any adhesions between the ovary and other peritoneal structures to ensure that no residual ovarian cells remain attached to the peritoneal surface. The ovarian vessels should be clamped and transected at least 2 cm proximal to the ovary, and preferably at the pelvic brim, to avoid leaving any ovarian tissue behind [69–71].

The fallopian tube should also be removed in its entirety with the exception of the interstitial portion of the fallopian tube as there have been no reports of malignant transformation in the tubal remnant after rrBSO [72]. However, clinicopathologic studies have found evidence of malignancy in all remaining portions of the fallopian tube [73]. In a series of 122 *BRCA*-positive women undergoing rrBSO, five of the seven occult malignancies diagnosed originated in the fimbrial portion of the fallopian tube [32]. Furthermore, the tubal fimbria is the site where noninvasive tubal intraepithelial carcinomas have been discovered, and these lesions may represent a precursor of invasive ovarian cancer [32]. Because of this, a special pathologic evaluation should be requested, which involves the entire specimen being serially micro-sectioned into 2- to 3-mm segments with particular attention being paid to the fimbria [14, 32].

Is There a Role for Hysterectomy at the Time of Risk-Reducing Salpingo-Oophorectomy?

There are limited data to inform the risk of uterine carcinoma among *BRCA* carriers. However, these data suggest that they are not at an overall increased risk. Shu et al. reported on a cohort of 1083 women with a known deleterious *BRCA* mutation (627 *BRCA1* and 453 *BRCA2*) who underwent rrBSO without hysterectomy between 1995 and 2011 [74]. With follow-up collected through October 2014, and compared to data generated from the SEER database, there was no significantly increased risk of uterine cancer identified with an incidence among *BRCA* carriers of 0.8% (observed to expected [O:E] ratio,1.9; 95% CI 0.8–3.7).

Although the risk of uterine cancer overall is not increased, some have raised concerns that the risk of uterine serous carcinoma (USC) is increased. However, the data on this topic are conflicting. Two case series including only women of Ashkenazi Jewish descent reported a 27 and 15% rate of *BRCA* mutations in patients with USC compared with a 2.0% rate in the general Ashkenazi Jewish population [75, 76]. In addition, a case series of the general population found a 2% rate of *BRCA* mutation in USC patients compared with a background rate of 0.6% [77]. Finally, five of the *BRCA* mutation carriers in the report by Shu et al. had serous (or serous-like) endometrial carcinomas, for an overall incidence of 0.4%. Compared to the SEER database, however, the risk was found to be elevated significantly among *BRCA1* carriers (O:E ratio, 22.2; 95% CI 6.1–56.9) but not *BRCA2* carriers (O:E ratio, 6.4; 95% CI 0.2–35.5) [74]. However, other data that reported the genetic testing results of 56 women with serous endometrial carcinoma found no women participants with *BRCA* mutations [78]. Taken together, while the overall risk of endometrial carcinoma is not increased, and while some data suggest the risk of USC is greatly increased among *BRCA1* carriers specifically, we note that the overall incidence of USC is low even among this subgroup. Therefore, we conclude that a hysterectomy at the time of rrBSO should not be recommended as a way to prevent the development of uterine carcinoma.

Nevertheless, there are specific situations in which performing a concomitant hysterectomy at the time of rrBSO may offer some benefit to individual patients. Women with Lynch syndrome and those who will take tamoxifen or hormone replacement therapy following their rrBSO should be offered a hysterectomy as part of their risk-reducing surgery. Women with Lynch syndrome are at increased risk of developing endometrial cancer, and a hysterectomy would decrease such risk [79]. Tamoxifen use is associated with a small risk of developing endometrial cancer; therefore, women who are taking or will be taking tamoxifen as chemoprophylaxis for breast cancer are also good candidates for a concurrent hysterectomy. Lastly, women who are concerned about the genitourinary symptoms of menopause, which include significant concerns about sexual health, may express interest in hormone replacement therapy. For these patients, estrogen replacement should be strongly favored over combined estrogen–progesterone treatment, particularly given data that the latter increases the risk of breast cancer in the general population [80–83]. Given the increased risk of endometrial cancer in patients taking estrogen therapy, hysterectomy should be offered.

Outside of these scenarios, the concurrent performance of a hysterectomy can be individualized to each patient. Preferably, the hysterectomy should also be performed laparoscopically (with or without robotic assistance), and the patient should be extensively counseled about the risks of hysterectomy compared to a rrBSO. These risks include increased surgical morbidity as well as a more prolonged recovery and the need for a short hospital stay.

Alternatives to Risk-Reducing Surgery

Ovarian Cancer Surveillance

Transvaginal ultrasound (TVUS) and CA-125 levels have a limited ability to detect ovarian cancer at an early, more curable stage of disease. Therefore, patients should be informed that there is no evidence that screening with these tests reduces mortality or improves the survival associated with ovarian cancer in high-risk populations [84]. However, given the high lifetime risk of these patients developing ovarian cancer, the Society of Gynecologic Oncology (SGO) and the American Congress of Obstetricians and Gynecologists (ACOG) recommend periodic screening with CA-125 and TVUS every six months starting at age 30–35 years old or 5–10 years before the age of first diagnosis in the family [84]. The NCCN guidelines, however, do not endorse routine screening with these interventions, but state that they may be considered starting at age 30–35 in women who choose to delay rrBSO [24]. The consensus opinion is that rrBSO is the best option for these patients and neither TVUS nor CA-125 should be considered reasonable substitutes.

Salpingectomy Alone

The possibility that the fallopian tubes are the primary site of carcinogenesis in *BRCA* mutation carriers has raised the question of whether bilateral salpingectomy with delayed oophorectomy may be an option for pre-menopausal women wanting to delay surgical menopause. Some trials have demonstrated that bilateral salpingectomies are safe and feasible [85], and from the patient's perspective, salpingectomy appears to be an acceptable option [86]. Among more than 200 *BRCA* carriers who had declined rrBSO, 33% reported an interest in salpingectomy, and most were willing to accept the need for interval oophorectomy [86]. However, more data are needed regarding its efficacy in reducing the risk for ovarian cancer before it can be recommend as an equivalent option, especially in women with

BRCA mutations. In addition, *BRCA* carriers who undergo salpingectomy without oophorectomy may not get the 50% reduction in breast cancer risk that has been reported after rrBSO in this patient population. Because of this, the NCCN discourages the use of salpingectomy alone outside of a clinical trial [24] and the SGO position is that salpingectomy alone is not a substitute for oophorectomy; however, it can be used in women who are done with childbearing but refuse rrBSO. In these women, risk-reducing oophorectomy should still be performed as soon as the woman is willing to accept menopause [25].

Chemoprevention

The use of oral contraceptive pills (OCPs) provides a significant protective effect against the development of ovarian cancer in women with *BRCA* mutations. Two meta-analyses have shown that OCPs are associated with ovarian cancer risk reduction in *BRCA* carriers. Friebel et al. found this risk reduction to range from 33 to 80% for *BRCA1* and 58–63% for *BRCA2* carriers with at least 1 year of use [87]. Meanwhile, Iodice et al. also showed a significant reduced risk of ovarian cancer (RR 0.50, 95% CI 0.33–0.75); however, this study found the effect to be similar in both *BRCA1* and *BRCA2* mutation carriers [88]. In addition, the degree of reduction appears to correlate with duration of OCPs use [88, 89].

Despite being clearly beneficial for ovarian cancer, multiple case–control studies have reported conflicting outcomes, from decreased risk, to no change, to increased risk of breast cancer with OCPs use [14]. The most concern stems from use of more than 5 years' therapy duration with formulations prior to 1975 [90]. However, in the meta-analysis described above, there was no evidence of a significantly increased breast cancer risk in oral contraceptive users overall, for users of current formulations of oral contraceptives, or in the first 10 years after cessation of use [88]. These risks should be discussed with patients; however, given the available data, OCPs use is a reasonable option for women with a *BRCA* mutation who have not undergone rrBSO and who are not trying to conceive.

Postoperative Care and Follow-up

Management of Symptoms and Long-Term Effects of Surgical Menopause

The use of systemic hormone replacement therapy (HRT) in women with *BRCA* mutations is controversial. On the one hand, menopausal symptoms that occur as a result of rrBSO can be alleviated with HRT. In addition, HRT has been shown to mitigate the risk of osteoporosis and fracture in women undergoing rrBSO [91, 92]. In contrast, given the increased risk of breast cancer in women with *BRCA* mutations, it is uncertain whether the data regarding the use of HRT in the general menopausal population are applicable to this patient population. Nevertheless, although high-quality data are not available, no studies to date have demonstrated an increased risk of breast cancer with HRT in *BRCA* carriers who undergo rrBSO prior to menopause [93–95]. In a prospective cohort of 462 women with germline *BRCA1/2* mutations, HRT of any type after rrBSO did not significantly alter the reduction in breast cancer risk associated with the rrBSO. The hazard ratio was 0.38 (95% CI, 0.09–1.59) for those who did not receive HRT versus 0.37 (95% CI, 0.14–0.96) in those patients who received HRT after rrBSO [93].

The decision to treat women with HRT after a rrBSO should be individualized and the patients made aware of the possible increased risk of breast cancer as well as the lack of quality data. This conversation should be had in the preoperative period because women who choose to be treated with HRT may benefit from undergoing a hysterectomy at the time of the rrBSO. A hysterectomy would allow for the patient to receive unopposed estrogen instead of combination estrogen/progesterone therapy. As previously described, unopposed estrogen is preferred because in postmenopausal women, it carries a smaller risk than combination HRT for the development of breast cancer. Hormonal therapy should also be targeted to the specific symptom that the patient is experiencing, and even women who choose to not undergo systemic HRT should be offered vaginal estrogen if they are experiencing symptoms of vaginal atrophy such as vaginal dryness and dyspareunia. There are no data regarding the risk of breast cancer in high-risk women treated with vaginal estrogen therapy. However, no increase in the risk of recurrence was found in 69 women with a history of breast cancer treated with vaginal estrogen for an average of one year [96]. Of note, in patients who do not undergo hysterectomy at the time of rrBSO and are treated with vaginal estrogen, a progestin may not be necessary with a low dose of vaginal estrogen. However, a course of progesterone once a month could be considered as a way to decrease the risk of endometrial hyperplasia/cancer.

Lastly, given the increased risk of osteoporosis after rrBSO, women should be counseled on use of calcium and vitamin D and incorporating weight-bearing exercise into their routine. Screening with dual-energy X-ray absorptiometry (DEXA) scans 2–3 years out from the time of rrBSO should also be considered.

Risk of Primary Peritoneal Cancer in Women After Undergoing Risk-Reducing Surgery

Women who undergo rrBSO for a *BRCA* mutation should be counseled about the risk of developing primary peritoneal carcinoma. Carcinoma of the peritoneum refers to a neoplasm identical to serous carcinoma of the ovary, which involves the peritoneal surfaces and develops in the presence or absence of ovarian tissue [97]. Studies looking at the likelihood of developing primary peritoneal carcinoma after rrBSO have quoted ranges between 0.5 and 10.7% for an average risk of 1.7% [69]. A larger prospective study including 1045 patients estimated a 4.3% cumulative incidence of peritoneal carcinoma in *BRCA* carriers at 20 years after rrBSO [42].

The risk appears to be highest in *BRCA1* carriers with a recent prospective study of 5783 *BRCA* carriers demonstrating a higher annual risk for *BRCA1* (0.20%) versus *BRCA2* carriers (0.01%) [26]. Currently, there are no data to recommend routine surveillance for primary peritoneal cancer in *BRCA* carriers following rrBSO.

References

1. American Cancer Society. Cancer facts & figures 2016. 2016. Available from: http://www.cancer.org/research/cancerfactsstatistics/cancerfactsfigures2016/. 10 July 2016.
2. Berek JS, Crum C, Friedlander M. Cancer of the ovary, fallopian tube, and peritoneum. Int J Gynaecol Obstet. 2012;119(Suppl 2):S118–29.
3. Pal T, Permuth-Wey J, Betts JA, Krischer JP, Fiorica J, Arango H, et al. *BRCA1* and *BRCA2* mutations account for a large proportion of ovarian carcinoma cases. Cancer. 2005;104 (12):2807–16.
4. Risch HA, McLaughlin JR, Cole DE, Rosen B, Bradley L, Kwan E, et al. Prevalence and penetrance of germline *BRCA1* and *BRCA2* mutations in a population series of 649 women with ovarian cancer. Am J Hum Genet. 2001;68(3):700–10.
5. Li AJ, Karlan BY. Genetic factors in ovarian carcinoma. Curr Oncol Rep. 2001;3(1):27–32.
6. Pennington KP, Swisher EM. Hereditary ovarian cancer: beyond the usual suspects. Gynecol Oncol. 2012;124(2):347–53.
7. Zhang S, Royer R, Li S, McLaughlin JR, Rosen B, Risch HA, et al. Frequencies of *BRCA1* and *BRCA2* mutations among 1,342 unselected patients with invasive ovarian cancer. Gynecol Oncol. 2011;121(2):353–7.
8. Berek JS, Chalas E, Edelson M, Moore DH, Burke WM, Cliby WA, et al. Prophylactic and risk-reducing bilateral salpingo-oophorectomy: recommendations based on risk of ovarian cancer. Obstet Gynecol. 2010;116(3):733–43.
9. Hartmann LC, Lindor NM. The role of risk-reducing surgery in hereditary breast and ovarian cancer. N Engl J Med. 2016;374(5):454–68.
10. Yoshida K, Miki Y. Role of *BRCA1* and *BRCA2* as regulators of DNA repair, transcription, and cell cycle in response to DNA damage. Cancer Sci. 2004;95(11):866–71.
11. Tutt A, Ashworth A. The relationship between the roles of *BRCA* genes in DNA repair and cancer predisposition. Trends Mol Med. 2002;8(12):571–6.
12. Huen MS, Sy SM, Chen J. *BRCA1* and its toolbox for the maintenance of genome integrity. Nat Rev Mol Cell Biol. 2010;11(2):138–48.
13. O'Donovan PJ, Livingston DM. *BRCA1* and *BRCA2*: breast/ovarian cancer susceptibility gene products and participants in DNA double-strand break repair. Carcinogenesis. 2010;31 (6):961–7.
14. Garcia C, Powell CB. A comprehensive approach to the identification and management of the *BRCA* patient. Obstet Gynecol Surv. 2015;70(2):131–43.
15. Oddoux C, Struewing JP, Clayton CM, Neuhausen S, Brody LC, Kaback M, et al. The carrier frequency of the *BRCA2* 6174delT mutation among Ashkenazi Jewish individuals is approximately 1%. Nat Genet. 1996;14(2):188–90.
16. Struewing JP, Abeliovich D, Peretz T, Avishai N, Kaback MM, Collins FS, et al. The carrier frequency of the *BRCA1* 185delAG mutation is approximately 1 percent in Ashkenazi Jewish individuals. Nat Genet. 1995;11(2):198–200.
17. Rebbeck TR, Mitra N, Wan F, Sinilnikova OM, Healey S, McGuffog L, et al. Association of type and location of *BRCA1* and *BRCA2* mutations with risk of breast and ovarian cancer. JAMA. 2015;313(13):1347–61.
18. Haber D. Prophylactic oophorectomy to reduce the risk of ovarian and breast cancer in carriers of *BRCA* mutations. N Engl J Med. 2002;346(21):1660–2.

19. Mavaddat N, Barrowdale D, Andrulis IL, Domchek SM, Eccles D, Nevanlinna H, et al. Pathology of breast and ovarian cancers among *BRCA1* and *BRCA2* mutation carriers: results from the Consortium of Investigators of Modifiers of *BRCA1/2* (CIMBA). Cancer Epidemiol Biomark Prev. 2012;21(1):134–47.
20. Chen S, Parmigiani G. Meta-analysis of *BRCA1* and *BRCA2* penetrance. J Clin Oncol. 2007;25(11):1329–33.
21. Mavaddat N, Peock S, Frost D, Ellis S, Platte R, Fineberg E, et al. Cancer risks for *BRCA1* and *BRCA2* mutation carriers: results from prospective analysis of EMBRACE. J Natl Cancer Inst. 2013;105(11):812–22.
22. Rebbeck TR, Lynch HT, Neuhausen SL, Narod SA, Van't Veer L, Garber JE, et al. Prophylactic oophorectomy in carriers of *BRCA1* or *BRCA2* mutations. N Engl J Med. 2002;346(21):1616–22.
23. Couch FJ, Wang X, McGuffog L, Lee A, Olswold C, Kuchenbaecker KB, et al. Genome-wide association study in *BRCA1* mutation carriers identifies novel loci associated with breast and ovarian cancer risk. PLoS Genet. 2013;9(3):e1003212.
24. NCCN. Genetic/familial high-risk assessment: breast and ovarian (Version 2.2016). 2016. Available from: https://www.nccn.org/professionals/physician_gls/pdf/genetics_screening.pdf. 6 Aug 2016.
25. Walker JL, Powell CB, Chen LM, Carter J, Bae Jump VL, Parker LP, et al. Society of gynecologic oncology recommendations for the prevention of ovarian cancer. Cancer. 2015;121(13):2108–20.
26. Finch AP, Lubinski J, Moller P, Singer CF, Karlan B, Senter L, et al. Impact of oophorectomy on cancer incidence and mortality in women with a *BRCA1* or *BRCA2* mutation. J Clin Oncol. 2014;32(15):1547–53.
27. Kurman RJ, Shih IM. The origin and pathogenesis of epithelial ovarian cancer: a proposed unifying theory. Am J Surg Pathol. 2010;34(3):433–43.
28. Risch HA. Hormonal etiology of epithelial ovarian cancer, with a hypothesis concerning the role of androgens and progesterone. J Natl Cancer Inst. 1998;90(23):1774–86.
29. Ness RB, Cottreau C. Possible role of ovarian epithelial inflammation in ovarian cancer. J Natl Cancer Inst. 1999;91(17):1459–67.
30. Piek JM, van Diest PJ, Zweemer RP, Jansen JW, Poort-Keesom RJ, Menko FH, et al. Dysplastic changes in prophylactically removed Fallopian tubes of women predisposed to developing ovarian cancer. J Pathol. 2001;195(4):451–6.
31. Cass I, Holschneider C, Datta N, Barbuto D, Walts AE, Karlan BY. *BRCA*-mutation-associated fallopian tube carcinoma: a distinct clinical phenotype? Obstet Gynecol. 2005;106(6):1327–34.
32. Callahan MJ, Crum CP, Medeiros F, Kindelberger DW, Elvin JA, Garber JE, et al. Primary fallopian tube malignancies in *BRCA*-positive women undergoing surgery for ovarian cancer risk reduction. J Clin Oncol. 2007;25(25):3985–90.
33. Leeper K, Garcia R, Swisher E, Goff B, Greer B, Paley P. Pathologic findings in prophylactic oophorectomy specimens in high-risk women. Gynecol Oncol. 2002;87(1):52–6.
34. Piek JM, Verheijen RH, Kenemans P, Massuger LF, Bulten H, van Diest PJ. *BRCA1/2*-related ovarian cancers are of tubal origin: a hypothesis. Gynecol Oncol. 2003;90(2):491.
35. Kindelberger DW, Lee Y, Miron A, Hirsch MS, Feltmate C, Medeiros F, et al. Intraepithelial carcinoma of the fimbria and pelvic serous carcinoma: evidence for a causal relationship. Am J Surg Pathol. 2007;31(2):161–9.
36. Przybycin CG, Kurman RJ, Ronnett BM. Shih Ie M, Vang R. Are all pelvic (nonuterine) serous carcinomas of tubal origin? Am J Surg Pathol. 2010;34(10):1407–16.
37. Colgan TJ, Murphy J, Cole DE, Narod S, Rosen B. Occult carcinoma in prophylactic oophorectomy specimens: prevalence and association with *BRCA* germline mutation status. Am J Surg Pathol. 2001;25(10):1283–9.

38. Finch A, Shaw P, Rosen B, Murphy J, Narod SA, Colgan TJ. Clinical and pathologic findings of prophylactic salpingo-oophorectomies in 159 *BRCA1* and *BRCA2* carriers. Gynecol Oncol. 2006;100(1):58–64.
39. Kuhn E, Kurman RJ, Vang R, Sehdev AS, Han G, Soslow R, et al. TP53 mutations in serous tubal intraepithelial carcinoma and concurrent pelvic high-grade serous carcinoma—evidence supporting the clonal relationship of the two lesions. J Pathol. 2012;226(3):421–6.
40. Rutter JL, Wacholder S, Chetrit A, Lubin F, Menczer J, Ebbers S, et al. Gynecologic surgeries and risk of ovarian cancer in women with *BRCA1* and *BRCA2* Ashkenazi founder mutations: an Israeli population-based case-control study. J Natl Cancer Inst. 2003;95 (14):1072–8.
41. Kauff ND, Domchek SM, Friebel TM, Robson ME, Lee J, Garber JE, et al. Risk-reducing salpingo-oophorectomy for the prevention of *BRCA1*- and *BRCA2*-associated breast and gynecologic cancer: a multicenter, prospective study. J Clin Oncol. 2008;26(8):1331–7.
42. Finch A, Beiner M, Lubinski J, Lynch HT, Moller P, Rosen B, et al. Salpingo-oophorectomy and the risk of ovarian, fallopian tube, and peritoneal cancers in women with a *BRCA1* or *BRCA2* Mutation. JAMA. 2006;296(2):185–92.
43. Rebbeck TR, Kauff ND, Domchek SM. Meta-analysis of risk reduction estimates associated with risk-reducing salpingo-oophorectomy in *BRCA1* or *BRCA2* mutation carriers. J Natl Cancer Inst. 2009;101(2):80–7.
44. Marchetti C, De Felice F, Palaia I, Perniola G, Musella A, Musio D, et al. Risk-reducing salpingo-oophorectomy: a meta-analysis on impact on ovarian cancer risk and all cause mortality in *BRCA1* and *BRCA2* mutation carriers. BMC Womens Health. 2014;14:150.
45. Domchek SM, Friebel TM, Singer CF, Evans DG, Lynch HT, Isaacs C, et al. Association of risk-reducing surgery in *BRCA1* or *BRCA2* mutation carriers with cancer risk and mortality. JAMA. 2010;304(9):967–75.
46. Graeser MK, Engel C, Rhiem K, Gadzicki D, Bick U, Kast K, et al. Contralateral breast cancer risk in *BRCA1* and *BRCA2* mutation carriers. J Clin Oncol. 2009;27(35):5887–92.
47. Domchek SM, Friebel TM, Neuhausen SL, Wagner T, Evans G, Isaacs C, et al. Mortality after bilateral salpingo-oophorectomy in *BRCA1* and *BRCA2* mutation carriers: a prospective cohort study. Lancet Oncol. 2006;7(3):223–9.
48. Rebbeck TR, Levin AM, Eisen A, Snyder C, Watson P, Cannon-Albright L, et al. Breast cancer risk after bilateral prophylactic oophorectomy in *BRCA1* mutation carriers. J Natl Cancer Inst. 1999;91(17):1475–9.
49. Kramer JL, Velazquez IA, Chen BE, Rosenberg PS, Struewing JP, Greene MH. Prophylactic oophorectomy reduces breast cancer penetrance during prospective, long-term follow-up of *BRCA1* mutation carriers. J Clin Oncol. 2005;23(34):8629–35.
50. Eisen A, Lubinski J, Klijn J, Moller P, Lynch HT, Offit K, et al. Breast cancer risk following bilateral oophorectomy in *BRCA1* and *BRCA2* mutation carriers: an international case-control study. J Clin Oncol. 2005;23(30):7491–6.
51. Heemskerk-Gerritsen BA, Seynaeve C, van Asperen CJ, Ausems MG, Collee JM, van Doorn HC, et al. Breast cancer risk after salpingo-oophorectomy in healthy *BRCA1/2* mutation carriers: revisiting the evidence for risk reduction. J Natl Cancer Inst. 2015;107(5).
52. Chai X, Domchek S, Kauff N, Rebbeck T, Chen J. RE. Breast cancer risk after salpingo-oophorectomy in healthy *BRCA1/2* mutation carriers: revisiting the evidence for risk reduction. J Natl Cancer Inst. 2015;107(9).
53. Kim J, Skrzynia C, Mersereau JE. A pilot study of *BRCA* mutation carriers' knowledge about the clinical impact of prophylactic-oophorectomy and views on fertility consultation: a single-center pilot study. J Genet Couns. 2015;24(1):149–57.
54. Gold EB, Colvin A, Avis N, Bromberger J, Greendale GA, Powell L, et al. Longitudinal analysis of the association between vasomotor symptoms and race/ethnicity across the menopausal transition: study of women's health across the nation. Am J Public Health. 2006;96(7):1226–35.

55. Woods NF, Mitchell ES. Symptoms during the perimenopause: prevalence, severity, trajectory, and significance in women's lives. Am J Med. 2005;118(Suppl 12B):14–24.
56. Yoshida T, Takahashi K, Yamatani H, Takata K, Kurachi H. Impact of surgical menopause on lipid and bone metabolism. Climacteric. 2011;14(4):445–52.
57. Finch A, Evans G, Narod SA. *BRCA* carriers, prophylactic salpingo-oophorectomy and menopause: clinical management considerations and recommendations. Womens Health (London). 2012;8(5):543–55.
58. Tuppurainen M, Kroger H, Honkanen R, Puntila E, Huopio J, Saarikoski S, et al. Risks of perimenopausal fractures—a prospective population-based study. Acta Obstet Gynecol Scand. 1995;74(8):624–8.
59. Colditz GA, Willett WC, Stampfer MJ, Rosner B, Speizer FE, Hennekens CH. Menopause and the risk of coronary heart disease in women. N Engl J Med. 1987;316(18):1105–10.
60. Rivera CM, Grossardt BR, Rhodes DJ, Brown RD Jr, Roger VL, Melton LJ 3rd, et al. Increased cardiovascular mortality after early bilateral oophorectomy. Menopause. 2009;16 (1):15–23.
61. Derby CA, Crawford SL, Pasternak RC, Sowers M, Sternfeld B, Matthews KA. Lipid changes during the menopause transition in relation to age and weight: the study of women's health across the nation. Am J Epidemiol. 2009;169(11):1352–61.
62. Woodard GA, Brooks MM, Barinas-Mitchell E, Mackey RH, Matthews KA, Sutton-Tyrrell K. Lipids, menopause, and early atherosclerosis in study of women's health across the nation heart women. Menopause. 2011;18(4):376–84.
63. Hidlebaugh DA, Vulgaropulos S, Orr RK. Treating adnexal masses. Operative laparoscopy vs. laparotomy. J Reprod Med. 1997;42(9):551–8.
64. Bisharah M, Tulandi T. Laparoscopic preservation of ovarian function: an underused procedure. Am J Obstet Gynecol. 2003;188(2):367–70.
65. Medeiros LR, Fachel JM, Garry R, Stein AT, Furness S. Laparoscopy versus laparotomy for benign ovarian tumours. Cochrane Database Syst Rev. 2005(3):CD004751.
66. Powell CB, Kenley E, Chen LM, Crawford B, McLennan J, Zaloudek C, et al. Risk-reducing salpingo-oophorectomy in *BRCA* mutation carriers: role of serial sectioning in the detection of occult malignancy. J Clin Oncol. 2005;23(1):127–32.
67. Medeiros F, Muto MG, Lee Y, Elvin JA, Callahan MJ, Feltmate C, et al. The tubal fimbria is a preferred site for early adenocarcinoma in women with familial ovarian cancer syndrome. Am J Surg Pathol. 2006;30(2):230–6.
68. Colgan TJ, Boerner SL, Murphy J, Cole DE, Narod S, Rosen B. Peritoneal lavage cytology: an assessment of its value during prophylactic oophorectomy. Gynecol Oncol. 2002;85 (3):397–403.
69. Dowdy SC, Stefanek M, Hartmann LC. Surgical risk reduction: prophylactic salpingo-oophorectomy and prophylactic mastectomy. Am J Obstet Gynecol. 2004;191 (4):1113–23.
70. Cass I, Walts A, Karlan BY. Does risk-reducing bilateral salpingo-oophorectomy leave behind residual tube? Gynecol Oncol. 2010;117(1):27–31.
71. Guillem JG, Wood WC, Moley JF, Berchuck A, Karlan BY, Mutch DG, et al. ASCO/SSO review of current role of risk-reducing surgery in common hereditary cancer syndromes. J Clin Oncol. 2006;24(28):4642–60.
72. Kauff ND, Barakat RR. Surgical risk-reduction in carriers of *BRCA* mutations: where do we go from here? Gynecol Oncol. 2004;93(2):277–9.
73. Alvarado-Cabrero I, Young RH, Vamvakas EC, Scully RE. Carcinoma of the fallopian tube: a clinicopathological study of 105 cases with observations on staging and prognostic factors. Gynecol Oncol. 1999;72(3):367–79.
74. Shu CA, Pike MC, Jotwani AR, Friebel TM, Soslow RA, Levine DA, et al. Uterine cancer after risk-reducing salpingo-oophorectomy without hysterectomy in women with *BRCA* mutations. JAMA Oncol. 2016.

75. Biron-Shental T, Drucker L, Altaras M, Bernheim J, Fishman A. High incidence of *BRCA1–2* germline mutations, previous breast cancer and familial cancer history in Jewish patients with uterine serous papillary carcinoma. Eur J Surg Oncol. 2006;32(10):1097–100.
76. Lavie O, Ben-Arie A, Segev Y, Faro J, Barak F, Haya N, et al. *BRCA* germline mutations in women with uterine serous carcinoma–still a debate. Int J Gynecol Cancer. 2010;20 (9):1531–4.
77. Pennington KP, Walsh T, Lee M, Pennil C, Novetsky AP, Agnew KJ, et al. *BRCA1*, TP53, and CHEK2 germline mutations in uterine serous carcinoma. Cancer. 2013;119(2):332–8.
78. Goshen R, Chu W, Elit L, Pal T, Hakimi J, Ackerman I, et al. Is uterine papillary serous adenocarcinoma a manifestation of the hereditary breast-ovarian cancer syndrome? Gynecol Oncol. 2000;79(3):477–81.
79. Committee on Practice B-G, Society of Gynecologic O. ACOG Practice Bulletin No. 147. Lynch syndrome. Obstet Gynecol. 2014;124(5):1042–54.
80. Anderson GL, Chlebowski RT, Aragaki AK, Kuller LH, Manson JE, Gass M, et al. Conjugated equine oestrogen and breast cancer incidence and mortality in postmenopausal women with hysterectomy: extended follow-up of the women's health initiative randomised placebo-controlled trial. Lancet Oncol. 2012;13(5):476–86.
81. Chlebowski RT, Anderson GL, Gass M, Lane DS, Aragaki AK, Kuller LH, et al. Estrogen plus progestin and breast cancer incidence and mortality in postmenopausal women. JAMA. 2010;304(15):1684–92.
82. Chlebowski RT, Rohan TE, Manson JE, Aragaki AK, Kaunitz A, Stefanick ML, et al. Breast cancer after use of estrogen plus progestin and estrogen alone: analyses of data from 2 women's health initiative randomized clinical trials. JAMA Oncol. 2015;1(3):296–305.
83. Collaborative Group on Hormonal Factors in Breast Cancer. Breast cancer and hormone replacement therapy: collaborative reanalysis of data from 51 epidemiological studies of 52,705 women with breast cancer and 108,411 women without breast cancer. Lancet. 1997;350(9084):1047–59.
84. ACOG Practice Bulletin No. 103. Hereditary breast and ovarian cancer syndrome. Obstet Gynecol. 2009;113(4):957–66.
85. McAlpine JN, Hanley GE, Woo MM, Tone AA, Rozenberg N, Swenerton KD, et al. Opportunistic salpingectomy: uptake, risks, and complications of a regional initiative for ovarian cancer prevention. Am J Obstet Gynecol. 2014;210(5):471 e1–11.
86. Holman LL, Friedman S, Daniels MS, Sun CC, Lu KH. Acceptability of prophylactic salpingectomy with delayed oophorectomy as risk-reducing surgery among *BRCA* mutation carriers. Gynecol Oncol. 2014;133(2):283–6.
87. Friebel TM, Domchek SM, Rebbeck TR. Modifiers of cancer risk in *BRCA1* and *BRCA2* mutation carriers: systematic review and meta-analysis. J Natl Cancer Inst. 2014;106(6): dju091.
88. Iodice S, Barile M, Rotmensz N, Feroce I, Bonanni B, Radice P, et al. Oral contraceptive use and breast or ovarian cancer risk in *BRCA1/2* carriers: a meta-analysis. Eur J Cancer. 2010;46(12):2275–84.
89. Narod SA, Risch H, Moslehi R, Dorum A, Neuhausen S, Olsson H, et al. Oral contraceptives and the risk of hereditary ovarian cancer. Hereditary ovarian cancer clinical study group. N Engl J Med. 1998;339(7):424–8.
90. Narod SA, Dube MP, Klijn J, Lubinski J, Lynch HT, Ghadirian P, et al. Oral contraceptives and the risk of breast cancer in *BRCA1* and *BRCA2* mutation carriers. J Natl Cancer Inst. 2002;94(23):1773–9.
91. Castelo-Branco C, Figueras F, Sanjuan A, Pons F, Vicente JJ, Vanrell JA. Long-term postmenopausal hormone replacement therapy effects on bone mass: differences between surgical and spontaneous patients. Eur J Obstet Gynecol Reprod Biol. 1999;83(2):207–11.
92. Challberg J, Ashcroft L, Lalloo F, Eckersley B, Clayton R, Hopwood P, et al. Menopausal symptoms and bone health in women undertaking risk reducing bilateral

salpingo-oophorectomy: significant bone health issues in those not taking HRT. Br J Cancer. 2011;105(1):22–7.

93. Rebbeck TR, Friebel T, Wagner T, Lynch HT, Garber JE, Daly MB, et al. Effect of short-term hormone replacement therapy on breast cancer risk reduction after bilateral prophylactic oophorectomy in *BRCA1* and *BRCA2* mutation carriers: the PROSE Study Group. J Clin Oncol. 2005;23(31):7804–10.
94. Eisen A, Lubinski J, Gronwald J, Moller P, Lynch HT, Klijn J, et al. Hormone therapy and the risk of breast cancer in *BRCA1* mutation carriers. J Natl Cancer Inst. 2008;100(19):1361–7.
95. Armstrong K, Schwartz JS, Randall T, Rubin SC, Weber B. Hormone replacement therapy and life expectancy after prophylactic oophorectomy in women with *BRCA1/2* mutations: a decision analysis. J Clin Oncol. 2004;22(6):1045–54.
96. Dew JE, Wren BG, Eden JA. A cohort study of topical vaginal estrogen therapy in women previously treated for breast cancer. Climacteric. 2003;6(1):45–52.
97. Bandera CA, Muto MG, Schorge JO, Berkowitz RS, Rubin SC, Mok SC. *BRCA1* gene mutations in women with papillary serous carcinoma of the peritoneum. Obstet Gynecol. 1998;92(4 Pt 1):596–600.
98. Ness RB, Cramer DW, Goodman MT, Kjaer SK, Mallin K, Mosgaard BJ, et al. Infertility, fertility drugs, and ovarian cancer: a pooled analysis of case-control studies. Am J Epidemiol. 2002;155(3):217–24.
99. Chittenden BG, Fullerton G, Maheshwari A, Bhattacharya S. Polycystic ovary syndrome and the risk of gynaecological cancer: a systematic review. Reprod Biomed Online. 2009;19 (3):398–405.
100. Pearce CL, Templeman C, Rossing MA, Lee A, Near AM, Webb PM, et al. Association between endometriosis and risk of histological subtypes of ovarian cancer: a pooled analysis of case-control studies. Lancet Oncol. 2012;13(4):385–94.
101. Jordan SJ, Whiteman DC, Purdie DM, Green AC, Webb PM. Does smoking increase risk of ovarian cancer? A systematic review. Gynecol Oncol. 2006;103(3):1122–9.
102. Collaborative Group on Epidemiological Studies of Ovarian Cancer, Beral V, Doll R, Hermon C, Peto R, Reeves G. Ovarian cancer and oral contraceptives: collaborative reanalysis of data from 45 epidemiological studies including 23,257 women with ovarian cancer and 87,303 controls. Lancet. 2008;371(9609):303–14.
103. Ip S, Chung M, Raman G, Trikalinos TA, Lau J. A summary of the Agency for Healthcare Research and Quality's evidence report on breastfeeding in developed countries. Breastfeed Med. 2009;4(Suppl 1):S17–30.
104. Cibula D, Widschwendter M, Majek O, Dusek L. Tubal ligation and the risk of ovarian cancer: review and meta-analysis. Hum Reprod Update. 2011;17(1):55–67.
105. Tsilidis KK, Allen NE, Key TJ, Dossus L, Lukanova A, Bakken K, et al. Oral contraceptive use and reproductive factors and risk of ovarian cancer in the European prospective investigation into cancer and nutrition. Br J Cancer. 2011;105(9):1436–42.
106. Kauff ND, Satagopan JM, Robson ME, Scheuer L, Hensley M, Hudis CA, et al. Risk-reducing salpingo-oophorectomy in women with a *BRCA1* or *BRCA2* mutation. N Engl J Med. 2002;346(21):1609–15.

Pathology of *BRCA* Tumors

6

Julio A. Ibarra

Introduction

Breast cancer is the most common malignancy in women. The American Cancer Society (ACS) estimates that 246,660 women will develop invasive breast cancer and there will be 40,450 deaths in the USA in 2016 [1]; worldwide, it is estimated that a million women will develop breast cancer every year. Miki et al. estimated that 5–10% of breast cancers are hereditary and are due to mutations in the high-penetrance susceptibility genes *BRCA1* and *BRCA2* [2]. Similar findings have been reported by other authors [3, 4]. Other genes associated with high risk of breast cancer include *p53* (in Li-Fraumeni syndrome) and *PTEN* (in Cowden's syndrome). Several other genes are associated with a small-to-moderate risk and include *CHEK2*, *ATM*, *NBS1*, *RAD51*, *BRIP1*, and *PALB2* [5–9].

BRCA1 mutation is associated with breast and ovarian cancer, while *BRCA2* mutations are associated with male breast cancer. *BRCA1* cancers usually occur in young women. Approximately 48% of *BRCA1* cancers are diagnosed before age 40 compared to 23% for *BRCA2* and 34% for controls. 82% of *BRCA1*-associated cancers occur before age 50, compared to 65% for *BRCA2* and 60% for controls [10]. The proportion of *BRCA1*-related cancer ranges from 5.3% for women less than 40 years of age, to 1.1% for women between ages 50 and 70 [11]. Women with *BRCA1* and *BRCA2* mutations carry an 80–87% risk of developing breast

J.A. Ibarra (✉)
Orange Coast Memorial Medical Center, 9920 Talbert Avenue, Fountain Valley, CA 92708, USA
e-mail: jibarra@memorialcare.org

J.A. Ibarra
Department of Pathology, University of California at Irvine, Irvine, CA 92868, USA

A.B. Chagpar (ed.), *Managing BRCA Mutation Carriers*,
DOI 10.1007/978-3-319-59198-8_6

cancer by age 70 [12–14]. Likewise, they carry a risk of developing ovarian cancer of 60 and 27%, respectively, by age 70 [14–16].

Identifying carriers by genetic analysis is the standard of care. Some patients develop breast cancers of specific types at younger ages than the control population; it is important to analyze the clinical and pathological characteristics to determine whether some of these carriers can be identified based on these criteria.

Hwang et al. reported that familial breast cancer accounts for up to 20% of breast cancer diagnosed in the USA, but fewer than 5% are attributable to mutations in *BRCA1* or *BRCA2* [17].

Treatment patterns of breast cancer, whether genetic or not, have strong tendency toward targeted therapy. For example, patients with hormone receptor-positive tumors will be treated with hormonal blockade (tamoxifen or aromatase inhibitors) and those with overexpression/amplification of ERBB2 (Her2neu) with Her2-targeted agents such as trastuzumab and pertuzumab. To this end, clinicians look for pathology information that will guide them on the best treatment schemes. In this chapter, we will explore the pathology characteristics of tumors occurring in patients that harbor the *BRCA1/2* mutation.

Histopathology of Invasive Breast Carcinoma

Some tumor types have been reported with more frequency in association with hereditary breast cancer such as invasive ductal carcinoma (IDC) of NST, lobular carcinoma, medullary carcinoma, and tubular carcinoma [18–23]. The basic pathologic features of these tumors will be described below.

The classification of breast cancers used in clinical practice today is based on histopathologic features. Molecular classification through gene profiling is being used more frequently but not routinely in clinical practice; the molecular groups include luminal A, luminal B, normal breast-like, Her2-like, and Basal-like [24–28]. Gene expression profiling is not practical for daily use; however, the use of immunohistochemistry has allowed us to classify most lesions and place them into the groups mentioned above. For example, luminal A breast cancers are estrogen receptor (ER)-positive and most are progesterone receptor (PR)-positive (97 and 68%, respectively) [29] and Her2neu-negative with a low proliferative rate (Ki67 < 14%). Likewise, luminal B tumors are ER- and PR-positive (97 and 61%, respectively) [29] and Her2neu-negative or Her2neu-positive but with a high proliferative rate (Ki67 > 14%). Basal cancers are generally ER-, PR-, and Her2neu-negative and express basal cytokeratins (CK5/6, CK14), smooth muscle actin, caveolin-1, P-cadherin, and EGFR among other markers. The triple-negative group has significant overlap with the basal group. Medullary carcinomas and salivary gland-like tumors such as acinic cell carcinoma and adenoid cystic carcinomas fall in this category. Not all basal-like cancers determined by gene expression profiling are triple-negative by IHC, and not all triple-negative cancers are basal-like by gene expression. Bertucci et al. [30] showed that only 71% of

Table 6.1 Histologic grading of invasive carcinoma

	One point	Two points	Three points
Tubules	>75%	10–75%	<10%
Nuclear pleomorphism	Uniform, small nuclei with uniform staining	Moderate pleomorphism	Marked pleomorphism, hyperchromasia, and distinct nucleoli
Mitosis	0–8/10 hpf's	9–17/10 hpf's	>18/10 hpf's

hpf's: high power fields. Number of mitosis depends on microscopic field size. The ranges listed here are for a field diameter of 0.55 mm (Olympus BX series). For a detailed list of mitotic counts from field diameter of 0.40 to 0.69 see the WHO Classification of Tumours of the Breast 4th Ed. [34]

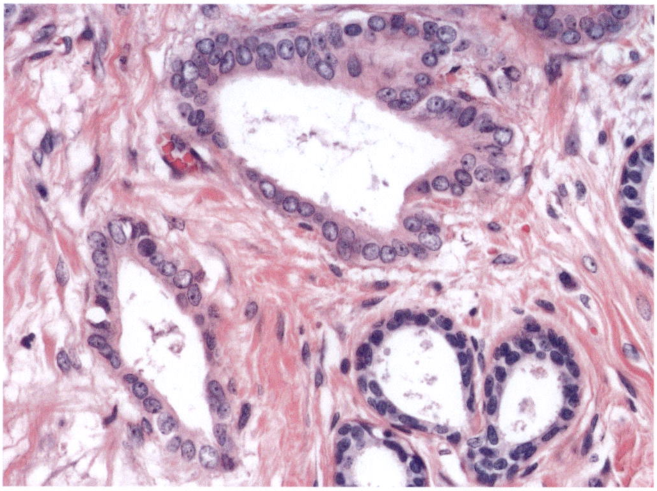

Fig. 6.1 40x magnification of a low-grade invasive ductal carcinoma with >75% tubule formation seen as tumor nests with open lumina, relatively uniform nuclei, and lack of mitotic activity (0–8 per 10 high power fields (hpf's)). Basement membrane is absent in these structures as proven by smooth muscle myosin heavy chain, p63, etc. A specific marker for determining the presence or absence of this structure

triple-negative cancers were basal-like by gene expression profiling and conversely only 77% basal-like tumors by gene profiling were triple-negative by IHC. Others have shown similar results [31, 32]. Lastly, Her2neu enriched are those lesions that overexpress ERBB2 (Her2neu) and do not express luminal markers.

Histologic grading of invasive carcinomas is done using the Nottingham combined histologic grade system [33], also known as modified Bloom Richardson scoring system. Three factors are considered: tubule formation, nuclear pleomorphism, and mitotic activity (Table 6.1).

Each category is given a point value of 1–3 with 1 being the better differentiated/lower grade and 3 the less differentiated/higher grade. The total number of points is added up to arrive at the final score, which can be from a minimum of 3 to a maximum of 9 points. A tumor with a score of 3–5 points is

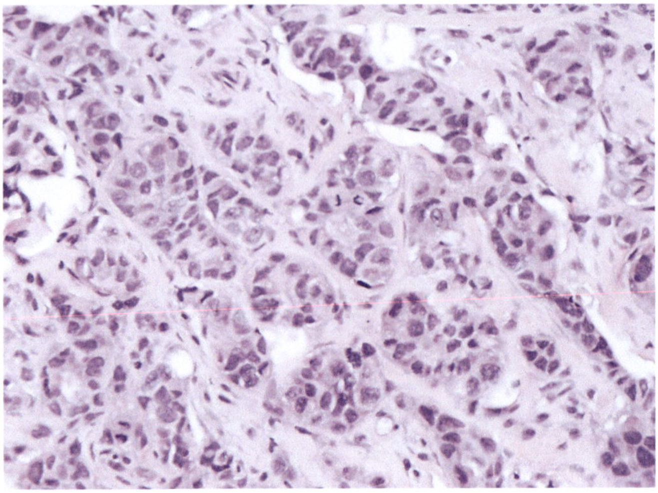

Fig. 6.2 20x magnification of an invasive ductal carcinoma of intermediate grade. Tubule formation between 10 and 75%, nuclei are grade 2 with moderate pleomorphism and 9–17 mitotic figures in 10 high power fields (hpfs)

considered low grade or well differentiated (Fig. 6.1), one with a score of 6 or 7 falls in the intermediate-grade group or moderately differentiated (Fig. 6.2), and those with a score of 8–9 in the high-grade group or poorly differentiated (Fig. 6.3). Histologic grade has been demonstrated to have prognostic significance [35]. *BRCA1* cancers tend to be of higher grade with grade 3 cancers reported in 66–100% of patients [10, 19, 36–42].

Regarding hormone receptors and Her2neu expression, breast cancers in *BRCA1* carriers are more frequently ER-negative [36, 38, 43–46]. The same is true for PR [36, 38, 43, 45, 46].

Invasive Ductal Carcinoma

The most frequent cancer seen in the female breast is IDC of NST, which accounts for approximately 75% of all breast cancers. These tumors represent a heterogeneous population of cancers that lack specific features and therefore are designated as no special type. Invasive ductal carcinoma of no special type have a large diversity of histologic and cytologic patterns ranging from well-formed tubules with open lumina to solid nests, from small to large nests, from low to high-grade nuclei, from infiltrating and spiculated borders to pushing borders, from low to high mitotic activity, from poor to rich lymphocytic infiltration, from rare to extensive lymphovascular invasion, from limited to extensive ductal carcinoma in situ in the background (Fig. 6.4a–h). Likewise, the prognostic markers (ER, PR, and Her2neu) vary significantly; approximately 80% of IDC NST express some degree of estrogen receptor and approximately 15% overexpress Her2neu. In essence, the

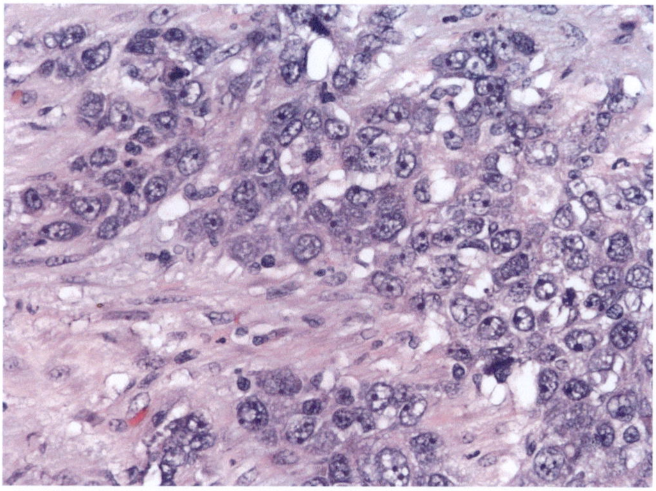

Fig. 6.3 40x magnification of a high grade invasive ductal carcinoma. Tumor grows in irregular nests, and cells have high-grade nuclei with prominent nucleoli. Mitosis amount to more than 18 in 10 high power fields (hpfs)

diagnosis of IDC NST comprises a large number of morphologic and molecular patterns that have few things in common. There are many other invasive carcinomas that are classified, by convention, under the umbrella of the "ductal" lesions but have special morphologic and molecular characteristics and are called IDC of special type. Some of the lesions included in this group are tubular carcinoma, medullary carcinoma, cribriform carcinoma, mucinous or colloid carcinoma, invasive micropapillary carcinoma, adenoid cystic carcinoma, invasive papillary carcinoma, apocrine carcinoma, inflammatory carcinoma, secretory carcinoma, and others. These carcinomas of special types have a wide variety of molecular changes, and their prognosis is significantly different. For example, patients with tubular carcinomas will have a "favorable" type of prognosis with 20-year survival that exceeds 95%, while micropapillary carcinomas have a tendency for early lymphovascular involvement and lymph node metastasis which results in a more guarded prognosis. The special types are rarely encountered in the *BRCA1*-related cancers, except for medullary cancers which are discussed later in this chapter. Some of the special types have been reported in *BRCA2* carriers.

Invasive Lobular Carcinoma

Invasive lobular carcinoma (ILC) accounts for 5–15% of all invasive breast cancers [47]. These tumors can have different patterns of growth such as classic (Fig. 6.5), alveolar (Fig. 6.6), solid (Fig. 6.7), pleomorphic (Fig. 6.8), signet ring (Fig. 6.9), histiocytoid (Figs. 6.10 and 6.11). ILC is often multifocal or diffuse, can affect both

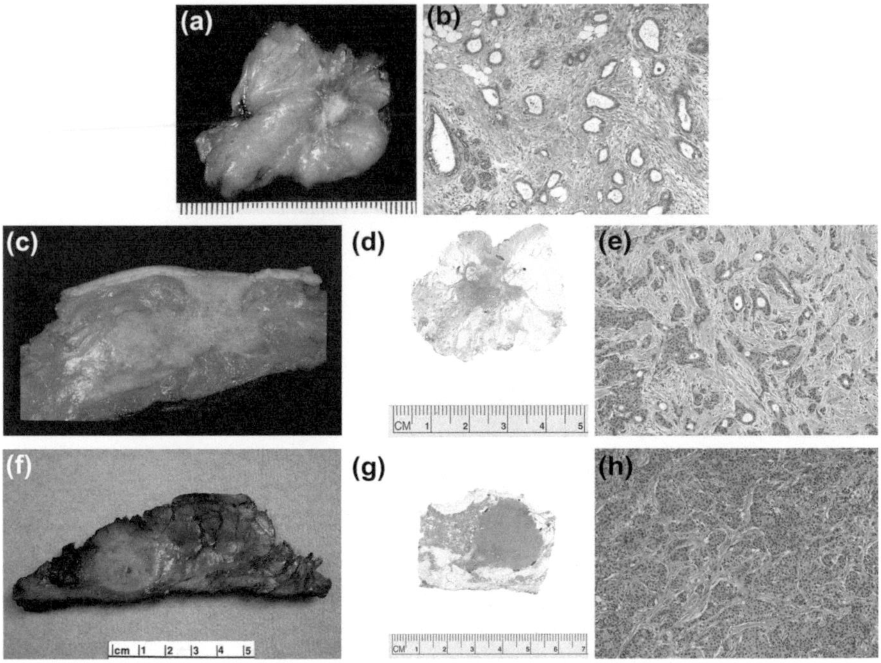

Fig. 6.4 **a** Macroscopic appearance of a low-grade invasive carcinoma with spiculated borders. This lesion is firm when compared to the surrounding tissues. This is an example of a low-grade invasive ductal carcinoma of tubular type. **b** 10x magnification of the lesion is shown in Fig. 6.4a. It represents a low-grade invasive ductal carcinoma. At 8 o'clock, there is an entrapped benign terminal duct lobular unit (TDLU). The tumor grows by forming glands or tubules with open lumina, and the stroma has a significant proliferation of fibroblasts that gives this tumor a firm consistency on palpation. **c** An invasive ductal carcinoma with infiltrative and irregular borders and involving the skin which has been retracted by the fibrosis created by the tumor. This lesion is firm when compared to the adjacent adipose tissue. This is an example of an infiltrating ductal cancer of no special type (NST). **d** This is a whole mount of the case shown in Fig. 6.4c. The advantage of these large format preparations is the ability to see margins and their relationship to the tumor. It allows for better measurement of tumor size and excellent radiology correlation. **e** 10x magnification of a photomicrograph from an invasive ductal carcinoma of intermediate grade. The tumor is growing in solid irregular nests and only a few open tubules, and the nuclei are more pleomorphic and have more mitoses. **f** An invasive ductal carcinoma with round and pushing borders. This tumor is also firm when compared to the surrounding fat but has very well-defined borders. This is an example of a cancer with medullary features. **g** Whole mount (large format histopathology) demonstrating the well-defined borders of the lesion. This is an infiltrating carcinoma; however, it invades by pushing the surrounding tissues with a well-demarcated border. Please note at 7 o'clock that there is a "tongue" of tumor wondering into the surrounding fat. This area has features of IDC NST, while the rest has medullary features. **h** 10x magnification microphotograph of an invasive ductal carcinoma of high grade with solid nests of tumor, high-grade nuclei and more mitoses

breasts, and is often associated with lobular neoplasia in the background. The classic variant, the most frequently seen, is characterized by small cells with round nuclei, which sometimes can have signet ring differentiation. A dense eosinophilic

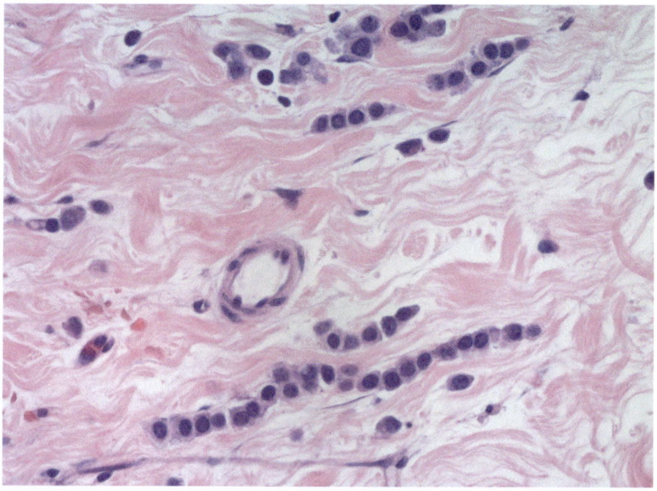

Fig. 6.5 40x magnification of an invasive lobular carcinoma (ILC) with a classic pattern of infiltration and low nuclear grade. The cells are small and arranged in a "single-file" style. Note that the stroma has very little fibroblastic reaction and is mostly composed of dense collagenous fibers

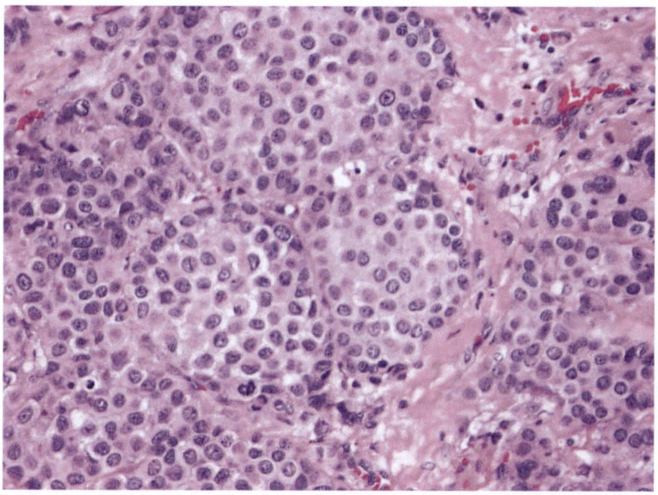

Fig. 6.6 40x magnification microphotograph of an "alveolar" variant of invasive lobular carcinoma (ILC). The cells are very similar to those seen in the "classic" type, and they demonstrate the cell separation (discohesion) that is characteristic of lobular lesions

droplet is sometimes present in the cytoplasm of these cells (Fig. 6.9). They form single files of tumor cells that infiltrate the surrounding breast parenchyma without eliciting a significant desmoplastic response; in fact, the background is often seen as

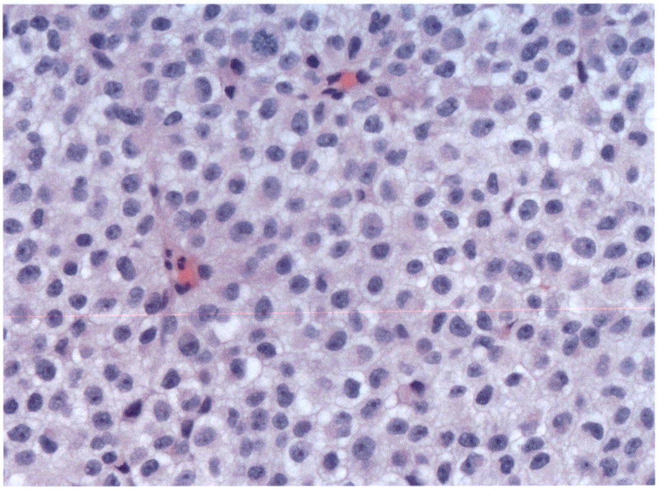

Fig. 6.7 40x magnification microphotograph of a "solid" variant of invasive lobular carcinoma (ILC). The cells are uniform and small, and they show the classic cell discohesion

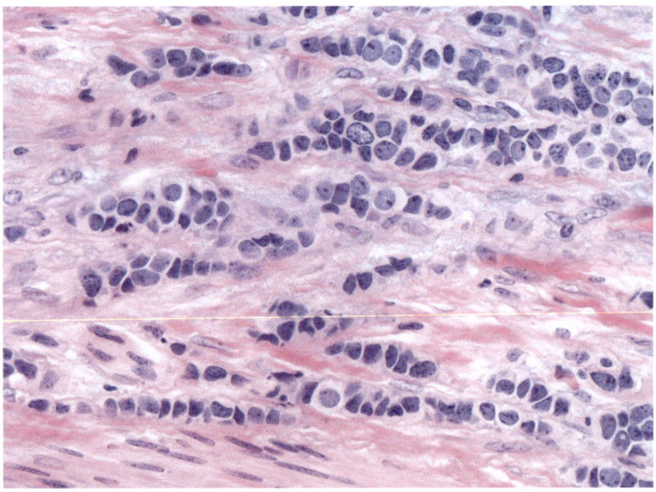

Fig. 6.8 40x magnification photomicrograph of a "pleomorphic" variant of invasive lobular carcinoma. The architectural pattern of infiltration can be classic, alveolar, or solid, but the nuclei are larger and irregular (intermediate to high grade)

dense collagen with little fibroblastic reaction. Frequently, the tumor cells infiltrate discreetly around adipocytes and can form concentric rings around preexisting acini (onion skin pattern). Mitotic activity is often low, and lymphovascular invasion is rare. It tends to grow in a diffuse pattern, and often, it is difficult to determine whether it is one lesion connected by thin strands or multiple lesions diffusely

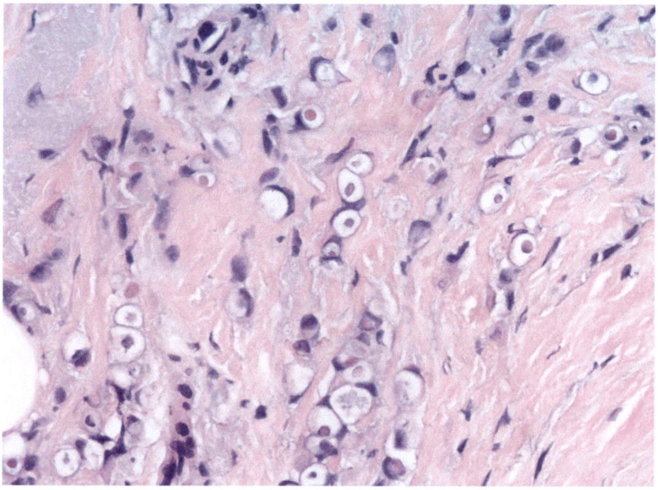

Fig. 6.9 40x magnification microphotograph of an invasive lobular carcinoma with cytoplasmic vacuoles giving them a "signet ring" appearance. These cells also have the classic dense eosinophilic droplets than can be seen some times in these lesions

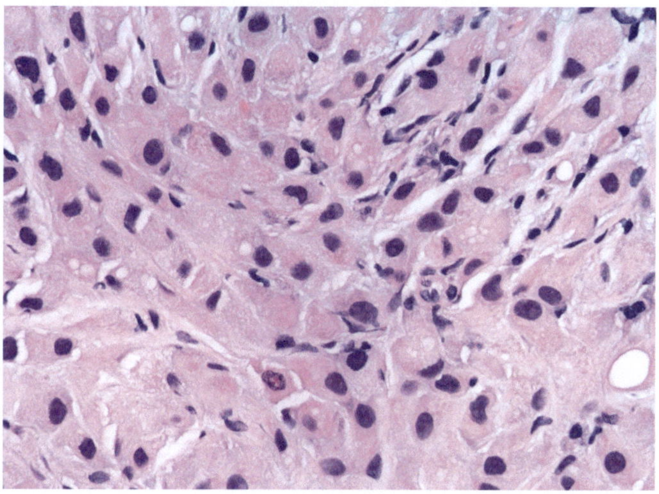

Fig. 6.10 40x magnification microphotograph of an invasive lobular carcinoma that demonstrates large amount of eosinophilic and somewhat "foamy" cytoplasm seen in the "histiocytoid" variant

involving the breast and not connecting with each other. Sometimes, they are also seen as a single solid spiculated mass. At the molecular level, lobular lesions have loss of expression of the adhesive molecule epithelial cadherin (E-Cadherin), which results in the classic non-cohesive appearance microscopically. The E-Cadherin gene is located in chromosome 16; lack of expression is due to loss of

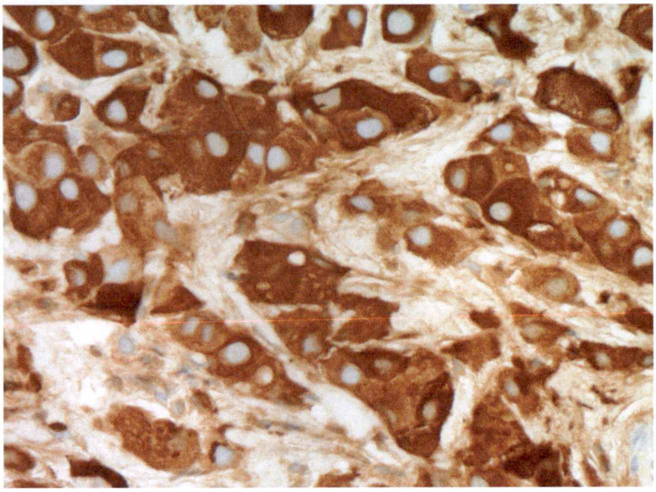

Fig. 6.11 40x magnification. Gross cystic disease fluid protein (GCDFP-15) immunostain. Histiocytoid and apocrine lesions usually stain very intensely with this marker

heterozygosity in 16q22.1 and mutation or methylation of the *CDH1* promoter. Loss of E-Cadherin is easily identified by immunohistochemistry (IHC). However, approximately 15% of lobular carcinomas express E-Cadherin [48, 49]. In cases where the histology is classic for lobular carcinoma and the E-Cadherin is positive, histology overrules IHC and the case is considered ILC with aberrant E-Cadherin staining. The solid and alveolar variants have similar cells to the classic type. The architecture is different. The alveolar variant (Fig. 6.6) grows in small rounded nests of cells surrounded by a thin fibrovascular layer. Lack of cell cohesion is also a feature in this variant. Solid variant (Fig. 6.7) grows in a confluent solid sheet of tumor cells with no stroma; in addition to having the same cytology as the classic variant, poor cell cohesion is also a feature. The pleomorphic variant (Fig. 6.8) of lobular carcinoma grows mostly in a "classic" pattern, but the cells have nuclear enlargement and variability in shape and staining quality. These tumors tend to be of higher grade, and while they have most of the genetic alteration seen in classic ILC, they can have a larger number of changes. Some pleomorphic lesions can overexpress Her2neu. The literature is confusing because many authors have combined immunohistochemical results of 2+ and 3+ as positive for overexpression leading to a positive rate between 40 and 80% [50–52]. We know, however, that the majority of Her2neu cases with a score of 2+ by IHC will not be amplified by FISH (64%) [53]. Classic forms of lobular carcinoma are rarely Her2neu-positive. However, Her2neu overexpression/amplification can be associated with pleomorphic lesions [54–56]. When only a score of 3+ by immunohistochemistry is considered a positive result, the number of cases classified as having overexpression drops significantly with some series reporting as low as 8% [57]. The histiocytoid variant (Fig. 6.10) falls in the category of pleomorphic lesions and has cells with abundant eosinophilic cytoplasm which stain strongly with GCDFP-15 (Fig. 6.11), a marker for apocrine/histiocytoid cell differentiation.

Medullary Carcinoma

Medullary carcinomas are uncommon tumors in the general population accounting for less than 1% of all invasive breast carcinomas. The original criteria were established by Ridolfi et al. [58] who showed better clinical outcomes at 10 years. These tumors have a well-defined rounded appearance and soft consistency on gross examination (Fig. 6.12). The classic histopathologic features include prominent lymphoplasmacytic infiltration, syncytial growth pattern in more than 75% of the tumor, pushing borders, and grade 2–3 nuclei (Fig. 6.13a, b). Tumors with some but not all of the features described in the original paper have been called atypical medullary carcinomas or invasive carcinoma with medullary features (Fig. 6.14a, b). Metaplastic changes may be seen, particularly squamous metaplasia. Because reproducibility has been a significant problem with these lesions, the WHO proposed to combine all "variants" into one category called carcinoma with medullary features [59]. These tumors are seen in younger patients and have been strongly associated with patients under age 50 that have *BRCA1* germline mutation. They are usually ER- and PR-negative and do not have overexpression/amplification of Her2neu. By molecular studies, they fall into the basal group and express basal cytokeratins (CK5/6, CK14, and CK17) as well as epidermal growth factor receptor (EGFR). The good prognosis of medullary carcinoma has been associated in part to the prominent lymphoplasmacytic infiltration.

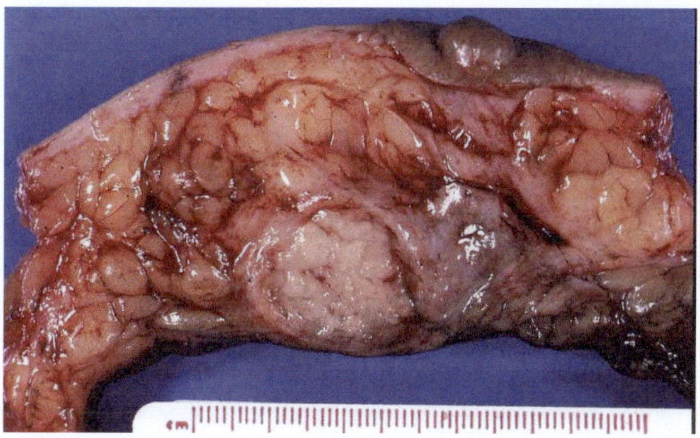

Fig. 6.12 Macroscopic image of a medullary cancer. These lesions are soft and well circumscribed

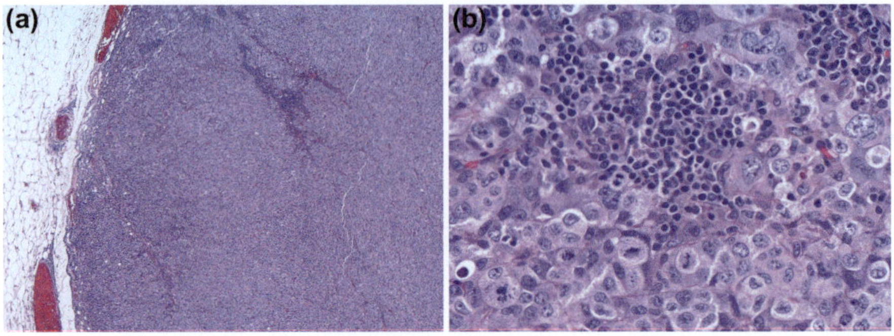

Fig. 6.13 **a** 4x magnification of a medullary cancer with well-defined, pushing borders, a syncytial pattern of growth and prominent lymphoid component. **b** 40x magnification microphotograph of a medullary carcinoma. These tumors have high-grade nuclei with prominent nucleoli and a significant lymphocytic component

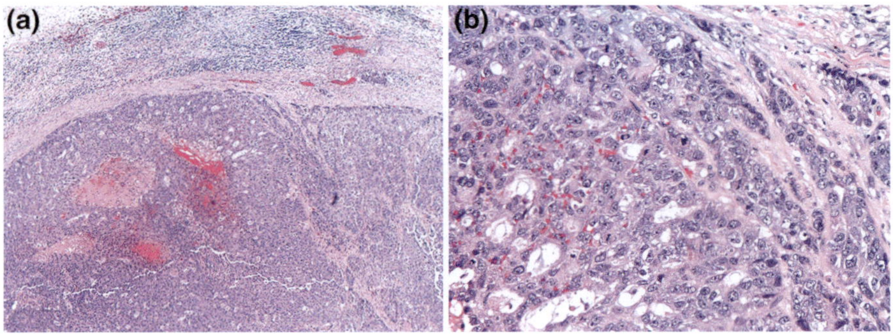

Fig. 6.14 **a** 4x magnification of an invasive ductal carcinoma with medullary features. Note the well-circumscribed border and the prominent lymphoid component. **b** This lesion has also high-grade nuclei but has tongues of infiltrating ductal carcinoma like the ones seen in IDC NST. The new WHO classification includes all these lesions under the category of invasive carcinoma with medullary features [59]

Tubular Carcinoma

Tubular carcinoma accounted for less than 4% of invasive breast carcinomas before the mammographic era but has become more prevalent with the use of screening mammography [60]. They are a low-grade invasive ductal lesion characterized by the formation of small glands with low- to intermediate-grade nuclei and open lumina (Fig. 6.4a, b). The glands are usually oval and often exhibit angulated shapes and grow in a haphazard fashion; the cells can have apocrine snouts. In the past, these classic glands with open lumina were required to be present in >75% of the tumor to be classified as pure tubular carcinomas [61, 62]. However, more recently the WHO classification suggests that they have to be present in >90% of the tumor in order to use the term of tubular carcinoma [60]. Tubular carcinomas are almost always ER- and PR-positive and Her2neu-negative and fall within the luminal A group. The prognosis of this tumor is excellent and thus is important to apply the recommended criteria to distinguish them from an IDC NST of low grade which will not have as good a prognosis. Lymph node metastases are rarely seen in tubular carcinomas (average of 10%, range 0–22%), [60], and when they occur, lymph node metastases are usually limited to one or two nodes [63].

The remainder of the special types of invasive carcinoma include colloid, adenoid cystic, cribriform, papillary. These do not have a significant association with *BRCA1/2* and therefore will not be discussed in this chapter.

Triple-Negative Breast Cancer

Triple-negative breast cancer (TNBC) is defined as a tumor that does not express estrogen receptor, progesterone receptor, or Her2neu protein. Approximately 10–20% of all breast cancers are TNBC [64, 65]. TNBC occur more frequently in younger women and are usually higher grade [66, 67]. Patients with TNBC have a worse prognosis [65, 68]. Tumors that have a triple-negative phenotype by IHC represent a heterogeneous group of breast cancers; when they are studied by gene profiling, only 71% fall into the basal-like group [30]. Conversely, only 77% of the basal-like tumors classified by gene profiling are triple-negative by IHC. Other authors have shown similar results [31, 32].

Histopathology of *BRCA1* Tumors

General Comments

The most frequent cancer seen in this group of patients is invasive ductal carcinoma (74%). Some but not all *BRCA1* tumors have distinct pathologic features such as more triple-negative breast cancers (TNBC) and higher nuclear grade tumors than the general population. In contrast, *BRCA2* tumors have pathologic features that are similar to cancers seen in non-carriers. *BRCA1*-related cancers occur in younger patients; according to Lakhani et al., 48% occur before the age of 40 [10]. Lakhani compared some histologic features seen in *BRCA1* and *BRCA2* and found that *BRCA1* carriers had significantly greater scores for mitotic count, lymphocytic infiltrate, continuous pushing margins, solid sheets of cells, and necrosis [21]. Other authors have reported well-defined borders, a prominent lymphoplasmacytic infiltrate, and more lymphovascular involvement [69]. Eerola et al. [70] reported that the classic features associated with *BRCA1*-related cancers (high grade, negative ER/PR, and positive *P53*) are limited to patients under age 50. Cancers found in *BRCA1* patients older than 50 differed only in grade from tumors in non-*BRCA1/2* patients.

Table 6.2 Major histologic types seen in *BRCA* carriers [10, 70, 71, 74]

	IDC NST (%)			ILC (%)			Medullary (%)		
	BRCA1	*BRCA2*	Control	*BRCA1*	*BRCA2*	Control	*BRCA1*	*BRCA2*	Control
Lakhani et al. [10]	74	76	74	3	10	10	13	3	2
Mavaddat et al. [71]	80	83		2.2	8.4		9.4	2.2	
Lakhani et al. [74]	78	77		3	9		11	2	
Eerola et al. [70]	72.5	62.7	67.1	15.7	28.8	19.7	9.8	–	2.9

IDC NST Invasive ductal carcinoma of no specific type; *ILC* Invasive lobular carcinoma
Series showing similar numbers of IDC NST in both *BRCA1* and *BRCA2* and an increase in the number of lobular carcinomas in *BRCA2* carrier cancers and an increase in medullary carcinomas in *BRCA1* carriers. When broken by age, the medullary cancers occur in the group under age 50 (see Table 6.3)

Table 6.3 Relationship of age to histopathologic features in *BRCA1* and *BRCA2* carriers

	BRCA1		*BRCA2*		Non-*BRCA1/2*	
	Age <50 (%)	Age ≥ 50 (%)	Age <50 (%)	Age ≥ 50 (%)	Age <50 (%)	Age ≥ 50 (%)
Ductal	70.6	76.5	54.3	75	58.6	72.3
Lobular	14.7	17.6	34.3	20.8	25.9	16.0
Medullary	14.7				3.4	1.1
Grade 3	84.4	47.1	23.3	31.8	17.3	23.3

Relationship of patient's age to major histologic types and grade. Not a significant difference in the incidence of invasive ductal cancers. There appears to be a slight increase incidence of lobular carcinomas in *BRCA2* under age 50. Medullary carcinomas appear to have their higher frequency in patient's under age 50 in the *BRCA1* group. Modified from Eerola et al. [70]

Histologic Types

The majority of cancers occurring in *BRCA1* carriers are IDC NST (74–80%), similar to *BRCA2* cancers (76–83%) [10, 71]. However, the proportion of ILC is higher in *BRCA2* carriers (8.4–10%) than in *BRCA1* (2.2–3%) (Table 6.2).

There are many reports in the literature indicating that there are more medullary cancers in *BRCA1* than *BRCA2* carriers (Table 6.2). These are overrepresented in *BRCA1*-related cancers [19, 21, 72–75]. Using multivariate analysis, the features that characterize *BRCA1* cancers include pushing borders, lymphocytic infiltration, and high mitotic activity; many of these cancers are not necessarily the classic medullary but medullary-like as defined by the WHO. On the other hand, *BRCA2* cancers were shown to have pushing borders and poor tubule formation [21]. When age is analyzed, the increased incidence of medullary cancers is only seen in patients <50 years of age and not in those older than 50 [70]. Table 6.3 shows the relationship between histopathologic features and age for *BRCA1* and *BRCA2* carriers as reported by Eerola et al. [70]. ER/Her2neu-negative phenotype is associated with the presence of necrosis [76]. Areas of necrosis are more frequent in *BRCA1*-associated cancers than in sporadic breast cancers [21, 77]. This phenotype has also been associated with medullary-like histologic patterns [78, 79].

Histologic Grade

Both *BRCA1* and *BRCA2* cancers tend to be higher grade than sporadic breast cancers. *BRCA1* tumors grow in solid nests and do not form tubules with open lumina, have high-grade nuclei with prominent pleomorphism, more frequently have high mitotic count and necrosis, all suggesting a more aggressive lesion [19, 21, 22, 75]. The CIMBA report [71], an international collaboration with 37 groups from 20 countries including 3797 *BRCA1* carriers, also shows that *BRCA1*-related tumors are of higher grade and with higher mitotic counts than age-matched controls. This group has also demonstrated that grade decreases with increasing age [71].

When the three elements of the grading system were analyzed, pleomorphism and increased mitotic activity were seen more often in *BRCA1* than *BRCA2* tumors [10].

High-grade lesions are also associated with ER-/Her2neu-negative phenotype, a pattern seen frequently in *BRCA1* cancers [80]. Tumor size and lymph node status are well-established parameters to judge prognosis in patients with breast cancer. The correlation between size of tumor and positive lymph nodes has been well documented [81–86]. This correlation is seen in *BRCA2* carrier cancers but not in *BRCA1* [87]. This lack of correlation is not due to small *BRCA1* cancers with positive nodes but rather due to large *BRCA1* cancers with negative nodes.

Gene Profiling

Basal phenotype is found in approximately 15% of breast cancers [88]. Basal phenotypes by IHC are ER- and Her2neu-negative. A distinct finding in these lesions is the expression of complex keratins (CK 5/6, CK 14, CK 15, and CK 17) which are present in the basal epithelium of the breast and other organs. A basal phenotype is often seen in *BRCA1*-associated cancers [89]. In fact, most *BRCA1*-associated carcinomas express basal or myoepithelial markers including caveolin 1 [90], CK5/6 and CK14 [91], vimentin and laminin [92], P-cadherin [93], and EGFR [94, 95]. Other basal markers include smooth muscle actin, glial fibrillary acidic protein, and calponin. Foulkes et al. [91] reported a basal-like phenotype (positive CK5/6 by IHC) in 45% of non-*BRCA1/2* carrier patients with ER-/Her2neu-negative tumors. In contrast, they found an 88% positivity in patients with ER-/Her2neu-negative tumors and *BRCA1/2* mutation.

Table 6.4 Relationship of patient's age to receptors and Her2neu status

	BRCA1 Age-groups		*BRCA2* Age-groups		Non-*BRCA* Age-groups	
	<50	>50	<50	>50	<50	>50
ER-negative	83.3	25.0	20.6	52.6	29.3	25.6
PR-negative	90.3	69.2	35.3	80.0	31.0	54.4
Her2neu-negative	76.7	92.3	83.3	84.2	81.8	83.1

Expression of receptors are significantly different in patients diagnosed before or after age 50. ER- and PR-negative cancers are seen more frequently in patients younger than 50 in the *BRCA1* group. Interestingly, there is an increase in PR-negative tumors in patients over age 50 in the *BRCA2* group. There is no significant change in the Her2neu status by age in any of the groups. Modified from Eerola et al. [70]

Immunophenotype (ER/PR/Her2neu)

BRCA1 and *BRCA2* cancers are more likely to be ER-, PR-, and Her2neu-negative and to be high grade [10, 21, 46, 70, 91, 96–100]. Some have estimated that 70–90% of *BRCA1*-associated tumors are ER-/PR-negative compared to a 30% rate for sporadic breast cancers [101], while *BRCA2* cancers have an immunophenotype similar to sporadic cancers (see below).

Eerola et al. [70] reported a significant difference in ER and PR expression between patients diagnosed before or after age 50 both in *BRCA1*- and in *BRCA2*-related cancers. They found a higher incidence of negative results in those younger than 50 in the *BRCA1* group and a higher incidence of negative results in those older than age 50 in the *BRCA2* group (see Table 6.4). There were no significant differences with age in the non-*BRCA1/2* cancer group. Her2neu-negative results do not appear to be associated with age. Her2neu overexpression rates are lower in *BRCA*-related tumors [44, 73, 75, 77, 97]. Lakhani et al. [97] reported that 3% of *BRCA1* and *BRCA2* cases are Her2neu-positive, compared with 15% of sporadic breast cancers.

Triple-Negative Breast Cancers/Basal-like Features

TNBC have been reported by several authors to account for 50–88% of *BRCA1*-related cancers [102–105]. TNBC are more common in *BRCA1* than *BRCA2* carriers. In a population-based study with 1460 patients (ages 20–49), Lee et al. reported that 48% of *BRCA* carriers had TNBC, compared with 12% in non-carriers [67]. They found that 69% of Ashkenazi women with *BRCA1* mutation had TNBC, while non-carrier Ashkenazi women had a similar proportion of TNBC (8%) as other ethnic/religious groups (13%). A number of investigators have similarly found that *BRCA1* carriers have a higher incidence of triple-negative breast cancers and a basal phenotype than *BRCA2* carriers and non-mutation controls [71, 91, 99, 106].

Non-*BRCA* carriers have similar percentages of TNBC to *BRCA2* carriers. Atchley reported TNBC in 57.1% of *BRCA1*, 23.3% of *BRCA2,* and 13.8% of non-*BRCA* patients [46]. *BRCA1* tumors fall into the basal-like subgroup with ER-negative, poor prognosis, and expression of basal markers. IHC studies have shown that *BRCA1* cancers expressed basal markers, whereas *BRCA2* tumors rarely do this [39, 91, 93, 107–110]. P-Cadherin, also a basal marker, has been reported to be increased in *BRCA1* cancers more frequently than in *BRCA2* and sporadic cancers [39, 93, 108]. SPARC, caveolin-1, and fascin are other basal markers that have been reported more frequently in *BRCA1* cancers [90, 109]. *BRCA1* cancers have basal phenotype, while *BRCA2* carcinomas have immunohistochemical features of luminal phenotype [109]. There is more EGFR overexpression in *BRCA1* tumors [45, 91, 108, 110, 111], a feature that has been reported in basal-like cancers.

Tumor Protein 53 (*P53*)

TP53 (tumor protein 53) is a tumor suppressor gene located on chromosome 17p13.1. It is mutated in the Li-Fraumeni syndrome, an autosomal dominant condition that predisposes to breast and other forms of cancer. *P53* inhibits cell cycle progression and facilitates apoptosis [112]. *P53* mutations are seen in 20–40% of sporadic breast cancers. This mutation has been reported more often in *BRCA1* cancers [37, 39, 97, 113] than in *BRCA2* cancers. Approximately 30–77% of *BRCA1* cancers have *P53* protein expression by immunohistochemistry compared to 20–63% of *BRCA2* cancers [37, 39, 40, 75, 97, 99, 113, 114] and 20% of the control population. *P53* mutations are often associated with higher histologic grade, increased mitotic activity, and worse clinical behavior [112, 115].

Premalignant Lesions/Prophylactic Mastectomy

Premalignant lesions in prophylactic mastectomies of *BRCA1* mutation carriers include an increased incidence of DCIS [116–118], LCIS [116], and atypical ductal (ADH) and lobular hyperplasia (ALH) [116–119]. The lymphoplasmacytic infiltrate seen in invasive tumors of *BRCA1* carriers is also seen in association with DCIS and within normal terminal ductal lobular units with T-cell lobulitis [119]. Hoogerbrugge et al. [116] reported a 57% incidence of atypical lesions (ALH 37%, LCIS 25%, ADH 39%, and DCIS 15%) in *BRCA* carriers. These lesions were seen more frequently in women older than 40 years and less frequently in patients who had oophorectomy. Kauff et al. [118] reported similar findings (ALH 13%, LCIS 4%, ADH 38%, and DCIS 13%); they also reported columnar cell change in 33% and sclerosing adenosis in 38%. Benign and non-atypical lesions such as cyst formation (83%), apocrine metaplasia (63%), and fibroadenomatoid change (46%) were prevalent but did not occur in frequencies that were significantly different than controls. Benign and non-atypical changes such as usual ductal hyperplasia, fibroadenomas, and lobulitis have also been described in increased numbers in carriers [119]. Lobulitis was defined as 100 or more lymphocytes or plasma cells in a lobule and was seen in 51% of carriers versus 10% of controls; other atypical lesions were found in 12% of carriers' prophylactic mastectomies versus 1% for non-carriers.

Other authors have reported fewer atypical and non-atypical lesions in prophylactic mastectomies of *BRCA* carriers [120] than controls (proliferative fibrocystic change: 7% in carriers vs. 22% in controls, ADH: 4% in carriers vs. 2% in controls, and lobular neoplasia: 0% in carriers vs. 7% in controls).

Claus et al. [121] has reported an increased prevalence of *BRCA1* (0.8%) and *BRCA2* (2.4%) mutation in patients diagnosed with DCIS.

Presence of DCIS in the Background

DCIS was reported as rare in *BRCA1* by some [17]. Lakhani et al. [10] in the Breast Cancer Linkage Consortium reported that DCIS within invasive carcinoma was less frequently seen in *BRCA1* and *BRCA2* cancers than in the control group (41% vs. 52% vs. 56%, respectively). Similar results were reported later by Lakhani et al. [21].

Hwang et al. [17] found DCIS alone or with invasive tumor in 28% of *BRCA* carriers versus 34% in high risk but non-carrier families. There was no difference in the numbers of DCIS-only events in the carriers versus non-carriers. They also found that high-grade DCIS was more frequent in *BRCA1* than *BRCA2* carriers or non-carriers (69, 40, 46%, respectively); furthermore, they found no low-grade DCIS in mutation carriers.

Other authors have reported DCIS is less frequently seen around the invasive cancer in *BRCA1* carriers than in controls [10, 122].

Proliferation, Apoptosis and Cell Cycle

BRCA1 tumors tend to have high levels of Ki67 [39, 97] and caspase 3, but low levels of BCL2 [39]. Caspase 3 activation is higher in high-grade and ER-negative tumors [123] which have expression of ER similar to sporadic breast cancers. BCL2 and Cyclin D1 are ER-associated genes and therefore are expected to be and usually are negative in *BRCA1* cancers [124, 125].

Cyclin D1 regulates progression from G1 to S phase in the cell cycle. Patients with high levels of estrogen show overexpression of Cyclin D1. Osin et al. [126] found 14% overexpression of Cyclin D1 in *BRCA1/2* cancers versus 35–36% in sporadic invasive and in situ ductal cancers.

p21 is a cyclin-dependent kinase inhibitor whose function is to block transition from G1 to S phase and suppresses cell proliferation. Data from a number of clinical studies are conflicting regarding p21 [112]. More recent studies suggest that p21 has an "antagonistic duality" inhibiting apoptosis (procancer), while having anti-proliferative effects (anticancer) [127]. Proteins that promote cell cycle progression such as Cyclins E, A, and B1 are increased in *BRCA1* cancers [128–130].

The transcription factor HIF-1 plays an important role in cellular response to oxygen levels [131, 132]. Hypoxia induces overexpression of this protein in patients with *BRCA1* tumors [133].

Molecular Genetics

BRCA1 cancers have genetic changes very similar to sporadic basal-like cancers [99, 134–136].

Clinical Implications

The histopathologic features combined with age at diagnosis can be strong indicators for genetic testing. Therefore, the finding of a high grade an ER-negative tumor in a patient younger than 35 years of age, suggests that she is a candidate for *BRCA* mutation analysis [137].

Similarly, Young et al. [138] concluded that young women with a high-grade triple-negative cancer and no family history of cancer may be candidates for genetic testing. However, women with other histologic forms of cancer (e.g., ER-positive or HER2-positive) and with no family history are unlikely to carry a mutation.

Even though it is mentioned in several portions of the text, it is important to emphasize that the "classic" features described for *BRCA1* tumors are age related and some of them are not seen as frequently in patients over the age of 55 [70, 99, 139].

Histopathology of *BRCA2* Tumors

General Comments

BRCA2-associated tumors represent a more heterogeneous group of cancers than *BRCA1* without a specific immunophenotype. The most frequent cancer seen in this group of patients is invasive ductal carcinoma (74%). While *BRCA1* tumors have distinct pathologic features such as more TNBC and higher nuclear grade tumors than the control population, *BRCA2* tumors have pathologic features that are similar to cancers seen in non-carriers.

Histologic Types

The majority of *BRCA2* cancers are invasive ductal carcinomas [10]. An increase incidence of lobular carcinomas, particularly invasive pleomorphic lesions, as well as tubular and cribriform carcinomas has been reported by some [18, 19, 21–23, 75]. The Consortium of Investigators of Modifiers of *BRCA1/2* [71] also found more lobular carcinomas in *BRCA2* than *BRCA1* carriers (8.4% in *BRCA2* vs. 2.2% in *BRCA1*).

Several authors have failed to demonstrate a specific histologic subtype in this group [10, 18, 21].

Eerola et al. [70] reported that invasive lobular carcinomas were slightly more frequently seen in the younger *BRCA2* (34.3% vs. 20.8%) (see Table 6.3).

On multivariate analysis, the features associated with *BRCA2* cancers were decreased tubule formation, lower mitotic rates, and pushing borders [21]. Because of the histologic features described, *BRCA* cancers tend to be moderately to poorly differentiated [10, 18, 37, 39].

Histologic Grade

BRCA2 tumors are more frequently intermediate- or high-grade tumors [18, 21, 37, 39].

Both *BRCA1* and *BRCA2* cancers tend to be higher grade than sporadic breast cancers. While both *BRCA1* and *BRCA2* tumors usually have no tubules, *BRCA1* tumors more often have high-grade nuclei (prominent pleomorphism) and increased mitotic activity, while *BRCA2* cancers usually have less nuclear pleomorphism or mitotic activity [21].

The Consortium of Investigators of Modifiers of *BRCA1* [71] reported that *BRCA2* tumors were of higher grade than age-matched controls, but the grade did not decrease with increasing age like it did in *BRCA1*. Similar to *BRCA1* cancers, *BRCA2* tumors more often have pushing borders when compared with controls [20, 21].

Gene Profiling

When using gene expression profiles, most *BRCA2* tumors are classified as luminal.

In contrast to *BRCA1* cancers, the *BRCA2* tend to express luminal cytokeratins (CK8, CK18) and not the basal ones expressed by *BRCA1* (CK5/6, CK14) [130].

Immunophenotype (ER/PR/Her2neu)

The immunophenotype of *BRCA2* cancers is similar to sporadic breast cancers. When compared to *BRCA1* cancers, they express more luminal markers and consequently are more often positive for ER alpha and PR [18–20, 70, 75, 77, 91, 97]; *BRCA2* cancers have ER- and PR-negative rates similar to sporadic breast cancers [39, 77, 97, 108, 140, 141]. Mavaddat et al. [71] reported a decrease in the percentage of ER-/PR-positive tumors in *BRCA2* carriers with increasing age, a phenomenon that is opposite to that seen in *BRCA1* carriers. They observed that the frequency of ER-negative tumors increased with age at cancer diagnosis; Her2 status did not vary with age at cancer diagnosis. The proportion of TNBC increased with age at cancer diagnosis. Grade 3 tumors were less likely to be ER-positive than grade 1 tumors. Regarding morphology and receptors, Mavaddat et al. found more ER-positive tumors in all morphologic types of *BRCA2* carriers than *BRCA1* (IDC 77% vs. 22%, ILC 88% vs. 57%, Medullary 48% vs. 11%).

For *BRCA2* cancers, Eerola et al. [70] reported that ER and PR negativity varied with age (20.6% vs. 52.6% for ER and 35.3% vs. 80.0% for PR in the <50 and the $\geq$ 50 year old group, respectively). There were no significant differences with age in the non-*BRCA1/2* cancer group (see Table 6.4).

BRCA2 tumors are usually Her2neu-negative [39, 97, 142], and Her2neu expression is equally low in *BRCA1* than *BRCA2*-related cancers [75, 77, 97].

Her2neu overexpression (3+) in *BRCA1/2* cancers is lower than that seen in the general population; the rates are 0–3.7% [39, 97].

Tumor Protein 53 (*P53*)

In *BRCA2* cancers, the rates of positive *P53* differ significantly between studies with some reporting 20% positive IHC [39] and others up to 50% [97].

Crook et al. reported a positive *P53* in 45% in *BRCA2* patients compared with 77% in *BRCA1* patients and 35% in sporadic breast cancers [113].

Premalignant Lesions

The incidence of DCIS and LCIS in *BRCA2* patients is similar to the control population; 52% versus 56% for DCIS and 3% versus 6% for LCIS, respectively [10]; others have found similar results [143, 144]. Bane et al. [20] found DCIS in 71% of *BRCA2* versus 69% in the control group.

Van der Groep et al. reported that the immunophenotype of accompanying DCIS in *BRCA2* patients is the same as that seen in the invasive component [145].

Proliferation, Apoptosis, and Cell Cycle

According to some, *BRCA2* cancers have similar expression of cell cycle proteins to that seen in sporadic breast cancers [130]. However, Osin et al. [126] found 14% overexpression of Cyclin D1 in *BRCA1/2* cancers versus 35–36% in sporadic invasive and in situ ductal cancers. Cyclin D1, a protein upregulated by estrogen, is seen more often in *BRCA2* than *BRCA1* cancers in series reported by Armes et al. [77] and Palacios et al. [130].

BRCA2 tumors show overexpression of apoptosis markers such as BAX and BCL2 [123, 146] confirming the correlation with ER-positive tumors. In one paper, there was more amplification of the c-myc gene in *BRCA2* than *BRCA1* cancers (62% vs. 18%) [39].

Molecular Genetics

The chromosomal gains and losses seen in *BRCA1* are different to those seen in sporadic cancers, and the gains and losses seen in *BRCA2* cancers are similar to the sporadic ones [147–149].

DCIS in *BRCA*

Hwang et al. reported an equal prevalence of DCIS in *BRCA* carriers as in women with a high family risk who are non-carriers (37% vs. 34%), but the age of presentation is earlier for carriers [17]. Also, they reported a higher incidence of high-grade DCIS for carriers than non-carriers.

Yang et al. [150] reported the presence of DCIS in 80.2% of all invasive cancers. The percentages of pure DCIS versus DCIS with the invasive component were very similar for *BRCA1* and *BRCA2* patients (21% vs. 63% for *BRCA1* and 23% vs. 61% for *BRCA2*).They found the DCIS to be high grade and mostly solid with necrosis and cribriform types. They found that the majority of the invasive and DCIS were triple-negative in *BRCA1* cancers and ER-/PR-positive and Her2neu-negative for *BRCA2* cancers. Even though the majority of *BRCA1* cancers were triple-negative, they found tumors were ER- and PR-positive in 20.8% of DCIS-associated cases.

Non-*BRCA1/2* Breast Cancers

These are believed to be associated with multiple low-penetrance genes. Even though they are described to be of lower histologic grade than sporadic cancers, their immunophenotype is very similar to sporadic cancers [39, 74, 75, 140].

Approximately 67–78% are invasive ductal carcinomas [39, 40, 74]. Invasive lobular carcinoma was found in 15% of non-*BRCA1/2* cancers [39, 74]. In comparison, ILC was seen only in 3% of *BRCA1* and 9% of *BRCA2* cancers. Non-*BRCA1/2* cancers are lower grade with more tubule formation, less nuclear pleomorphism, and lower mitotic activity [39, 74]. Honrado's group [140] demonstrated by IHC that non-*BRCA1/2* cancers may be classified into the 5 subgroups originally described by expression profiling analysis [88, 89].

Acknowledgements This work would not have been possible without the help of Mrs. Elizabeth Mason-Renteria from the MemorialCare library at Long Beach who helped me in obtaining most of the papers used as references for this chapter.

References

1. Siegel RL, Miller KD, Jemal A. Cancer statistics, 2016. CA Cancer J Clin. 2016;66:7–30.
2. Miki Y, Swensen J, Shattuck-Eidens D, et al. A strong candidate for the breast and ovarian cancer susceptibility gene *BRCA1*. Science. 1994;266:66–71.
3. Wooster R, Bignell G, Lancaster J, et al. Identification of the breast cancer susceptibility gene *BRCA2*. Nature. 1995;378:789–92.
4. Rahman N, Scott RH. Cancer genes associated with phenotypes in monoallelic and biallelic mutation carriers: new lessons from old players. Hum Mol Genet. 2007;16(Spec No. 1): R60–6.
5. McKinnon PJ. *ATM* and ataxia telangiectasia. EMBO Rep. 2004;5:772–6.
6. Da Silva L, Lakhani SR. Pathology of hereditary breast cancer S49. Mod Pathol. 2010;23: S46–51.
7. Nevanlinna H, Bartek J. The *CHEK2* gene and inherited breast cancer susceptibility. Oncogene. 2006;25:5912–9.
8. Rahman N, Seal S, Thompson D, et al. *PALB2*, which encodes a *BRCA2*-interacting protein, is a breast cancer susceptibility gene. Nat Genet. 2007;39:165–7.
9. Stratton MR, Rahman N. The emerging landscape of breast cancer susceptibility. Nat Genet. 2008;40:17–22.
10. Lakhani SR, Easton DF, Stratton FR, et al. Pathology of familial breast cancer: differences between breast cancers in carriers of *BRCA1* or *BRCA2* mutations and sporadic cases. Breast cancer linkage consortium. Lancet. 1997;349:1505–10.
11. Ford D, Easton DF, Peto J. Estimates of the gene frequency of *BRCA1* and its contribution to breast and ovarian cancer incidence. Am J Hum Genet. 1995;57:1457–62.
12. Easton DF, Narod SA, Ford D, et al. Letter to the editor; The genetic epidemiology of *BRCA1*. Breast cancer linkage consortium. Lancet. 1994;344:761.
13. Ford D, Easton DF, Bishop DT, et al. Risks of cancer in *BRCA1*-mutation carriers: breast cancer linkage consortium. Lancet. 1994;343:692–5.
14. Ford D, Easton DF, Stratton M, et al. Genetic heterogeneity and penetrance analysis of the *BRCA1* and *BRCA2* genes in breast cancer families: the breast cancer linkage consortium. Am J Hum Genet. 1998;62:676–89.
15. Marroni F, Aretini P, D'Andrea E, et al. Penetrances of breast and ovarian cancer in a large series of families tested for *BRCA1/2* mutations. Eur J Hum Genet. 2004;12:899–906.
16. Easton DF, Ford D, Bishop DT. Breast and ovarian cancer incidence in *BRCA1*-mutation carriers. Breast cancer linkage consortium. Am J Hum Genet. 1995;56:265–71.
17. Hwang ES, Mclennan JL, Moore DH, et al. Ductal carcinoma in situ in *BRCA* mutation carriers. J Clin Oncol. 2007;25:642–7.
18. Agnarsson BA, Jonasson JG, Bjornsdottir IB, et al. Inherited *BRCA2* mutation associated with high grade breast cancer. Breast Cancer Res Treat. 1998;47:121–7.
19. Armes JE, Egan AJ, Southey MC, et al. The histologic phenotypes of breast carcinoma occurring before age 40 years in women with and without *BRCA1* or *BRCA2* germline mutations: a population-based study. Cancer. 1998;83:2335–45.
20. Bane AL, Beck JC, Bleiweiss I, et al. *BRCA2* mutation-associated breast cancers exhibit a distinguishing phenotype based on morphology and molecular profiles from tissue microarrays. Am J Surg Pathol. 2007;31:121–8.
21. Lakhani SR, Jacquemier J, Sloane JP, et al. Multifactorial analysis of differences between sporadic breast cancers and cancers involving *BRCA1* and *BRCA2* mutations. J Natl Cancer Inst. 1998;90:1138–45.
22. Marcus JN, Watson P, Page DL, et al. Hereditary breast cancer: pathobiology, prognosis, and *BRCA1* and *BRCA2* gene linkage. Cancer. 1996;77:697–709.
23. Marcus JN, Watson P, Page DL, et al. *BRCA2* hereditary breast cancer pathophenotype. Breast Cancer Res Treat. 1997;44:275–7.

24. Sorlie T, Perou CM, Tibshirani R, et al. Gene expression patterns of breast carcinomas distinguishing tumor subclasses with clinical implications. Proc Natl Acad Sci USA. 2001;98(19):10869–74.
25. Brenton JD, Carey LA, Ahmed AA, et al. Molecular classification and molecular forecasting of breast cancer: ready for clinical application? J Clin Oncol. 2005;23(29):7350–60.
26. Rakha EA, El Sayed ME, Reis-Filho JS, et al. Expression profiling technology: its contribution to our understanding of breast cancer. Histopathol. 2008;52(1):67–81.
27. Parker JS, Mullins M, Cheang MC, et al. Supervised risk predictor of breast cancer based on intrinsic subtypes. J Clin Oncol. 2009;27(8):1160–7.
28. Jones RL, Constantinidou A, Reis-Filho JS. Molecular classification of breast cancer. Surg Pathol Clin. 2012;5(3):701–17.
29. Cheang MCU, Chia SK, Voduc D, et al. Ki67 index, HER2 status, and prognosis of patients with luminal B breast cancer. J Natl Cancer Inst. 2009;101:736–50.
30. Bertucci F, Finetti P, Cervera N, et al. How basal are triple-negative breast cancers? Int J Cancer. 2008;123:236–40.
31. de Ronde JJ, Hannemann J, Halfwerk H, et al. Concordance of clinical and molecular breast cancer subtyping in the context of preoperative chemotherapy response. Breast Cancer Res Treat. 2010;119:119–26.
32. Parker JS, Mullins M, Cheang MC, et al. Supervised risk predictor of breast cancer based on intrinsic subtypes. J Clin Oncol. 2009;27:1160–7.
33. Elston CW, Ellis IO. Pathological prognostic factors in breast cancer. The value of histological grade in breast cancer: experience from a large study with long-term follow up. Histopathology. 1991;19(5):403–10.
34. Ellis IO, Reis-Filho JS, Simpson JF, et al. Grading chapter. In: Lakhani S, Ellis IO, Schnitt et al., editors. WHO classification of tumours of the breast. Lyon: IARC Press; 2012. p. 19–20.
35. Rakha EA, Reis-Filho JS, Baehner, et al. Breast cancer prognostic classification in the molecular era: the role of histological grade. Breast Cancer Res. 2010;12(4):207–218.
36. Johannsson OT, Idvall I, Anderson C, et al. Tumour biological features of *BRCA1*-induced breast and ovarian cancer. Eur J Cancer. 1997;33:362–71.
37. Lynch BJ, Holden JA, Buys SS, et al. Pathobiologic characteristics of hereditary breast cancer. Hum Pathol. 1998;29:1140–4.
38. Karp SE, Tonin PN, Begin LR, et al. Influence of *BRCA1* mutations on nuclear grade and estrogen receptor status of breast carcinoma in Ashkenazi Jewish women. Cancer. 1997;80:435–41.
39. Palacios J, Honrado E, Osorio A, et al. Immunohistochemical characteristics defined by tissue microarray of hereditary breast cancer not attributable to *BRCA1* or *BRCA2* mutations: differences from breast carcinomas arising in *BRCA1* and *BRCA2* mutation carriers. Clin Cancer Res. 2003;9:3606–14.
40. Eerola H, Heikkila P, Tamminen A, et al. Histopathological features of breast tumours in *BRCA1*, *BRCA2* and mutation-negative breast cancer families. Breast Cancer Res. 2005;7: R93–100.
41. Eisinger F, Nogues C, Guinebretiere JM, et al. Novel indications for *BRCA1* screening using individual clinical and morphological features. Int J Cancer. 1999;84:263–7.
42. Chappuis PO, Nethercot V, Foulkes WD. Clinico-pathological characteristics of *BRCA1*- and *BRCA2*-related breast cancer. Semin Surg Oncol. 2000;18:287–95.
43. Loman N, Johannsson O, Bendahl PO, et al. Steroid receptors in hereditary breast carcinomas associated with *BRCA1* or *BRCA2* mutations or unknown susceptibility genes. Cancer. 1998;83:310–9.
44. Noguchi S, Kasugai T, Miki Y, et al. Clinicopathologic analysis of *BRCA1*- or *BRCA2*-associated hereditary breast carcinoma in Japanese women. Cancer. 1999;85:2200–5.
45. van der Groep P, Bouter A, van der Zanden R, et al. Distinction between hereditary and sporadic breast cancer on the basis of clinicopathological data. J Clin Pathol. 2006;59:611–7.

46. Atchley DA. Clinical and pathologic characteristics of patients With *BRCA*-positive and *BRCA*-negative breast cancer. J Clin Oncol. 2008;26:4282–8.
47. Lakhani S, Rakha E, Simpson PT. Invasive lobular carcinoma. In: Lakhani S, Ellis IO, Schnitt, et al., editors. WHO classification of tumours of the breast. Lyon: IARC Press; 2012. p. 40–2.
48. DaSilva L, Parry S, Reid L, et al. Aberrant expression of E-Cadherin in lobular carcinomas of the breast. Am J Surg Pathol. 2008;32(5):773–83.
49. Rakha EA, Patel A, Powe DG, et al. Clinical and biological significance of E-Cadherin protein expression in invasive lobular carcinoma of the breast. Am J Surg Pathol. 2010;34 (10):1472–9.
50. Manucha V, Khilko N, Reilly K, Zhang X. Invasive pleomorphic lobular carcinoma, negative for ER, PR and Her/2neu–a case report. Int J Clin Exp Pathol. 2011;4(2):200–5.
51. Palacios J, Sarrio D, Garcia-Macias MC, et al. Frequent E-cadherin gene inactivation by loss of heterozygosity in pleomorphic lobular carcinoma of the breast. Mod Pathol. 2003;16 (7):674–8.
52. Moe RE, Anderson BO. Distinctive biology of pleomorphic lobular carcinoma of the breast. J Surg Oncol. 2005;90(2):47–50.
53. Ridolfi RL, Jamehdor MR, Arber JM. HER-2/neu testing in breast carcinoma: a combined immunohistochemical and fluorescence in situ hybridization approach. Mod Pathol. 2000;13 (8):866–73.
54. Hoff ER, Tubbs RR, Myles JL, et al. HER2/neu amplification in breast cancer: stratification by tumor type and grade. Am J Clin Pathol. 2002;117:916–21.
55. Lal P, Tan LK, Chen B. Correlation of HER-2 status with estrogen and progesterone receptors and histologic features in 3655 invasive breast carcinomas. Am J Clin Pathol. 2005;123:541–6.
56. Porter PL, Garcia R, Moe R, et al. *C-erbB-2* oncogene protein in in situ and invasive lobular breast neoplasia. Cancer. 1991;68:331–4.
57. Frolik D, Caduff R, Varga Z. Pleomorphic lobular carcinoma of the breast: its cell kinetics, expression of oncogenes and tumour suppressor genes compared with invasive ductal carcinomas and classical infiltrating lobular carcinomas. Histopathology. 2001;39(5):503–13.
58. Ridolfi RL, Rosen PP, Port A, et al. Medullary carcinoma of the breast: a clinicopathologic study with 10 year follow up. Cancer. 1977;40:1365–85.
59. Jacquemier J, Reis-Filho, Lakhani SR, et al. Carcinomas with medullary features. In Lakhani S, Ellis IO, Schnitt, et al. WHO classification of tumors of the breast. Lyon: IARC Press; 2012. p. 46–7.
60. Rakha E, Pinder SE, Shin SJ, et al. Invasive tubular carcinoma. In: Lakhani S, Ellis IO, Schnitt, et al., editors. WHO classification of tumors of the breast. Lyon: IARC Press; 2012. p. 43–5.
61. Peters GN, Wolff M, Haagensen CD. Tubular carcinoma of the breast. Clinical pathologic correlations based on 100 cases. Ann Surg. 1981;193:138–49.
62. Carstens PH, Greenberg RA, Francis D, Lyon H. Tubular carcinoma of the breast. A long term follow-up. Histopathology. 1985;9:271–80.
63. Sullivan T, Raad RA, Goldberg S, et al. Tubular carcinoma of the breast: a retrospective analysis and review of the literature. Breast Cancer Res Treat. 2005;93:199–205.
64. Reis-Filho JS, Tutt AN. Triple negative tumors: a critical review. Histopathology. 2008;52:108–18.
65. Bauer KR, Brown M, Cress RD, et al. Descriptive Analysis of estrogen receptor (ER)-negative, progesterone receptor (PR)-negative and Her2-negative invasive breast cancer, the so call triple negative phenotype: a population based study from the California Cancer Registry. Cancer. 2007;109:1721–8.
66. Anders C, Carey LA. Understanding and treating triple-negative breast cancer. Oncology (Williston Park) 2008;22:1233–9; discussion 1239–40, 1243.

67. Lee E, McKean-Cowdin R, Ma H, et al. Characteristics of triple-negative breast cancer in patients with a *BRCA1* mutation: results from a population-based study of young women. J Clin Oncol. 2011;29:4373–80.
68. Carey LA, Dees EC, Sawyer L, et al. The triple negative paradox: primary tumor chemosensitivity of breast cancer subtypes. Clin Cancer Res. 2007;13:2329–34.
69. Heerma van Voss MR, van der Groep P, Bart J, et al. Lympho-vascular invasion in *BRCA* related breast cancer compared to sporadic controls. BMC Cancer. 2010;10:145–51.
70. Eerola H, Heikkila P, Tamminen A, et al. Relationship of patients' age to histopathological features of breast tumours in *BRCA1* and *BRCA2* and mutation-negative breast cancer families. Breast Cancer Res. 2005;7:R465–9.
71. Mavaddat N, Barrowdale D, Andrulis IL, et al. Pathology of breast and ovarian cancers among *BRCA1* and *BRCA2* mutation carriers: results from the Consortium of investigators of modifiers of *BRCA1/2* (CIMBA). Cancer Epidemiol Biomarkers Prev. 2012;21(1):134–47.
72. Eisinger F, Nogues C, Bimbaum D, et al. *BRCA1* and medullary breast cancer. JAMA. 1998;280:1227–8.
73. Phillips K, Andulis IL, Goodwin PJ, et al. Breast carcinomas arising in carriers of mutations in *BRCA1* or *BRCA2*: are they prognostically different? J Clin Oncol. 1999;17:3653–63.
74. Lakhani SR, Gusterson BA, Jacquemier J, et al. The pathology of familial breast cancer: histological features of cancers in families not attributable to mutations in *BRCA1* or *BRCA2*. Clin Cancer Res. 2000;6:782–9.
75. Honrado E, Benitez J, Palacios J. The molecular pathology of hereditary breast cancer: genetic testing and therapeutic implications. Mod Pathol. 2005;18:1305–20.
76. Tsuda H, Takarabe T, Hasegawa F, et al. Large, central acellular zones indicating myoepithelial tumor differentiation in high-grade invasive ductal carcinomas as markers of predisposition to lung and brain metastases. Am J Surg Pathol. 2000;24:197–202.
77. Armes JE, Trute L, White D, et al. Distinct molecular pathogeneses of early-onset breast cancers in *BRCA1* and *BRCA2* mutation carriers: a population-based study. Cancer Res. 1999;59:2011–7.
78. Tot T. The cytokeratin profile of medullary carcinoma of the breast. Histopathology. 2000;37:175–81.
79. Takei H, Iino Y, Horiguchi J, et al. Low and high molecular weight cytokeratins in invasive breast carcinoma. Oncol Rep. 1997;4:33–8.
80. Jones C, Nonni AV, Fulford L, et al. CGH analysis of ductal carcinoma of the breast with basaloid/myoepithelial cell differentiation. Br J Cancer. 2001;85:422–7.
81. Fisher B, Slack NH, Bross ID. Cancer of the breast: size of neoplasm and prognosis. Cancer. 1969;24:1071–80.
82. Smart CR, Myers MH, Gloeckler LA. Implications from SEER data on breast cancer management. Cancer. 1978;41:787–9.
83. Nemoto T, Vana J, Bedwani RN, et al. Management and survival of female breast cancer: results of a national survey by the American College of Surgeons. Cancer. 1980;45:2917–24.
84. Reger V, Beito G, Jolly PC. Factors affecting the incidence of lymph node metastases in small cancers of the breast. Am J Surg. 1989;157:501–2.
85. Carter CL, Allen C, Henson DE. Relation of tumor size, lymph node status, and survival in 24,740 breast cancer cases. Cancer. 1989;63:181–7.
86. Olivotto IA, Jackson JS, Mates D, et al. Prediction of axillary lymph node involvement of women with invasive breast carcinoma: a multivariate analysis. Cancer. 1998;83:948–55.
87. Foulkes WD, Metcalfe K, Hanna W, et al. Disruption of the expected positive correlation between breast tumor size and lymph node status in *BRCA1*-related breast carcinoma. Cancer. 2003;98(8):1569–77.
88. Perou CM, Sørlie T, Eisen MB, et al. Molecular portraits of human breast tumours. Nature. 2000;406:747–52.
89. Sørlie T, Tibshirani R, Parker J, et al. Repeated observation of breast tumor subtypes in independent gene expression data sets. Proc Natl Acad Sci USA. 2003;100:8418–23.

90. Pinilla SM, Honrado E, Hardisson D, et al. Caveolin-1 expression is associated with a basal-like phenotype in sporadic and hereditary breast cancer. Breast Cancer Res Treat. 2006;99:85–90.
91. Foulkes WD, Stefansson IM, Chappuis PO, et al. Germline *BRCA1* mutations and a basal epithelial phenotype in breast cancer. J Natl Cancer Inst. 2003;95:1482–5.
92. Rodriguez-Pinilla SM, Sarrio D, Honrado E, et al. Vimentin and laminin expression is associated with basal-like phenotype in both sporadic and *BRCA1*-associated breast carcinomas. J Clin Pathol. 2007;60:1006–12.
93. Arnes JB, Brunet JS, Stefansson I, et al. Placental cadherin and the basal epithelial phenotype of *BRCA1*-related breast cancer. Clin Cancer Res. 2005;11:4003–11.
94. Cheang MC, Voduc D. Bajdik C, et al. Basal-like breast cancer defined by five biomarkers has superior prognostic value than triple negative phenotype. Clin Cancer Res. 2008;14:1368–76.
95. Collins LC, Martyniak A, Kandel MJ. Basal cytokeratin and epidermal growth factor receptor expression are not predictive of *BRCA1* mutation status in women with triple negative breast cancers. Am J Surg Pathol. 2009;33:1093–7.
96. Lakhani SR. The pathology of familial breast cancer: morphological aspects. Breast Cancer Res. 1999;1:31–5.
97. Lakhani SR, Van De Vijver MJ, Jacquemier J, et al. The pathology of familial breast cancer: predictive value of immunohistochemical markers estrogen receptor, progesterone receptor, HER-2, and *p53* in patients with mutations in *BRCA1* and *BRCA2*. J Clin Oncol. 2002;20:2310–8.
98. Foulkes WD, Metcalfe K, Sun P, et al. Estrogen receptor status in *BRCA1*- and *BRCA2*-related breast cancer: the influence of age, grade, and histological type. Clin Cancer Res. 2004;10:2029–34.
99. Turner NC, Reis-Filho JS. Basal-like breast cancer and the *BRCA1* phenotype. Oncogene. 2006;25:5846–53.
100. Rakha EA, Reis-Filho JS, Ellis IO. Basal-like breast cancer: a critical review. J Clin Oncol. 2008;26:2568–81.
101. Reis-Filho JS, Richardson AL, van de Vijver MJ, et al. Genetic alterations and gene expression profiles of *BRCA1* and *BRCA2* associated breast cancers. In: Lakhani, SL, editor. WHO classification of tumors of the breast, 4th ed. WHO Press; 2012. p. 181–2.
102. Haffty BG, Yang Q, Reiss M, et al. Locoregional relapse and distant metastasis in conservatively managed triple negative early-stage breast cancer. J Clin Oncol. 2006;24:5652–7.
103. Musolino A, Bella MA, Bortesi B, et al. *BRCA* mutations, molecular markers, and clinical variables in early-onset breast cancer: a population-based study. Breast. 2007;16:280–92.
104. Li WF, Hu Z, Rao NY, et al. The prevalence of *BRCA1* and *BRCA2* germline mutations in high-risk breast cancer patients of Chinese Han nationality: two recurrent mutations were identified. Breast Cancer Res Treat. 2008;110(1):99–109.
105. Byrski T, Gronwald J, Huzarski T, et al. Response to neo-adjuvant chemotherapy in women with *BRCA1*-positive breast cancers. Breast Cancer Res Treat. 2008;108:289–96.
106. Lakhani SR, Reis-Filho JS, Fulford L, et al. Prediction of *BRCA1* status in patients with breast cancer using estrogen receptor and basal phenotype. Clin Cancer Res. 2005;11:5175–80.
107. Nielsen TO, Hsu FD, Jensen K, et al. Immunohistochemical and clinical characterization of the basal-like subtype of invasive breast carcinoma. Clin Cancer Res. 2004;10:5367–74.
108. Laakso M, Loman N, Borg A, et al. Cytokeratin 5/14-positive breast cancer: true basal phenotype confined to *BRCA1* tumors. Mod Pathol. 2005;18:1321–8.
109. Rodriguez-Pinilla SM, Sarrio D, Honrado E, et al. Prognostic significance of basal-like phenotype and fascin expression in node-negative invasive breast carcinomas. Clin Cancer Res. 2006;12:1533–9.
110. van der Groep P, Bouter A, van der Zanden R, Re. Germline *BRCA1* mutations and a basal epithelial phenotype in breast cancer. J Natl Cancer Inst. 2004;96:712–3; author reply 714.

111. van Diest PJ, van der Groep P, van der Wall E. EGFR expression predicts *BRCA1* status in patients with breast cancer. Clin Cancer Res. 2006;12:670, author reply 671.
112. Elledge RM, Allred DC. Prognostic and predictive value of *p53* and p21 in breast cancer. Breast Cancer Res Treat. 1998;52:79–98.
113. Crook T, Brooks LA, Crossland S, et al. *p53* mutation with frequent novel condons but not a mutator phenotype in *BRCA1*- and *BRCA2*-associated breast tumours. Oncogene. 1998;17:1681–9.
114. Phillips KA, Nichol K, Ozcelik H, et al. Frequency of *p53* mutations in breast carcinomas from Ashkenazi Jewish carriers of *BRCA1* mutations. J Natl Cancer Inst. 1999;91:469–73.
115. Rudolph P, Olsson H, Bonatz G, et al. Correlation between *p53*, *cerbB-2*, and topoisomerase II alpha expression, DNA ploidy, hormonal receptor status and proliferation in 356 node-negative breast carcinomas: prognostic implications. J Pathol. 1999;187:207–16.
116. Hoogerbrugge N, Bult P, de Widt-Levert LM, et al. High prevalence of premalignant lesions in prophylactically removed breasts from women at hereditary risk for breast cancer. J Clin Oncol. 2003;21:41–5.
117. Isern AE, Loman N, Malina J, et al. Histopathological findings and follow-up after prophylactic mastectomy and immediate breast reconstruction in 100 women from families with hereditary breast cancer. Eur J Surg Oncol. 2008;34:1148–54.
118. Kauff ND, Brogi E, Scheuer L, et al. Epithelial lesions in prophylactic mastectomy specimens from women with *BRCA* mutations. Cancer. 2003;97:1601–8.
119. Hermsen BB, von Mensdorff-Pouilly S, Fabry HF, et al. Lobulitis is a frequent finding in prophylactically removed breast tissue from women at hereditary high risk of breast cancer. J Pathol. 2005;206:220–3.
120. Adem C, Reynolds C, Soderberg CL, et al. Pathologic characteristics of breast parenchyma in patients with hereditary breast carcinoma, including *BRCA1* and *BRCA2* mutation carriers. Cancer. 2003;97:1–11.
121. Claus EB, Petruzella S, Matloff E, et al. Prevalence of *BRCA1* and *BRCA2* mutations in women diagnosed with ductal carcinoma in situ. JAMA. 2005;293:964–9.
122. Jacquemier J, Eisinger F, Guinebretiere JM, et al. Intraductal component and *BRCA1*-associated breast cancer. Letter to the editor. Lancet. 1996;348:1098.
123. Hadjiloucas I, Gilmore AP, Bundred NJ, et al. Assessment of apoptosis in human breast tissue using an antibody against the active form of caspase 3: relation to tumour histopathological characteristics. Br J Cancer. 2001;85:1522–6.
124. Simpson PT, Reis-Filho JS, Lambros MB, et al. Molecular profiling pleomorphic lobular carcinomas of the breast: evidence for a common molecular genetic pathway with classic lobular carcinomas. J Pathol. 2008;215:231–44.
125. Freneaux P, Stoppa-Lyonnet D, Mouret E, et al. Low expression of BCL-2 in *BRCA1*-associated breast cancers. Br J Cancer. 2000;83:1318–22.
126. Osin P, Gusterson BA, Philp E, et al. Predicted anti-oestrogen resistance in *BRCA*-associated familial breast cancers. Eur J Cancer. 1998;34:1683–6.
127. Gartel AL, Tyner AL. The role of the cyclin-dependent kinase inhibitor p 21 in apoptosis 1. Mol Cancer Therap. 2002;639–49.
128. Foulkes WD, Brunet JS, Stefansson IM, et al. The prognostic implication of the basal-like (cyclin E high/p27 low/*p53*+/glomeruloid-microvascular-proliferation+) phenotype of *BRCA1* related breast cancer. Cancer Res. 2004;64:830–5.
129. Chappuis PO, Donato E, Goffin JR, et al. Cyclin E expression in breast cancer: predicting germline *BRCA1* mutations, prognosis and response to treatment. Ann Oncol. 2005;16:735–42.
130. Palacios J, Honrado E, Osorio A, et al. Phenotypic characterization of *BRCA1* and *BRCA2* tumors based in a tissue microarray study with 37 immunohistochemical markers. Breast Cancer Res Treat. 2005;90:5–14.
131. Semenza GL. Regulation of mammalian O2 homeostasis by hypoxia-inducible factor 1. Ann Rev Cell Dev Biol. 1999;15:551–78.

132. Semenza GL. HIF-1: mediator of physiological and pathophysiological responses to hypoxia. J Appl Physiol. 2000;88(4):1474–80.
133. van der Groep P, Bouter A, Menko FH, et al. High frequency of HIF-1alpha overexpression in *BRCA1* related breast cancer. Breast Cancer Res Treat. 2008;111:475–80.
134. Turner N, Tutt A, Ashworth A. Hallmarks of "*BRCA*ness" in sporadic cancers. Nat Rev Cancer. 2004;4:814–9.
135. Chin K, DeVries S, Fridlyand J, et al. Genomic and transcriptional aberrations linked to breast cancer pathophysiologies. Cancer Cell. 2006;10:529–41.
136. Bergamaschi A, Kim YH, Wang P, et al. Distinct patterns of DNA copy number alteration are associated with different clinicopathological features and gene expression subtypes of breast cancer. Genes Chromosom Cancer. 2006;45:1033–40.
137. Lidereau R, Eisinger F, Champeme MH, et al. Major improvement in the efficacy of *BRCA1* mutation screening using morphoclinical features of breast cancer. Cancer Res. 2000;60:1206–10.
138. Young SR, Pilarski RT, Donenberg T, et al. The prevalence of *BRCA1* mutations among young women with triple-negative breast cancer. BMC Cancer. 2009;9:86–90.
139. Vaziri SA, Krumroy LM, Elson P, et al. Breast tumor immunophenotype of *BRCA1*-mutation carriers is influenced by age at diagnosis. Clin Cancer Res. 2001;7:1937–45.
140. Honrado E, Osorio A, Milne RL, et al. Immunohistochemical classification of non-*BRCA1/2* tumors identifies different groups that demonstrate the heterogeneity of *BRCAX* families. Mod Pathol. 2007;20:1298–306.
141. Robson ME, Chappuis PO, Satagopan J, et al. A combined analysis of outcome following breast cancer: differences in survival based on *BRCA1/BRCA2* mutation status and administration of adjuvant treatment. Breast Cancer Res. 2004;6:R8–17.
142. Quenneville LA, Phillips KA, Ozcelik H, et al. HER-2/ neu status and tumor morphology of invasive breast carcinomas in Ashkenazi women with known *BRCA1* mutation status in the Ontario Familial Breast Cancer Registry. Cancer. 2002;95:2068–75.
143. Arun B, Vogel KJ, Lopez A, et al. High prevalence of preinvasive lesions adjacent to *BRCA1/2*-associated breast cancers. Cancer Prev Res (Phila. PA). 2009;2:122–7.
144. Hoogerbrugge N, Bult P, Bonenkamp JJ, et al. Numerous high-risk epithelial lesions in familial breast cancer. Eur J Cancer. 2006;42:2492–8.
145. van der Groep P, van Diest PJ, Menko FH, et al. Molecular profile of ductal carcinoma in situ of the breast in *BRCA1* and *BRCA2* germline mutation carriers. J Clin Pathol. 2009;62:926–30.
146. Schuyer M, van der Burg ME, Henzen-Logmans SC, et al. Reduced expression of BAX is associated with poor prognosis in patients with epithelial ovarian cancer: a multifactorial analysis of *TP53*, p21, BAX and BCL-2. Br J Cancer. 2001;85:1359–67.
147. Jonsson G, Naylor TL, Vallon-Christersson J, et al. Distinct genomic profiles in hereditary breast tumors identified by array-based comparative genomic hybridization. Cancer Res. 2005;65:7612–21.
148. van Beers EH, van Welsem T, Wessels LF, et al. Comparative genomic hybridization profiles in human *BRCA1* and *BRCA2* breast tumors highlight differential sets of genomic aberrations. Cancer Res. 2005;65:822–7.
149. Wessels LF, van Welsem T, Hart AA, et al. Molecular classification of breast carcinomas by comparative genomic hybridization: a specific somatic genetic profile for *BRCA1* tumors. Cancer Res. 2002;62:7110–7.
150. Yang RL, Mick R, Lee K. DCIA in *BRCA1* and *BRCA2* mutation carriers: prevalence, phenotype, and expression of oncodrivers C-Met and HER3. J Transl Med. 2015;13:335–42.

Surgical Management of Breast Cancer in *BRCA* Mutation Carriers

7

Patricia A. Cronin and Hiram S. Cody III

Introduction

Women with germline *BRCA1/2* mutations have a breast cancer prevalence of 46–71% by age 70 [1]. Preventive strategies that may significantly reduce this risk include bilateral prophylactic mastectomy [2–4] or hormonal interventions such as bilateral oophorectomy and tamoxifen[5–7]. For mutation carriers diagnosed with breast cancer, the best surgical option for local therapy is still debated. For women without a mutation, breast-conserving therapy (BCT) with adjuvant radiation therapy (RT) is often the treatment of choice as it is less invasive than mastectomy but with comparable survival [8]. However, mutation carriers must also consider the risks of ipsilateral breast tumor recurrence (IBTR) and of contralateral breast cancer (CBC), both higher than in non-carriers, and whether other patient and/or treatment factors (such as systemic adjuvant therapy and risk-reducing contralateral prophylactic mastectomy) could alter the risk for IBTR and/or CBC. The potential survival benefit of an intervention should also be included in decision making. This chapter will explore the options available for the surgical management of breast cancer in mutation carriers.

P.A. Cronin
Memorial Sloan Kettering Cancer Center, Breast Service, Evelyn Lauder Breast and Imaging Center, 300 East 66th Street, New York, NY 10065, USA
e-mail: patcronin@rcsi.ie

H.S. Cody III (✉)
Breast Service, Department of Surgery, Memorial Sloan Kettering Cancer Center, Weill Cornell Medical College, 300 East 66th Street, #831, New York, NY 10065, USA
e-mail: codyh@mskcc.org

A.B. Chagpar (ed.), *Managing BRCA Mutation Carriers*,
DOI 10.1007/978-3-319-59198-8_7

Surgical Management

Primary Tumor Surgical Management

BCT, combining breast-conserving surgery (lumpectomy) with adjuvant RT, is the treatment of choice for most non-mutation carriers as it is a minimally invasive and cosmetically acceptable approach for suitable tumors with survival comparable to mastectomy. For mutation carriers, the possibility of increased risk of local recurrence compared to non-mutation carriers has made the option of BCT controversial. There were also concerns about the potentially harmful effects of RT in patients with *BRCA* mutations, but these have not been substantiated in the literature [9, 10].

Many studies have compared the risk of ipsilateral breast tumor recurrence (IBTR) following BCT in mutation versus non-mutation carriers, but have produced conflicting results and are limited by retrospective designs and small sample size. Several have concluded that BCT was not associated with an increased risk of IBTR in mutation carriers compared to non-mutation carriers, but in studies with longer follow-up, there was some evidence for an increased risk of IBTR in mutation carriers.

A recent meta-analysis by Valachis et al. [11] compiled results from ten studies that addressed this question. Results from six cohort [12–17] and four case–control studies [18–21] allowed comparison of 526 *BRCA* mutation carriers and 2320 controls with respect to IBTR following BCT. The pooled rates of IBTR for mutation carriers were 17.3% (95% confidence interval (CI) 11.4–24.2%) compared to 11% (95% CI 6.5–15.4%) in non-mutation carrier controls. There was no significant difference in IBTR between carriers and controls (risk ratio (RR) 1.45, 95% CI 0.98–2.14, $p = 0.07$) (Table 7.1).

In six studies (1212 patients) with a median follow-up less than 7 years, there was no significant difference in IBTR between mutation carriers and non-mutation carrier controls (11.7 vs. 8.9%, $p = 0.51$). Among five studies (1,634 patients) with median follow-up greater than 7 years, there was a significantly higher risk for IBTR in *BRCA* mutation carriers (23.7 vs. 15.9%, $p = 0.003$) [11].

Cases of IBTR beyond 7 years could be new primary breast cancers given that all of the residual breast tissue after BCT carries a persistent mutation-related risk. Two studies [20, 21] aimed to differentiate between IBTR events which were "true recurrence" as opposed to "new primary" breast cancers and found no significant increase in true recurrences (RR 1.37, 95% CI 0.44–4.21, $p = 0.59$) but a trend for a higher rate of new ipsilateral primary cancers (RR 2.07, 95% CI 0.99–4.36, $p = 0.05$) in *BRCA* mutation versus non-mutation carriers [11].

Another important consideration is whether there are differences in IBTR, and whether differential approaches should be adopted, according to mutation type. Four studies reported on IBTR by mutation type (405 patients with *BRCA1* mutations and 203 with *BRCA2* mutations), and in the meta-analysis there was no

Table 7.1 Risk for ipsilateral breast tumor recurrence following breast-conserving therapy in *BRCA* mutation carriers versus non-mutation carriers

		Risk ratio [95% CI]
Cohort studies		
	Brekelmans et al. [12]	0.61 [0.33–1.11]
	Chappuis et al. [13]	0.97 [0.22–4.15]
	El-Tamer et al. [14]	3.22 [1.15–9.01]
	Haffty et al. [15]	2.15 [1.13–4.07]
	Robson et al. [16]	0.46 [0.06–3.34]
	Robson et al. [17]	1.57 [0.73–3.36]
	Subtotal	1.32 [0.70–2.46]
Case–control studies		
	Eccles et al. [18]	0.69 [0.30–1.58]
	Garcia-Etienne et al. [19]	4.50 [1.32–15.35]
	Kirova et al. [20]	1.90 [1.22–2.97]
	Pierce et al. [21]	1.51 [0.89–2.56]
	Subtotal	1.60 [0.94–2.72]
Total		1.45 [0.98–2.14]

Adapted from [11]

difference in risk of IBTR between *BRCA1* and *BRCA2* mutation carriers (RR 0.76, 95% CI 0.49–1.16, $p = 0.20$) [11].

Some studies have examined the impact of BCT on survival in mutation carriers compared to non-mutation carriers. Two studies found no difference in overall survival (OS) following BCT in mutation carriers compared to non-mutation carrier controls. The first study [20] found no difference in OS at a follow-up of over 13 years. The second study [9], with a median follow-up of 5.3 years for mutation carriers, found an insignificant difference in 5-year OS between carriers and non-carriers: 86 versus 91%. In the same study, breast cancer-specific survival (BCSS) survival was also similar: 92 versus 91%. Another study observed a significantly higher risk for death due to breast cancer, at a median follow-up of 116 months, in patients with *BRCA* mutations (multivariable hazard ratio (HR): 2.39, 95% CI 1.20–4.75) [17].

Few studies address the risk for IBTR in *BRCA* mutation carriers with BCT compared to mastectomy. One study [22] compared the risk for IBTR in *BRCA* mutation carriers after mastectomy versus BCT, finding a higher risk for BCT: 23.5 versus 5.5% at 15 years, $p < 0.0001$). Of note, both BCSS and OS were similar between the two groups (91.7 vs. 92.8% ($p = 0.85$) and 87.3 vs. 89.8% ($p = 0.73$), respectively. Most IBTR events in the BCT group appeared to be new primary cancers and most in the mastectomy group to be true recurrences. The lack of survival difference between mastectomy and BCT may reflect a less biologically aggressive phenotype in the new primary cancers.

A recent study comparing BCT and mastectomy in *BRCA1*/ *BRCA2* mutation carriers reached similar conclusions. BCT was associated with an increased risk of IBTR in a multivariable analysis adjusting for tumor stage, age, and use of adjuvant

Table 7.2 Summary of risk factors for ipsilateral breast tumor recurrence following breast-conserving therapy in *BRCA* mutation carriers

Risk factors	Hazard ratio [95% CI]
Age (continuous)	0.96 [0.92–0.99]
Age > 50	0.69 [0.27–1.77]
Positive margins	0.76 [0.18–3.19]
ER positive	1.74 [0.71–4.25]
Grade 3	0.95 [0.35–2.59]
T stage ($\geq$ T2)	0.76 [0.37–1.53]
Stage II	0.69 [0.36–1.33]
Nodal metastases	0.86 [0.39–1.89]
Tamoxifen use	0.73 [0.39–1.39]
Tamoxifen use (ovaries intact)	0.39 [0.09–1.69]
Chemotherapy	0.51 [0.31–0.84]
Oophorectomy	0.42 [0.22–0.81]

Adapted from [11]

chemotherapy (HR 2.9; CI 1.1–7.8). The cumulative 15-year incidence of IBTR for BCT was 32% versus 9% for the mastectomy group, but there were no significant differences in OS, BCSS, or distant recurrence. IBTR post-mastectomy was seen in the first five years, whereas IBTR post-BCT continued to occur beyond 5 years, again likely reflecting more aggressive biology in "true recurrent" compared to "second primary" cancers [23].

In considering BCT in a *BRCA* mutation carrier, other treatment factors may mitigate the patient's risk for IBTR. Two studies [21, 24] have reviewed these risk modifiers and found that two (supported by a moderate level of evidence) were associated with a reduced risk of IBTR: the use of adjuvant chemotherapy (RR 0.51, 95% CI 0.31–0.84) and having undergone an oophorectomy (RR 0.42, 95% CI 0.22–0.81). The use of adjuvant tamoxifen was not significantly associated with IBTR in the meta-analysis [11] (Table 7.2).

Current data show that BCT does not increase the risk for "true recurrence" IBTR in *BRCA* mutation carriers compared to non-carriers, indicating that RT in mutation carriers is at least as effective as in non-mutation carriers. The increased risk for IBTR among mutation carriers in studies with longer follow-up reflects a higher risk for new primary cancers, since all of the residual breast tissue is at a higher mutation-related risk. No difference in OS has been demonstrated between mutation carriers and non-mutation carriers [17]. Taking into account the comparable rates of IBTR and similar survival outcomes after BCT between mutation carriers and non-mutation carriers, it would not be unreasonable to offer BCT to *BRCA* mutation carriers who can accept the likelihood of an increased risk of new ipsilateral primary breast cancer events and the need for ongoing surveillance. The risk of IBTR is modified by other treatment factors such as oophorectomy or adjuvant chemotherapy, and these should be included in surgeon–patient counseling and decision making.

Contralateral Prophylactic Mastectomy

The option of contralateral prophylactic mastectomy (CPM) in *BRCA* mutation carriers is another consideration in the surgical decision. *BRCA* mutation carriers have a higher risk for contralateral breast cancer (CBC) compared with non-mutation carriers, and the risk of CBC is higher for *BRCA1* than for *BRCA2*. Pooled rates of CBC in a meta-analysis of 11 studies (7 cohort [12–17, 25] and 4 case–control [18–21]), including 807 mutation carriers and 3163 non-mutation carrier controls, were 23.7% (95% CI 17.6–30.5%) and 6.8% (95% CI 4.2–10%), respectively. Patients with *BRCA* mutations had a higher risk for CBC compared with non-mutation carriers (RR 3.56, 95% CI 2.50–5.08, $p < 0.001$) [11] (Table 7.3).

Among seven studies comparing the risk of CBC between *BRCA1* mutation carriers (n = 1532) and *BRCA2* mutation carriers (n = 950), the rates of CBC were 21.1 versus 15.1%, respectively ($p < 0.04$) [26]. Another meta-analysis found that the risk of CBC at 5 years was 15 versus 9% and at 10 years was 27 versus 19%. The cumulative risk of CBC was substantially lower in non-mutation carriers, 3% at 5 years and 5% at 10 years [26].

The risk of CBC is potentially modifiable and may be dependent on other factors such as age. Six studies [17, 21, 27–30] have investigated potential risk factors for CBC in *BRCA* mutation carriers, and three studies [31–33] have investigated the role of age at first breast cancer diagnosis as a risk factor for CBC. Factors associated with a decreased risk for CBC include oophorectomy (RR 0.52, 95% CI 0.37–0.74), tamoxifen use (RR 0.57, 95% CI 0.43–0.75), and increasing age [11]

Table 7.3 Risk of CBC in *BRCA* mutation carriers versus non-mutation carriers

		Risk ratio [95% CI]
Cohort studies		
	Brekelmans et al. [12]	3.54 [2.28–5.49]
	Chappuis et al. [13]	7.97 [1.39–45.81]
	El-Tamer et al. [14]	1.74 [0.98–3.11]
	Haffty et al. [15]	4.77 [1.86–12.24]
	Robson et al. [16]	4.88 [1.89–12.58]
	Robson et al. [17]	3.51 [2.05–6.01]
	Stoppa-Lyonnet et al. [25]	0.89 [0.39–2.04]
	Subtotal	2.90 [1.85–4.53]
Case–control studies		
	Eccles et al. [18]	3.60 [2.15–6.03]
	Garcia-Etienne et al. [19]	15.0 [1.79–125.57]
	Kirova et al. [20]	3.67 [2.07–6.48]
	Pierce et al. [21]	8.34 [4.45–15.63]
	Subtotal	5.00 [2.97–8.40]
Total		3.56 [2.50–5.08]

Adapted from [11, 25]

Table 7.4 Summary of risk factors for contralateral breast cancer in *BRCA* mutation carriers

Risk factors	Hazard ratio [95% CI]
Age (continuous)	0.98 [0.95–1.02]
Age > 50	0.47 [0.27–0.82]
ER positive	1.02 [0.64–1.62]
Grade 3	0.84 [0.50–1.41]
Nodal metastases	0.76 [0.51–1.12]
Tamoxifen use	0.57 [0.43–0.75]
Tamoxifen use (ovaries intact)	0.42 [0.27–0.63]
Tamoxifen use (post-oophorectomy)	0.83 [0.24–2.89]
Chemotherapy	0.90 [0.66–1.22]
Oophorectomy	0.52 [0.37–0.74]

Adapted from [17, 21, 27–30]

(Table 7.4). The protective effect of tamoxifen seems to be stronger in patients who did not undergo oophorectomy (RR 0.42, 95% CI 0.27–0.63). Use of adjuvant chemotherapy did not alter the risk for CBC (RR 0.90, 95% CI 0.66–1.22). Although a cumulative HR for the impact of increasing age could not be calculated in the meta-analysis, Metcalfe et al. [27] demonstrated that age > 50 years at diagnosis was associated with decreased risk of CBC at 15 years compared to age at diagnosis < 50 years (16.8 vs. 37.6%, $p = 0.001$). Similarly, in an earlier study by Graeser et al. age > 50 years at diagnosis was associated with a decreased risk of CBC in *BRCA1* but not *BRCA2* families. The protective effect of tamoxifen and prophylactic oophorectomy in *BRCA2* mutation carriers may reflect the higher rate of estrogen receptor (ER) positive disease in this group [27].

The aim of CPM in mutation carriers is to reduce future CBC development and thereby confer a survival advantage to mutation carriers. Two studies [12, 34] have examined survival differences in *BRCA* mutation carriers after CPM versus therapeutic mastectomy alone (at a median follow-up of 4.3 and 3.4 years). There was no difference in BCSS between patients with *BRCA* mutation who had CPM and those who did not (HR 0.78, 95% CI 0.44–1.39, $p = 0.40$) [11]. Although van Sprundel et al. found that OS with CPM was 94% and without CPM was 77% ($p = 0.03$), there was no survival advantage for CPM after adjustment for prophylactic oophorectomy ($p = 0.14$).

A large study [35] compared 242 mutation carriers with primary unilateral breast cancer who had CPM with 341 mutation carriers who did not. At a median follow-up of 11.4 years, 4 (2%) patients in the CPM group developed CBC compared to 64 (19%) in the surveillance group ($p < 0.001$). Mortality was lower for CPM than for surveillance (9.6 and 21.6 per 1000 person-years of observation, respectively; adjusted HR 0.49, 95% CI 0.29–0.82). Survival benefit was most apparent in patients who were young (<40 years), or had low/intermediate grade and/or non-triple-negative cancers, or who were not treated with adjuvant chemotherapy. Chemotherapy was more frequent in the CPM group and may have contributed to improved OS.

The survival benefit of CPM in mutation carriers with breast cancer remains debatable. Certainly, CBC risk and potential modifying factors (age, oophorectomy, and/or tamoxifen use) need to be considered when counseling patients about CPM. An aggressive surgical approach may also be tempered by the stage of the index lesion and the patient's prognosis. *BRCA1* mutation carriers seem to have increased risk of CBC, and so perhaps a differential approach should be considered dependent on mutation type. Since *BRCA1* mutation carriers tend to develop breast cancers at a younger age and have a higher proportion of triple-negative cancers, they may not receive oophorectomy or tamoxifen; CPM may therefore be more beneficial in this setting.

One must also consider the potential psychosocial and emotional impact of CPM. Although studies have shown that psychosocial outcomes and quality of life are similar between women at increased risk of breast cancer who choose prophylactic mastectomy and those who do not [36–39], negative effects on body image and sexuality occur in a significant minority of patients [36, 38, 39]. Therefore, an adequate discussion and counseling of carriers with unilateral breast cancer should also include the psychosocial dimensions of each surgical option.

Patient Selection

Current evidence is insufficient to generate a clinical guideline for surgical management of unilateral breast cancer which encompasses all *BRCA* mutation carriers, and patients should be counseled on a case-by-case basis. There is evidence to identify subgroups of patients who may be at a lower risk for IBTR and/or CBC, and who may therefore benefit from less aggressive surgery. Adjuvant chemotherapy and oophorectomy are each associated with a 50% decreased risk for IBTR. *BRCA1-*/ *BRCA2*-related cancers may be more sensitive to chemotherapy, decreasing the risk for IBTR, and may lead to onset of premature menopause, further decreasing the risk for patients with hormone-sensitive tumors. The *proportional* reduction in risk of IBTR for mutation carriers who have had oophorectomy is similar to the reduction in risk for development of breast cancer in mutation carriers who have had prophylactic oophorectomy [40]. For mutation carriers who will not be receiving chemotherapy and/or oophorectomy, a more aggressive surgical approach (mastectomy with or without CPM) may be reasonable.

Oophorectomy reduces the risk of CBC by about 50%, as does adjuvant tamoxifen. Older age at first breast cancer diagnosis is also associated with a reduced risk of CBC. Indeed, the risk for CBC significantly decreased with increased age, with 50 years being the age cut-off used in most studies. Accordingly, younger patients who receive neither oophorectomy nor tamoxifen constitute another subgroup of patients who may benefit from a more aggressive surgical approach.

Nipple Sparing Mastectomy

Nipple sparing mastectomy (NSM) is a surgical technique in which the breast tissue is removed preserving the entire mastectomy skin envelope and nipple areola complex (NAC). For some women, the NAC plays an important role in their body image, and hence, NSM may enhance cosmetic outcome and offer psychological benefit [41]. Compared to skin-sparing mastectomy, patients report greater cosmetic satisfaction [42, 43] and improved psychosocial and sexual well-being [44].

Candidates for NSM include patients undergoing prophylactic mastectomy and those with ductal carcinoma in situ (DCIS) or invasive breast cancer of limited extent [45]. Appropriately selected, only 12% of patients will have tumor involvement at the NAC, precluding NSM [46, 47]. Factors associated with nipple involvement include tumors larger than 2–4 cm, a tumor–nipple distance of less than 2 cm, tumors involving more than one quadrant, and tumors with unfavorable biology (poorly differentiated, ER/PR negative, HER2 positive, or extensive intraductal component). With increasing experience, the selection boundaries for NSM are being extended to encompass more advanced disease [48–51], and a study from the National Cancer Institute's Surveillance, Epidemiology, and End Results (SEER) database has demonstrated a 202% increase in the use of NSM between 2005 and 2009 in the USA [52].

The oncologic safety of NSM in *BRCA* mutation carriers is controversial. One study of *BRCA* patients identified terminal ductal lobular units in 24% of the NACs and 8% of nipples, with occult tumor involving the NAC in 0% of risk-reducing specimens, and 10% in therapeutic specimens [53]. These rates are similar to those for non-BRCA mutation carriers. In a recent study of 177 NSMs with immediate breast reconstruction (IBR) in 89 *BRCA* mutation carriers, 26 patients had NSM with CPM for early-stage disease and 63 has prophylactic NSM, and at 26- to 28-month follow-up, there were no loco-regional recurrences or new breast cancers [54]. Five patients (6%) required subsequent excision of the nipple-areola complex for oncological or other reasons. Larger studies with longer follow-up are needed before the oncological safety of NSM in *BRCA* mutation carriers can be assured.

Breast Reconstruction Considerations

Planning for breast reconstruction, immediate or delayed, must consider the impact of post-mastectomy radiotherapy (PMRT). The two main issues which raise concern are compromised delivery of RT by the reconstructed breast and the impact of RT on the long-term cosmetic result of the reconstruction [55]. Cosmesis and symmetry issues may be more evident in the setting of bilateral mastectomy with immediate reconstruction, followed by unilateral PMRT.

Historically, patients requiring PMRT have been encouraged to have delayed breast reconstruction, based on concern that the reconstruction would compromise the delivery of PMRT [56–59]. Specific concerns include compromised delivery to the internal mammary nodes, non-uniform delivery, underdosing of the chest wall, and increased dose to normal tissues [55], but the evidence for these concerns is conflicting. Motwani et al. [59] reported compromised delivery of RT in 52% of patients who had undergone immediate reconstruction, compared with 7% of controls. Koutcher et al. [60] found no compromise in PMRT for most patients, with a 30-month loco-regional control rate of 97%. Surgeons at the University of Texas MD Anderson Cancer Center have advocated an "delayed–immediate" reconstruction algorithm for patients who need to receive PMRT [61]. A tissue expander is placed at the time of mastectomy, deflated during adjuvant RT, expanded post-RT, and followed by autologous flap reconstruction 4–6 months later [62]. They report low complication rates, tissue expander loss in 14% of patients, and local recurrence at 32 months' follow-up of 3% [63]. They suggest that the complication rate with a "delayed–immediate" approach and subsequent flap reconstruction may be lower than that for a standard delayed flap reconstruction (26 vs. 38%, $p = 0.4$) [62], but many others have reported acceptable cosmetic and oncologic outcomes with immediate reconstruction followed by PMRT [60]. In an analysis of 191 patients requiring PMRT who underwent TRAM flap reconstruction, the risk of loco-regional recurrence at 40 months' follow-up did not significantly differ between immediate and delayed procedures (3.7 vs. 1.8%, $p = 0.65$) [64]. In a more recent report from the same authors, among 492 patients with stage II–III disease who received mastectomy, chemotherapy, and PMRT, at a median follow-up of 7.2 years, there was no difference in local recurrence, disease-free survival, or overall survival, between immediate and delayed flap reconstruction [65]. Similarly, Wright et al. [66] reported on 104 patients with tissue expander reconstruction who underwent exchange to a permanent implant prior to PMRT. Local control rates were excellent, and immediate breast reconstruction was not associated with increased risk of distant metastases or death. In contrast, Nahabedian et al. [67] retrospectively analyzed 146 patients who underwent immediate or delayed reconstruction after PMRT, finding that loco-regional recurrence rates were higher in patients who underwent immediate versus delayed reconstruction (27 vs. 15%, $p = 0.04$). These data should be interpreted with caution based on the surprisingly high rates of recurrence [67, 68]. Since randomized trials are unlikely, the safety of breast reconstruction prior to PMRT remains controversial.

Regarding cosmesis and PMRT, the data favor delayed breast reconstruction. PMRT to the reconstructed breast is associated with fat necrosis, impaired wound healing, contracture, fibrosis, volume loss, and architectural distortion [69]. In a systematic review [69] of 10 published reports of patients undergoing immediate and delayed reconstruction and PMRT, the authors found a higher incidence of breast fibrosis and contracture with immediate reconstruction. Adesiyun et al. [70] compared immediate versus delayed breast reconstruction in 113 patients receiving PMRT and found a trend toward fewer complications in the delayed reconstruction group (32 vs. 44%, $p = 0.18$), but with a comparable proportion of patients satisfied with their cosmetic outcomes (68%) [70]. Another group found no significant difference in complication rates with immediate versus delayed flap reconstruction in patients who received PMRT, but the authors ultimately recommended delayed reconstruction [71]. In contrast, others have reported acceptable cosmesis and complication rates with immediate reconstruction. A meta-analysis of 11 studies by Barry et al. [72] concluded that postoperative outcomes did not differ by the sequencing of reconstruction and PMRT, but that autologous flaps appeared to have superior outcomes, with lower rates of fibrosis, contracture, infection, fat necrosis, and reoperation for flap- versus implant-based reconstruction [73]. The cosmetic results of autologous flap reconstruction appear to be superior to those of tissue expander/implant reconstruction, especially in the setting of PMRT [74], but successful outcomes have been achieved with implant-based reconstruction. For example, Cordeiro et al. [75, 76] reported satisfactory aesthetic results with immediate tissue expander placement followed by exchange for a permanent implant prior to PMRT, with aesthetic results categorized as "good to excellent" in 90% of patients, and with an implant loss rate of 9.1% [76].

Another important issue is post-mastectomy reconstruction of the previously irradiated breast. Mutation carriers who are considering BCT must understand that if they later develop IBTR then mastectomy is the standard treatment. Others may have discovered their mutation status belatedly after completing BCT and are now considering completion mastectomy and CPM. In a recent systematic review [77] of patients with prior RT having mastectomy and immediate breast reconstruction, the rates of reoperation and of reconstructive failure were 37 and 17%, respectively. Prior breast RT is not an absolute contraindication to NSM, and although complication rates are higher in this setting, reconstruction failure and/or nipple/areola necrosis is infrequent [78]. Prior RT to one breast may affect the symmetry of a bilateral reconstruction; this should be considered in *BRCA* mutation carriers who are planning surgery for the initial breast cancer and, for those who previously elected BCT, in managing a subsequent IBTR.

Breast reconstruction could delay the start of adjuvant chemotherapy, although in a systematic review of four studies of women with versus without immediate breast reconstruction followed by chemotherapy, Xavier Harmeling et al. [79] found delays after immediate breast reconstruction of only 6.6–16.8 days. One could hypothesize that free flap reconstruction might lead to the greatest delay, but Kontos et al. [80] compared 27 women with free flap reconstruction to non-reconstructed controls and found that the mean time to chemotherapy was 55 versus 40 days, with delays past 6 weeks in 67% of flap patients versus 29% of controls. The most common reasons for delays were flap and donor-site complications. Since current guidelines recommend the initiation of adjuvant chemotherapy 4–12 weeks post-mastectomy [81], an adverse effect from immediate breast reconstruction is unlikely.

Conclusions

All surgical options suitable for the management of a unilateral breast cancer are open to *BRCA* mutation carriers. Regarding BCT, studies with short follow-up suggest no difference in the rate of "true" IBTR compared to non-mutation carriers, while longer follow-up studies suggest higher rates of "second primary" IBTR. In comparing BCT with mastectomy, there may be differences in BCSS but not in OS. Present evidence for the oncological safety of BCT in mutation carriers will continue to evolve and mature. Mutation carriers have a 3.5-fold increased risk of CBC, with the risk for *BRCA1* being higher than for *BRCA2* carriers. CPM can reduce this risk but should be considered in the context of patient age, comorbidities, the stage of the index lesion, increased surgical morbidity, and psychosocial aspects. A survival benefit for CPM has yet to be proven. Chemotherapy, endocrine therapy, and/or oophorectomy can each alter the risks of IBTR and of CBC, and need to be included in risk/benefit discussions.

References

1. Nelson HD, Fu R, Goddard K, Mitchell JP, Okinaka-Hu L, Pappas M, Zakher B. Risk assessment, genetic counseling, and genetic testing for BRCA-related cancer: systematic review to update the U.S. Preventive Services task force recommendation. Rockville (MD): Agency for Healthcare Research and Quality (US); 2013 Dec. Report No.: 12-05164-EF-1. U. S. Preventive Services Task Force Evidence Syntheses, formerly Systematic Evidence Reviews.
2. Hartmann LC, Sellers TA, Schaid DJ, Frank TS, Soderberg CL, Sitta DL, et al. Efficacy of bilateral prophylactic mastectomy in BRCA1 and BRCA2 gene mutation carriers. J Natl Cancer Inst. 2001;93(21):1633–7.
3. Meijers-Heijboer H, van Geel B, van Putten WL, Henzen-Logmans SC, Seynaeve C, Menke-Pluymers MB, et al. Breast cancer after prophylactic bilateral mastectomy in women with a BRCA1 or BRCA2 mutation. N Engl J Med. 2001;345(3):159–64.
4. Rebbeck TR, Friebel T, Lynch HT, Neuhausen SL, van't Veer L, Garber JE, et al. Bilateral prophylactic mastectomy reduces breast cancer risk in BRCA1 and BRCA2 mutation carriers: the PROSE Study Group. J Clin Oncol. 2004;22(6):1055–62.
5. Rebbeck TR, Levin AM, Eisen A, Snyder C, Watson P, Cannon-Albright L, et al. Breast cancer risk after bilateral prophylactic oophorectomy in BRCA1 mutation carriers. J Natl Cancer Inst. 1999;91(17):1475–9.
6. Rebbeck TR, Lynch HT, Neuhausen SL, Narod SA, Van't Veer L, Garber JE, et al. Prevention and Observation of Surgical End Points Study Group. Prophylactic oophorectomy in carriers of BRCA1 or BRCA2 mutations. N Engl J Med. 2002;346(21):1616–22.
7. Narod SA, Brunet JS, Ghadirian P, Robson M, Heimdal K, Neuhausen SL, et al. Hereditary Breast Cancer Clinical Study Group. Tamoxifen and risk of contralateral breast cancer in BRCA1 and BRCA2 mutation carriers: a case-control study. Lancet. 2000;356(9245): 1876–81.
8. Clarke M, Collins R, Darby S, Davies C, Elphinstone P, Evans V, et al. Early Breast Cancer Trialists' Collaborative Group (EBCTCG). Effects of radiotherapy and of differences in the extent of surgery for early breast cancer on local recurrence and 15-year survival: an overview of the randomised trials. Lancet. 2005;366(9503):2087–106.
9. Pierce LJ, Strawderman M, Narod SA, Oliviotto I, Eisen A, Dawson L, et al. Effect of radiotherapy after breast-conserving treatment in women with breast cancer and germline BRCA1/2 mutations. J Clin Oncol. 2000;18(19):3360–9.
10. Drooger J, Akdeniz D, Pignol JP, Koppert LB, McCool D, Seynaeve CM, et al. Adjuvant radiotherapy for primary breast cancer in BRCA1 and BRCA2 mutation carriers and risk of contralateral breast cancer with special attention to patients irradiated at younger age. Breast Cancer Res Treat. 2015;154(1):171–80.

11. Valachis A, Nearchou AD, Lind P. Surgical management of breast cancer in BRCA-mutation carriers: a systematic review and meta-analysis. Breast Cancer Res Treat. 2014;144(3): 443–55.
12. Brekelmans CT, Tilanus-Linthorst MM, Seynaeve C, vd Ouweland A, Menke-Pluymers MB, Bartels CC, et al. Tumour characteristics, survival and prognostic factors of hereditary breast cancer from BRCA2-, BRCA1- and non-BRCA1/2 families as compared to sporadic breast cancer cases. Eur J Cancer. 2007;43(5):867–76.
13. Chappuis PO, Kapusta L, Bégin LR, Wong N, Brunet JS, Narod SA, et al. Germline BRCA1/2 mutations and p27(Kip1) protein levels independently predict outcome after breast cancer. J Clin Oncol. 2000;18(24):4045–52.
14. El-Tamer M, Russo D, Troxel A, Bernardino LP, Mazziotta R, Estabrook A, et al. Survival and recurrence after breast cancer in BRCA1/2 mutation carriers. Ann Surgical Oncol. 2004;11(2):157–64.
15. Haffty BG, Harrold E, Khan AJ, Pathare P, Smith TE, Turner BC, et al. Outcome of conservatively managed early-onset breast cancer by BRCA1/2 status. Lancet. 2002;359 (9316):1471–7.
16. Robson M, Gilewski T, Haas B, Levin D, Borgen P, Rajan P, et al. BRCA-associated breast cancer in young women. J Clin Oncol. 1998;16(5):1642–9.
17. Robson ME, Chappuis PO, Satagopan J, Wong N, Boyd J, Goffin JR, et al. A combined analysis of outcome following breast cancer: differences in survival based on BRCA1/BRCA2 mutation status and administration of adjuvant treatment. Breast Cancer Res. 2004;6(1):R8–17.
18. Eccles D, Simmonds P, Goddard J, Coultas M, Hodgson S, Lalloo F, et al. Familial breast cancer: an investigation into the outcome of treatment for early stage disease. Fam Cancer. 2001;1(2):65–72.
19. Garcia-Etienne CA, Barile M, Gentilini OD, Botteri E, Rotmensz N, Sagona A, et al. Breast-conserving surgery in BRCA1/2 mutation carriers: are we approaching an answer? Ann Surg Oncol. 2009;16(12):3380–7.
20. Kirova YM, Savignoni A, Sigal-Zafrani B, de La Rochefordiere A, Salmon RJ, This P, et al. Is the breast-conserving treatment with radiotherapy appropriate in BRCA1/2 mutation carriers? Long-term results and review of the literature. Breast Cancer Res Treat. 2010; 120(1):119–26.
21. Pierce LJ, Levin AM, Rebbeck TR, Ben-David MA, Friedman E, Solin LJ, et al. Ten-year multi-institutional results of breast-conserving surgery and radiotherapy in BRCA1/2-associated stage I/II breast cancer. J Clin Oncol. 2006;24(16):2437–43.
22. Pierce LJ, Phillips KA, Griffith KA, Buys S, Gaffney DK, Moran MS, et al. Local therapy in BRCA1 and BRCA2 mutation carriers with operable breast cancer: comparison of breast conservation and mastectomy. Breast Cancer Res Treat. 2010;121(2):389–98.
23. Nilsson MP, Hartman L, Kristoffersson U, Johannsson OT, Borg A, Henriksson K, et al. High risk of in-breast tumor recurrence after BRCA1/2-associated breast cancer. Breast Cancer Res Treat. 2014;147(3):571–8.
24. Metcalfe K, Lynch HT, Ghadirian P, Tung N, Kim-Sing C, Olopade OI, et al. Risk of ipsilateral breast cancer in BRCA1 and BRCA2 mutation carriers. Breast Cancer Res Treat. 2011;127(1):287–96.
25. Stoppa-Lyonnet D, Ansquer Y, Dreyfus H, Gautier C, Gauthier-Villars M, Bourstyn E, et al. Familial invasive breast cancers: worse outcome related to BRCA1 mutations. J Clin Oncol. 2000;18(24):4053–9.
26. Molina-Montes E, Pérez-Nevot B, Pollán M, Sánchez-Cantalejo E, Espín J, Sánchez MJ. Cumulative risk of second primary contralateral breast cancer in BRCA1/BRCA2 mutation carriers with a first breast cancer: a systematic review and meta-analysis. Breast. 2014;23(6):721–42.

27. Metcalfe K, Gershman S, Lynch HT, Ghadirian P, Tung N, Kim-Sing C, et al. Predictors of contralateral breast cancer in BRCA1 and BRCA2 mutation carriers. Br J Cancer. 2011; 104(9):1384–92.
28. Mavaddat N, Peock S, Frost D, Ellis S, Platte R, Fineberg E, et al. EMBRACE. Cancer risks for BRCA1 and BRCA2 mutation carriers: results from prospective analysis of EMBRACE. J Natl Cancer Inst. 2013;105(11):812–22.
29. Gronwald J, Tung N, Foulkes WD, Offit K, Gershoni R, Daly M, et al. Hereditary Breast Cancer Clinical Study Group. Tamoxifen and contralateral breast cancer in BRCA1 and BRCA2 carriers: an update. Int J Cancer. 2006;118(9):2281–4.
30. Reding KW, Bernstein JL, Langholz BM, Bernstein L, Haile RW, Begg CB, et al. WECARE Collaborative Study Group. Adjuvant systemic therapy for breast cancer in BRCA1/BRCA2 mutation carriers in a population-based study of risk of contralateral breast cancer. Breast Cancer Res Treat. 2010;123(2):491–8.
31. Graeser MK, Engel C, Rhiem K, Gadzicki D, Bick U, Kast K, et al. Contralateral breast cancer risk in BRCA1 and BRCA2 mutation carriers. J Clin Oncol. 2009;27(35):5887–92.
32. Malone KE, Begg CB, Haile RW, Borg A, Concannon P, Tellhed L, et al. Population-based study of the risk of second primary contralateral breast cancer associated with carrying a mutation in BRCA1 or BRCA2. J ClinOncol. 2010;28(14):2404–10.
33. Verhoog LC, Brekelmans CT, Seynaeve C, Meijers-Heijboer EJ, Klijn JG. Contralateral breast cancer risk is influenced by the age at onset in BRCA1-associated breast cancer. Br J Cancer. 2000;83(3):384–6.
34. van Sprundel TC, Schmidt MK, Rookus MA, Brohet R, van Asperen CJ, Rutgers EJ, et al. Risk reduction of contralateral breast cancer and survival after contralateral prophylactic mastectomy in BRCA1 or BRCA2 mutation carriers. Br J Cancer. 2005;93(3):287–92.
35. Heemskerk-Gerritsen BA, Rookus MA, Aalfs CM, Ausems MG, Collée JM, Jansen L, et al. Improved overall survival after contralateral risk-reducing mastectomy in BRCA1/2 mutation carriers with a history of unilateral breast cancer: a prospective analysis. Int J Cancer. 2015;136(3):668–77.
36. Brandberg Y, Sandelin K, Erikson S, Jurell G, Liljegren A, Lindblom A, et al. Psychological reactions, quality of life, and body image after bilateral prophylactic mastectomy in women at high risk for breast cancer: a prospective 1-year follow-up study. J Clin Oncol. 2008; 26(24):3943–9.
37. Tercyak KP, Peshkin BN, Brogan BM, DeMarco T, Pennanen MF, Willey SC, et al. Quality of life after contralateral prophylactic mastectomy in newly diagnosed high-risk breast cancer patients who underwent BRCA1/2 gene testing. J Clin Oncol. 2007;25(3):285–91.
38. Unukovych D, Sandelin K, Liljegren A, Arver B, Wickman M, Johansson H, Brandberg Y. Contralateral prophylactic mastectomy in breast cancer patients with a family history: a prospective 2-years follow-up study of health related quality of life, sexuality and body image. Eur J Cancer. 2012;48(17):3150–6.
39. Frost MH, Slezak JM, Tran NV, Williams CI, Johnson JL, Woods JE, et al. Satisfaction after contralateral prophylactic mastectomy: the significance of mastectomy type, reconstructive complications, and body appearance. J Clin Oncol. 2005;23(31):7849–56.
40. Rebbeck TR, Kauff ND, Domchek SM. Meta-analysis of risk reduction estimates associated with risk-reducing salpingo-oophorectomy in BRCA1 or BRCA2 mutation carriers. J Natl Cancer Inst. 2009;101(2):80–7.
41. Wellisch DK, Schain WS, Noone RB, Little JW 3rd. The psychological contribution of nipple addition in breast reconstruction. Plast Reconstr Surg. 1987;80(5):699–704.
42. Boneti C, Yuen J, Santiago C, Diaz Z, Robertson Y, Korourian S, et al. Oncologic safety of nipple skin-sparing or total skin-sparing mastectomies with immediate reconstruction. J Am Coll Surg. 2011;212(4):686–93; discussion 693–5.
43. Didier F, Radice D, Gandini S, Bedolis R, Rotmensz N, Maldifassi A, et al. Does nipple preservation in mastectomy improve satisfaction with cosmetic results, psychological adjustment, body image and sexuality? Breast Cancer Res Treat. 2009;118(3):623–33.

44. Wei CH, Scott AM, Price AN, Miller HC, Klassen AF, Jhanwar SM, et al. Psychosocial and sexual well-being following nipple-sparing mastectomy and reconstruction. Breast J. 2016; 22(1):10–7.
45. Chen CM, Disa JJ, Sacchini V, Pusic AL, Mehrara BJ, et al. Nipple-sparing mastectomy and immediate tissue expander/implant breast reconstruction. Plast Reconstr Surg. 2009; 124(6):1772–80.
46. Crowe JP, Patrick RJ, Yetman RJ, Djohan R. Nipple-sparing mastectomy update: one hundred forty-nine procedures and clinical outcomes. Arch Surg. 2008;143(11):1106–10; discussion 1110.
47. Paepke S, Schmid R, Fleckner S, Paepke D, Niemeyer M, Schmalfeldt B, et al. Subcutaneous mastectomy with conservation of the nipple-areola skin: broadening the indications. Ann Surg. 2009;250(2):288–92.
48. Burdge EC, Yuen J, Hardee M, Gadgil PV, Das C, Henry-Tillman R, et al. Nipple skin-sparing mastectomy is feasible for advanced disease. Ann Surg Oncol. 2013; 20(10):3294–302.
49. Fortunato L, Loreti A, Andrich R, Costarelli L, Amini M, Farina M, et al. When mastectomy is needed: is the nipple-sparing procedure a new standard with very few contraindications? J Surg Oncol. 2013;108(4):207–12.
50. Poruk KE, Ying J, Chidester JR, Olson JR, Matsen CB, et al. Breast cancer recurrence after nipple-sparing mastectomy: one institution's experience. Am J Surg. 2015;209(1):212–7.
51. Peled AW, Wang F, Foster RD, Alvarado M, Ewing CA, et al. Expanding the indications for total skin-sparing mastectomy: is it safe for patients with locally advanced disease? Ann Surg Oncol. 2016;23(1):87–91.
52. Agarwal S, Agarwal S, Neumayer L, Agarwal JP. Therapeutic nipple-sparing mastectomy: trends based on a national cancer database. Am J Surg. 2014;208(1):93–8.
53. Reynolds C, Davidson JA, Lindor NM, Glazebrook KN, Jakub JW, Degnim AC, et al. Prophylactic and therapeutic mastectomy in BRCA mutation carriers: can the nipple be preserved? Ann Surg Oncol. 2011;18(11):3102–9.
54. Manning AT, Wood C, Eaton A, Stempel M, Capko D, Pusic A, et al. Nipple-sparing mastectomy in patients with BRCA1/2 mutations and variants of uncertain significance. Br J Surg. 2015;102(11):1354–9.
55. Buchholz TA, Strom EA, Perkins GH, McNeese MD. Controversies regarding the use of radiation after mastectomy in breast cancer. Oncologist. 2002;7(6):539–46.
56. Bristol IJ, Woodward WA, Strom EA, Cristofanilli M, Domain D, Singletary SE, et al. Locoregional treatment outcomes after multimodality management of inflammatory breast cancer. Int J Radiat Oncol Biol Phys. 2008;72(2):474–84.
57. Schechter NR, Strom EA, Perkins GH, Arzu I, McNeese MD, Langstein HN, et al. Immediate breast reconstruction can impact postmastectomy irradiation. Am J Clin Oncol. 2005; 28(5):485–94.
58. Kronowitz SJ, Robb GL. Breast reconstruction and adjuvant therapies. Semin Plast Surg. 2004;18(2):105–15.
59. Motwani SB, Strom EA, Schechter NR, Butler CE, Lee GK, Langstein HN, et al. The impact of immediate breast reconstruction on the technical delivery of postmastectomy radiotherapy. Int J Radiat Oncol Biol Phys. 2006;66(1):76–82.
60. Koutcher L, Ballangrud A, Cordeiro PG, McCormick B, Hunt M, Van Zee KJ, et al. Postmastectomy intensity modulated radiation therapy following immediate expander-implant reconstruction. Radiother Oncol. 2010;94(3):319–23.
61. Kronowitz SJ, Hunt KK, Kuerer HM, Babiera G, McNeese MD, Buchholz TA, et al. Delayed-immediate breast reconstruction. Plast Reconstr Surg. 2004;113(6):1617–28.
62. Kronowitz SJ. Delayed-immediate breast reconstruction: technical and timing considerations. Plast Reconstr Surg. 2010;125(2):463–74.

63. Buchholz TA, Kronowitz SJ, Kuerer HM. Immediate breast reconstruction after skin-sparing mastectomy for the treatment of advanced breast cancer: radiation oncology considerations. Ann Surg Oncol. 2002;9(8):820–1.
64. Huang CJ, Hou MF, Lin SD, Chuang HY, Huang MY, Fu OY, Lian SL. Comparison of local recurrence and distant metastases between breast cancer patients after postmastectomy radiotherapy with and without immediate TRAM flap reconstruction. Plast Reconstr Surg. 2006;118(5):1079–86; discussion 1087–8.
65. Lee HH, Hou MF, Wei SY, Lin SD, Luo KH, Huang MY, et al. Comparison of long-term outcomes of postmastectomy radiotherapy between breast cancer patients with and without immediate flap reconstruction. PLoS ONE. 2016;11(2):e0148318.
66. Wright JL, Cordeiro PG, Ben-Porat L, Van Zee KJ, Hudis C, Beal K, McCormick B. Mastectomy with immediate expander-implant reconstruction, adjuvant chemotherapy, and radiation for stage II-III breast cancer: treatment intervals and clinical outcomes. Int J Radiat Oncol Biol Phys. 2008;70(1):43–50.
67. Nahabedian MY, Momen B. The impact of breast reconstruction on the oncologic efficacy of radiation therapy: a retrospective analysis. Ann Plast Surg. 2008;60(3):244–50.
68. Anavekar NS1, Rozen WM, Le Roux CM, Ashton MW. Achieving autologous breast reconstruction for breast cancer patients in the setting of post-mastectomy radiotherapy. J Cancer Surviv. 2011;5(1):1–7.
69. Javaid M, Song F, Leinster S, Dickson MG, James NK. Radiation effects on the cosmetic outcomes of immediate and delayed autologous breast reconstruction: an argument about timing. J Plast Reconstr Aesthet Surg. 2006;59(1):16–26.
70. Adesiyun TA, Lee BT, Yueh JH, Chen C, Colakoglu S, Anderson KE, et al. Impact of sequencing of postmastectomy radiotherapy and breast reconstruction on timing and rate of complications and patient satisfaction. Int J Radiat Oncol Biol Phys. 2011;80(2):392–7.
71. Spear SL, Ducic I, Low M, Cuoco F. The effect of radiation on pedicled TRAM flap breast reconstruction: outcomes and implications. Plast Reconstr Surg. 2005;115(1):84–95.
72. Barry M, Kell MR. Radiotherapy and breast reconstruction: a meta-analysis. Breast Cancer Res Treat. 2011;127(1):15–22.
73. Barry AS, Bacin F, Kodjikian L, Benbouzid F, Balmitgere T, Grange JD. Choroidal metastases of lung neoplasm treated with external radiotherapy and polychemotherapy: a study of four clinical cases. J Fr Ophtalmol. 2012;35(2):122 e1–8.
74. Newman LA, Kuerer HM, Hunt KK, Ames FC, Ross MI, Theriault R, et al. Feasibility of immediate breast reconstruction for locally advanced breast cancer. Ann Surg Oncol. 1999; 6(7):671–5.
75. Cordeiro PG, Pusic AL, Disa JJ, McCormick B, VanZee K. Irradiation after immediate tissue expander/implant breast reconstruction: outcomes, complications, aesthetic results, and satisfaction among 156 patients. Plast Reconstr Surg. 2004;113(3):877–81.
76. Cordeiro PG, Albornoz CR, McCormick B, Hu Q, Van Zee K. The impact of postmastectomy radiotherapy on two-stage implant breast reconstruction: an analysis of long-term surgical outcomes, aesthetic results, and satisfaction over 13 years. Plast Reconstr Surg. 2014; 134(4):588–95.
77. El-Sabawi B, Sosin M, Carey JN, Nahabedian MY, Patel KM. Breast reconstruction and adjuvant therapy: a systematic review of surgical outcomes. J Surg Oncol. 2015;112(5): 458–64.
78. Tang R, Tang R, Coopey SB, Colwell AS, Specht MC, Gadd MA, et al. Nipple-sparing mastectomy in irradiated breasts: selecting patients to minimize complications. Ann Surg Oncol. 2015;22(10):3331–7.
79. Harmeling JX, Kouwenberg CAE, Bijlard E, Burger KNJ, Jage A, Mureau MAM. The effect of immediate breast reconstruction on the timing of adjuvant chemotherapy: a systematic review. Breast Cancer Res Treat. 2015;153(2):241–51.

80. Kontos M, Lewis RS, Lüchtenborg M, Holmberg L, Hamed H. Does immediate breast reconstruction using free flaps lead to delay in the administration of adjuvant chemotherapy for breast cancer? Eur J Surg Oncol. 2010;36(8):745–9.
81. Lohrisch C, Paltiel C, Glemon K, Speers C, Taylor S, et al. Impact on survival of time from definitive surgery to initiation of adjuvant chemotherapy for early-stage breast cancer. J Clin Oncol. 2006;24(30):4888–94.

Medical Management of Breast Cancer in *BRCA* Mutation Carriers

8

Soley Bayraktar and Banu K. Arun

Role of *BRCA* Genes

BRCA1 and *BRCA2* function as tumor suppressor genes and are important in the maintenance of genomic stability through their role in DNA damage signaling and DNA repair. Both *BRCA1* and *BRCA2* have been implicated in mediating the repair of double-strand breaks by homologous recombination (HR) by interactions with *RAD51*. Upon DNA damage, *BRCA1* associates with *RAD51* and localizes to the damaged region; *BRCA1* is then phosphorylated. *BRCA2* functions downstream of *BRCA1* by forming a complex with *RAD51*. The primary function of *BRCA2* is to facilitate HR [1]. Cells deficient for *BRCA1* or *BRCA2* are unable to repair double-strand breaks via error-free HR, resulting in repair via the error-prone non-homologous end-joining (NHEJ) pathway, which introduces chromosomal instability [2, 3]. During S-phase, the expression levels of *BRCA1* and *BRCA2* increase, indicating a function in maintaining genomic stability during the DNA replication process [4]. In addition to its role in HR, *BRCA1* appears to have functions in DNA repair. *BRCA1* is part of the *BRCA1*-associated genome-surveillance complex (BASC), which includes ATM, *RAD50*, *MRE11*, *NBS1*, and the mismatch repair proteins MLH1, PMS2, MSH2, and MSH6 [5]. *BRCA1* is also involved in transcription-coupled excision repair, chromatin

S. Bayraktar
Department of Medical Oncology, Memorial Cancer Center,
Piyalepasa Bulvari, Sisli, 34340 Istanbul, Turkey
e-mail: soley.bayraktar@gmail.com

B.K. Arun (✉)
Department of Breast Medical Oncology, The University of Texas MD Anderson
Cancer Center, 1515 Holcombe Boulevard, Unit 1354, Houston, TX 77030-4009, USA
e-mail: barun@mdanderson.org

A.B. Chagpar (ed.), *Managing BRCA Mutation Carriers*,
DOI 10.1007/978-3-319-59198-8_8

remodeling, and, together with *BARD1*, the ubiquitination process, through which proteins are tagged for degradation by the proteasome [1, 6].

A germline mutation in *BRCA1* or *BRCA2* only represents the first hit in the classical Knudson's two-hit hypothesis, whereas the second inactivating somatic mutation often involves deletion of the wild-type allele, termed loss of heterozygosity (LOH). LOH is present in the majority (80%) of tumors arising from mutation carriers [4, 5]. By contrast, small somatic mutations involving a single or few bases are very rare [6]. Another somatic inactivation mechanism, epigenetic silencing by promoter methylation, has been reported for *BRCA1* in 9–13% of sporadic breast tumors and up to 42% of non-*BRCA1/2* hereditary breast tumors, leading to reduced *BRCA1* expression [7, 8]. By contrast, *BRCA1* promoter methylation is rare in tumors from *BRCA1* and *BRCA2* mutation carriers [9], and *BRCA2* promoter methylation is seldom observed in both sporadic and hereditary breast cancers [10]. Genetic testing for *BRCA* mutations is now widely available, and multiple professional societies have published guidelines for testing and management. Genetic testing trends include utilization of multi-gene panels that take advantage of next-generation sequencing and testing for low- and moderate-penetrance susceptibility genes [11].

BRCA1- and *BRCA2*-Associated Breast Cancers

Up to 10% of breast cancers result from specific genetic mutations in *BRCA1*, *BRCA2*, *CHEK2*, *TP53*, and *PTEN* [12]. Families carrying genetic mutations in the above-mentioned genes exhibit an apparently dominant inheritance pattern and are often characterized by early age of onset and over-representation of ovarian, bilateral breast, and male breast cancers [13].

Early reports suggested that germline mutations in the genes *BRCA1* and *BRCA2* were responsible for the majority of hereditary breast cancers, although more recent studies have demonstrated that mutations in the two genes only account for 25–28% of the family risk [14, 15]. However, additional *BRCA1/2* mutations likely remain undetected by the screening methods used today. Women carrying a *BRCA1* or *BRCA2* germline mutation also have increased risk of other cancer types, such as ovarian cancer and fallopian tube cancer, male breast cancer, prostate cancer, pancreas cancer, gastrointestinal cancers (e.g., gall bladder, bile duct, and stomach), and melanoma [16–18]. In a large study by the Consortium of Investigators of Modifiers of *BRCA1/2* (CIMBA), the median age of breast cancer diagnosis was 40 years among *BRCA1* and 43 years among *BRCA2* mutation carriers [19].

Although germline mutations in *BRCA1* and *BRCA2* confer high risk of breast and ovarian cancers, the penetrance of these genes is incomplete. The risk of developing breast cancer by the age of 70 is 45–87% in *BRCA1* and *BRCA2* mutation carriers. For ovarian cancer, the risk is 45–60% among *BRCA1* mutation carriers and 11–35% among *BRCA2* mutation carriers [20–22]. Typically, breast cancer presents earlier than ovarian cancer, especially in *BRCA1* carriers [23]; the risk of developing a metachronous ovarian cancer is greatest for women with early-onset breast cancer (younger than 40 years at diagnosis) or with a family history of breast or ovarian cancer [24]. A large-scale historical cohort study looked at the risk of metachronous ovarian cancer after *BRCA* breast cancer and determined that the 10-year actuarial risk for *BRCA1* carriers was 12.7% and that for *BRCA2* carriers was 6.8% [25].

The majority of invasive breast cancers that arise in *BRCA1* and *BRCA2* carriers are invasive ductal carcinomas (IDCs) (80%) [26]. A higher frequency of *BRCA1* tumors are classified as medullary carcinomas compared with sporadic tumors (9% vs. 2%, respectively) [19, 27]. Notably, 11% of medullary carcinomas carry *BRCA1* germline mutations [28]. By contrast, an excess of invasive lobular and tubular carcinomas has been reported for *BRCA2* tumors relative to *BRCA1* tumors [27]. *BRCA1* tumors are more frequently high grade compared with sporadic tumors [29]. Most *BRCA2* tumors are grade 2/3 with high mitotic rates.

A recent study examining pathology data from 4325 *BRCA1* to 2568 *BRCA2* mutation carriers reported that 78% of tumors arising in *BRCA1* carriers were estrogen receptor (ER)-negative, while only 23% of tumors arising in *BRCA2* mutation carriers were ER-negative. Furthermore, HER2 overexpression was only observed in approximately 10% of tumors from mutation carriers. Consequently, 69% of the *BRCA1* tumors were triple-negative (TN), which was true for only 16% of the *BRCA2* tumors [19]. In contrast to *BRCA1* tumors, *BRCA2* tumors seem to be more similar to sporadic tumors with respect to the expression of IHC markers. Most *BRCA2* breast tumors exhibit a luminal phenotype featuring overexpression of ER, progesterone receptor (PR), and cytokeratins CK8 and CK18 [30].

Recent studies have observed preinvasive lesions both in prophylactic mastectomy specimens from mutation carriers and in normal breast tissue adjacent to breast cancers [31]. Among *BRCA1/2*-associated breast cancers, 59% had at least one associated preinvasive lesion compared with 75% of controls. Preinvasive lesions were more prevalent in *BRCA2* mutation carriers than in *BRCA1* mutation carriers (70% vs. 52%, respectively). The most common preinvasive lesion in both groups was ductal carcinoma in situ (DCIS); 56% of *BRCA1/2*-associated breast cancers and 71% of the sporadic breast cancers had adjacent intraductal disease, respectively [31]. These findings suggest that *BRCA1/2*-associated breast cancers progress through the same intermediate steps as sporadic breast cancers and that DCIS should be considered part of the *BRCA1/2* tumor spectrum.

While most studies indicate a similar prognosis for women with hereditary breast cancers compared with age-matched women with sporadic breast cancers [32–38], other studies have reported worse survival outcomes [39–43]. Lee et al. [44] reported similar survival rates in *BRCA1* mutation carriers with TN disease compared with non-carriers. Confirming those findings, Bayraktar et al. [45] observed a 50% prevalence of deleterious *BRCA1/2* mutations in high-risk women diagnosed with TN breast cancer. Overall prognosis of TN breast cancer in *BRCA* carriers and non-carriers was not significantly different within the first 5 years following initial diagnosis. In other studies, the aggressive nature of these breast cancers has been demonstrated through a higher-median Oncotype DX recurrence score in ER-positive, node-negative BRCA-associated breast cancers as compared with controls [46] and inferior worse overall survival (OS) in *BRCA1* mutation carriers with breast and ovarian cancer as compared with non-carrier patients. Notably, this OS difference was not seen in *BRCA2* mutation carriers [47].

Systemic Therapy Options for Women with *BRCA* Mutation-Associated Breast Cancer

Traditionally, for those who develop breast or ovarian cancer, systemic therapy has been selected similarly to those with sporadic cancers, and the choice of chemotherapy (adjuvant or neoadjuvant as appropriate), endocrine therapy, and radiation has been based on ER/PR/HER2 status, lymph node involvement, and the size of the tumor. However, the approach to treatment is changing based on the recent data suggesting unique patterns of sensitivity and resistance to systemic therapies in *BRCA* mutation-associated breast cancers [48–53].

Platinum Agents

Due to the involvement of the *BRCA1/2* protein products in DNA-repair mechanisms, *BRCA* mutational status may impact sensitivity to different chemotherapeutic agents [54–56]. In vitro studies have demonstrated that *BRCA1*-defective cell lines are sensitive to DNA-damaging agents, such as platinum, and are relatively resistant to taxanes compared with BRCA-competent cell lines [57, 58]. Several subsequent clinical studies have supported these preclinical findings [48, 49, 51]. Byrski et al. reported a remarkable pathological complete response (pCR) rate of 80% in a small prospective trial evaluating neoadjuvant cisplatin in *BRCA1* mutation-associated breast cancer [48]. A subsequent study by the same group treated 107 women with stage 1–3 breast cancer with known *BRCA1* mutation with four cycles of neoadjuvant cisplatin and found a pCR rate of 61% (65/107 patients) [59]. The promising neoadjuvant data with cisplatin initiated a randomized phase III trial comparing carboplatin to docetaxel in metastatic *BRCA* mutation-associated breast cancer (NCT00321633) and a smaller phase II trial evaluating cisplatin for metastatic *BRCA1* mutation-associated breast cancer. Early results from the phase II trial have been encouraging, with 46% of women achieving a complete response and 26% of women achieving a partial response [60].

Antimicrotubule Agents

Several studies have evaluated the use of taxane agents in breast cancer patients with germline *BRCA* mutations; overall, these studies have found poorer outcomes in patients with *BRCA* mutations. A phase III randomized controlled trial comparing carboplatin with docetaxel in patients with metastatic breast cancer (MBC) or recurrent locally advanced BRCA-mutated breast cancer found that overall response rate (ORR) was 68% in those patients receiving carboplatin versus 33.3% in those who received docetaxel with progression free survival (PFS) of 6.8 months versus 3.1 months, respectively [61].

Interestingly, several studies have suggested that the lack of efficacy with taxane administration is limited to *BRCA1* carriers. A trial presented at the 2010 American Society of Clinical Oncology (ASCO) Annual Meeting showed improved ORR in *BRCA2* patients compared tonon-carriers treated with taxanes (ORR 75% vs. 36%); notably, however, the two groups had similar PFS (4.6 vs. 4.7 months). A more recent study presented at the 2015 San Antonio Breast Cancer Symposium (SABCS) evaluated the response of patients with stage I-III breast cancer, including 12 *BRCA* carriers, to neoadjuvant weekly taxane followed by adriamycin/cyclosphosphamide (AC) or 5-fluorouracil/epirubicin/cyclophosphamide (FEC). None of the four patients with *BRCA1* mutations had a radiographic complete response (CR) after taxanes (0%), while 2/8 *BRCA2* carriers had a radiographic CR after taxanes (25%); 16.1% (18/112) of non-carriers had a radiographic CR after taxane treatment alone. Notably, following taxane + AC/FEC treatment, pCR rate was 50% of *BRCA* carriers and 31.3% of non-carriers [62].Similarly, another study found that women with metastatic, hormone receptor-negative *BRCA1* mutation-associated breast cancer had lower response rates and shorter time to progression with a taxane-containing regimen compared with hormone receptor-negative sporadic breast cancer controls [63]. Overall, these findings suggest that *BRCA* mutation status predicts taxane resistance, with differential efficacy between *BRCA1* and *BRCA2* mutation carriers, and suggests that normal *BRCA1* may be required for clinical response to antimicrotubule agents.

While of limited use as monotherapy in *BRCA* patients, taxane administration in combination with DNA-damaging agents has been shown to have increased efficacy. A study from MD Anderson Cancer Center demonstrated that *BRCA1* carriers had a high pCR to neoadjuvant anthracycline-taxane-based chemotherapy (pCR in 46% of *BRCA1* carriers vs. 22% of non-carriers) [50]. The phase II Gepar Sixto study found in patients treated with weekly paclitaxel/non-pegylated liposomal doxorubicin, a pCR rate of 57.9% in *BRCA* carriers, compared to 40.2% pCR rate in non-carriers. The pCR rate increased by 25% with the addition of weekly carboplatin in *BRCA* carriers (as compared with a 14% increase in pCR rate in non-carriers) [64]. Another study presented at the 2014 ASCO meeting evaluated the efficacy of neoadjuvant combination therapy with carboplatin and docetaxel in sporadic and BRCA-associated TNBC. 86% (12/14) patients with deleterious mutations achieved a pCR in this trial compared to 50% in the 28 patients with sporadic TNBC [65].

In addition to taxanes, the antimicrotubule agent Eribulin has also been evaluated in BRCA-associated breast cancer patients. The neoadjuvant Gepar Quinto study evaluated 74 *BRCA* carriers with TNBC treated with epirubicin/cyclophosphamide followed by four cycles of docetaxel, with or without bevacizumab, and found a pCR rate of 50% (37/74) in *BRCA* carriers as compared with 31.1% pCR rate in non-carriers [66].

Trabectedin

Trabectedin has been shown to block DNA binding of the oncogenic transcription factor FUS-CHOP. In addition, preclinical and clinical data have suggested that trabectedin may have specific activity against nucleotide excision repair intact or HR repair-deficient metastatic breast cancer, suggesting potential efficacy in BRCA-mutated breast cancer. When evaluated in heavily pretreated MBC patients with germline *BRCA1* and two mutations, trabectedin treatment every 3 weeks was found to lead to partial response (PR) in 6/35 patients (17%), with median PFS of 3.9 months [67]. Similarly, a phase II trial of *BRCA1/2* mutation carrier patients with pretreated MBC, trabectedin resulted in PR in 4/29 (14%) patients, and median PFS of 3.3 months [68].

Lurbinectedin (PM01183)

Lurbinectedin (PM01183) binds covalently to DNA and induces the formation of double-strand breaks in a wide range of cancer cell lines, with particular activity against platinum resistant tumors and HR-deficient cell lines. In this setting, it was investigated in previously treated MBC patients with germline *BRCA1/2* mutations with an ORR of 41% (1 CR, 6 PR, 6 stable disease (SD), 4 progressive disease (PD) in 17 evaluable patients) with a median duration of response of 5 months. This was as compared with an ORR of 9% and median duration of response of 3.3 months in an unselected cohort. In an exploratory analysis, the ORR in the BRCA-mutated cohort was higher (64%, 7/11 patients) in PARP inhibitor naïve patients [69]. An ongoing trial is currently accruing to investigate this agent in *BRCA* mutation carriers with MBC (NCT01525589).

PARP Inhibitors

With advances in molecularly targeted therapy in solid tumors, an appealing targeted therapy for *BRCA1/2* carriers, poly (ADP-ribose) polymerase (PARP) inhibitors, has also been developed. PARP proteins play a role in single-strand DNA repair; when PARP is inhibited, single-strand breaks cannot be repaired leading to double-strand breaks at the replication fork [70, 71]. Because *BRCA1* and *BRCA2* proteins are critical in double-strand DNA repair, combining PARP inhibition with tumors that have defective *BRCA1* or *BRCA2* proteins exerts a synergistic lethal effect [72, 73]. This hypothesis has been supported by in vitro studies showing enhanced cytotoxicity in *BRCA1*- and *BRCA2*-deficient cells compared with cells with wild-type *BRCA* proteins [74, 75].

Olaparib

In a phase I study of 60 patients, of whom 22 were *BRCA* carriers, patients were treated with two different dose levels of the PARP inhibitor olaparib; the maximum tolerated dose (MTD) was determined to be 400 mg orally twice daily. The most common side effects were grade 1 or 2 nausea, vomiting, fatigue, dysgeusia, and anorexia. Myelosuppression (anemia or thrombocytopenia) was also observed in a few patients [53]. All patients had a PR, according to Response Evaluation Criteria in Solid Tumors (RECIST), with responses lasting 20–80 weeks in the 19 *BRCA* mutation carriers with ovarian, breast, or prostate cancer who could be evaluated for tumor response.

Phase II multicenter, multinational studies that examined breast and ovarian cancers independently were then conducted in mutation carriers. Both used the phase I MTD, 400 mg twice daily, and 100 mg twice daily because this dose was the lowest dose at which an antitumor effect was seen in the phase I trial. The primary endpoint for both studies was the ORR. Among breast cancer patients, the ORR for those in the 400-mg arm was 41% (11 of 27), with an additional 44% (12/27) of women achieving SD. For those with MBC on 100-mg olaparib, 22% (6/27) had PR and an additional 44% (12/27) of the patients achieved SD. These results were particularly impressive because the patients had undergone a median of three prior chemotherapy regimens. Similar results were observed in the ovarian cancer study, with an ORR of 33% (11/33) in the 400-mg arm. The most common side effects were nausea and fatigue [76, 77]. A subsequent multicenter phase II basket trial evaluated olaparib monotherapy in 298 patients with heavily pretreated recurrent cancers including ovarian, breast, pancreatic, and prostate cancers with *BRCA1/2* mutations. Of the 62 patients with BRCA-mutated breast cancer, ORR was 12.9% (8/62) and 47% of patients had disease stabilization for at least 8 weeks. This response rate was better for those patients without prior platinum exposure (20% vs. 9.5%). This lower objective response rate compared with previous studies was suggested to be due to the heavily pretreated nature of these patients, with a mean of 4.6 prior chemotherapy regimens in the metastatic setting [78]. These data resulted in a paradigm shift assuming that *BRCA1/2* carriers have differential susceptibility to systemic therapy compared with non-carriers. Ongoing studies include OlympiA, a phase III randomized trial of olaparib as adjuvant monotherapy in BRCA-mutated TNBC patients, which began enrolling in April 2014 with 1320 patients targeted across 550 sites and 25 countries worldwide (clinicaltrials.gov, NCT02032823).

Veliparib (ABT 888)

Veliparib is a potent, orally administered small molecule inhibitor of PARP1 and PARP2. Similar to olaparib, monotherapy with veliparib has been shown to be efficacious in BRCA-associated breast cancer patients. In a phase I study of single-agent veliparib, ORR was 29% (4/14 patients) in BRCA-mutated breast cancer patients as compared with ORR 5% (1/21 patients) in non-carriers [79]. In a subsequent phase II trial of BRCA-mutated MBC, single-agent veliparib was administered orally at 400 mg twice daily until progression, at which time therapy was transitioned to combination veliparib dosed at 150 mg orally twice daily in combination with carboplatin (AUC of 5) once every 3 weeks. Forty-one/forty-four patients enrolled were treated, all with *BRCA1/2* mutations. As of the time of presentation at the ASCO 2014 meeting, the rate of PR was 17% (2/12 patients) in *BRCA1* patients and 23% (3/13 patients) in *BRCA2* mutations who had at least four cycles of follow-up. The authors noted time to failure on veliparib of 2.0 months for *BRCA1* and 5.1 months for *BRCA2* carriers. Of the ten patients who had proceeded to combination treatment with veliparib and carboplatin, the authors noted one PR in a *BRCA1* carrier [80].

Rucaparib (AG-014699, PF-01367338)

Rucaparib, a potent selective PARP1 and PARP2 inhibitor, was evaluated in a multicenter, single arm phase II trial with 41 patients with BRCA-mutated breast (17 patients) and ovarian (24 patients) cancer. Rucaparib was given on days 1–5 of a 21-day cycle. Of these 41 patients, 38 had RECIST assessments with 5% ORR (2/38); 26% of patients achieved SD for at least 4 months (10/38). The intermittent dosing schedule was suspected to be the cause of the lower ORR found in this study [81].

Niraparib (MK4827)

Niraparib is a potent, selective, orally available PARP1 and PARP2 inhibitor shown in a phase I dose escalation study to have a 50% ORR (2/4 patients with PR) in patients with advanced BRCA-associated breast cancer. Of those patients with BRCA-associated ovarian cancer, 8/20 (40%) had PR [82]. Niraparib is also being evaluated in HER2-negative MBC patients with germline *BRCA1/2* mutations versus physician's choice (eribulin, vinorelbine, capecitabine, or gemcitabine) in an ongoing clinical trial (NCT01905592).

Talazoparib (BMN673)

Talazoparib is the most potent and specific inhibitor of PARP1 and 2 in clinical development with an IC_{50} of less than 1 nM; it also functions by trapping PARP on DNA. In the first in human dose escalation study of talazoparib, eight patients were included with breast cancer (6/8 with deleterious *BRCA* mutations). Objective responses occurred in 2/6 breast cancer patients with *BRCA* mutations (33%) [83]. In a subsequent study presented at 2013 SABCS, patients with solid tumors, including 18 with BRCA-associated breast cancer, were treated with talazoparib from 900 to 1000 μg/day. Of these 18 patients, one had CR, six PR, and five SD for at least 12 weeks. Notably, four of the BRCA-associated breast cancer patients enrolled in the trial had not responded to prior platinum-containing agents, none of these responded to talazoparibeither [83]. Multiple upcoming studies will evaluate talazoparib in patients with BRCA-associated breast cancer. These include:

- The phase II ABRAZO study randomizing patients with *BRCA* associated locally advanced or MBC with one of two cohorts: those previously responding to a platinum-containing regimen for MBC or those without prior platinum therapy but having previously received more than two prior chemotherapy regimens for MBC (NCT02034916) [84]

- The phase III EMBRACA trial in BRCA-mutated patients with locally advanced or MBC comparing talazoparib given 1 mg/day in 21-day cycles versus physician's choice (capecitabine, eribulin, gemcitabine, or vinorelbine) [85]
- The phase II single-center, non-randomized, multi-cohort trial evaluating the use of talazoparib in patients with advanced solid tumors without curative therapeutic options. This trial will specifically look at cohorts including: *BRCA* somatic mutations, *BRCA* somatic deletions, mutations, or homozygous deletions in other *BRCA* pathway genes including *ATM*, *PALB2*, *NBS1*, Fanconi Anemia genes, mutations, or homozygous deletions in *PTEN* and/or *PTEN* loss by IHC, HR defects, and germline *BRCA1/2* mutations [86].

Endocrine Therapy

While some studies have reported that tamoxifen reduces the risk of CBC by 50% to 70% in *BRCA* mutation carriers [87–89], other studies have not reported a significant reduction [90–94]. For example, two studies have demonstrated that adjuvant use of tamoxifen was associated with a CBC risk reduction of 50% for *BRCA1* carriers and 58% for *BRCA2* carriers, regardless of ER status [87, 95]. This result differed from that of a small retrospective study comparing outcomes in early-stage *BRCA* mutation-associated and sporadic breast cancer treated with tamoxifenwhich observed a lower OS in *BRCA* carriers, suggesting relative resistance to tamoxifen [96]. Similarly, Metcalfe et al. did not observe a statistically significant reduction in CBC risk associated with the use of tamoxifen [97]. Importantly, in none of the studies [87, 95] tamoxifen was associated with a risk reduction in women after oophorectomy. These results, however, require confirmation, and the use of adjuvant tamoxifen is recommended in patients with *BRCA* mutation-associated ER-positive breast cancer.

Currently, the role of aromatase inhibitors (AI) after RRSO or as an adjuvant endocrine therapy in *BRCA* mutation-associated breast cancer is unknown. The IBIS-II study is evaluating anastrozole versus placebo in high-risk women. In addition, there is an ongoing French study evaluating letrozole versus placebo in women with *BRCA* mutations [ClinicalTrials.gov identifier: NCT00673335].

Increased Surveillance

Current screening recommendations for the asymptomatic *BRCA* mutation carrier encompass examination, imaging, and laboratory evaluation. Surveillance for female carriers emphasizes screening techniques for breast and ovarian cancers.

There is general agreement that women with a higher lifetime risk of breast cancer, such as that conferred by a *BRCA* mutation, should undergo earlier and more frequent screening, with additional imaging modalities considered. A consolidated summary of breast cancer screening recommendations published by the National Comprehensive Cancer Network (NCCN), American Cancer Society (ACS), American College of Radiology (ACR), and other national organizations for the asymptomatic, female, *BRCA* mutation carrier includes the following [98, 99]:

- Monthly breast self-examination (BSE) beginning at the age of 18 years
- Semiannual clinical breast examination (CBE) beginning at the age of 25 years
- Alternating annual mammograms with annual breast magnetic resonance imaging (MRI) beginning at the age of 25–30 years or individualized based on the earliest age of cancer onset in the family [100].

While RRSO is more effective in preventing ovarian cancer in these women compared to general population, some may not opt to pursue this intervention until after their childbearing years. In the absence of more effective screening methods, transvaginal ultrasound (TVU) and CA-125 levels continue to be recommended and endorsed by national organizations for women who are at high risk for hereditary breast and ovarian cancer syndromes (HBOC) [101]. Current NCCN screening guidelines for *BRCA* mutation carriers who are not undergoing RRSO include the following:

- Semiannual concurrent pelvic examination, TVU, and CA-125 antigen determination beginning at the age of 35 years or 5–10 years earlier than the youngest age at which any family member was diagnosed with ovarian cancer.

Mutations in the tumor suppressor genes *BRCA1* and *BRCA2* place male and female carriers at increased risk for a number of other cancers, notably pancreatic, melanoma, colorectal, and other gastrointestinal tumors. No expert consensus or evidence-based guidelines exist regarding screening for these cancers. Some literature and investigational studies support considering the following additional surveillance modalities [102–104]:

- Pancreatic: annual endoscopic ultrasound, beginning at the age of 50 years or 10 years prior to the earliest pancreatic cancer diagnosis in the family
- Melanoma: annual full body skin and ocular examination
- Colorectal: population screening guidelines, beginning at the age of 50 years and continuing until 75 years old
- Annual fecal occult blood testing
- Sigmoidoscopy every 5 years or colonoscopy every 10 years.

References

1. Roy R, Chun J, Powell SN. *BRCA1* and *BRCA2*: different roles in a common pathway of genome protection. Nature Rev Cancer. 2012;12(1):68–78.
2. Yuan SS, Lee SY, Chen G, Song M, Tomlinson GE, Lee EY. *BRCA2* is required for ionizing radiation-induced assembly of *Rad51* complex in vivo. Cancer Res. 1999;59(15):3547–51.
3. Zhong Q, Chen CF, Li S, Chen Y, Wang CC, Xiao J, et al. Association of *BRCA1* with the h*Rad50*-h*Mre11*-p95 complex and the DNA damage response. Science. 1999;285(5428): 747–50.
4. Collins N, McManus R, Wooster R, Mangion J, Seal S, Lakhani SR, et al. Consistent loss of the wild type allele in breast cancers from a family linked to the *BRCA2* gene on chromosome 13q12-13. Oncogene. 1995;10(8):1673–5.
5. Cornelis RS, Neuhausen SL, Johansson O, Arason A, Kelsell D, Ponder BA, et al. High allele loss rates at 17q12-q21 in breast and ovarian tumors from BRCAl-linked families. The Breast Cancer Linkage Consortium. Genes Chromosomes Cancer. 1995;13(3):203–10.
6. Khoo US, Ozcelik H, Cheung AN, Chow LW, Ngan HY, Done SJ, et al. Somatic mutations in the *BRCA1* gene in Chinese sporadic breast and ovarian cancer. Oncogene. 1999;18(32): 4643–6.
7. Esteller M, Silva JM, Dominguez G, Bonilla F, Matias-Guiu X, Lerma E, et al. Promoter hypermethylation and *BRCA1* inactivation in sporadic breast and ovarian tumors. J Natl Cancer Inst. 2000;92(7):564–9.
8. Tapia T, Smalley SV, Kohen P, Munoz A, Solis LM, Corvalan A, et al. Promoter hypermethylation of *BRCA1* correlates with absence of expression in hereditary breast cancer tumors. Epigenetics. 2008;3(3):157–63.
9. Dworkin AM, Spearman AD, Tseng SY, Sweet K, Toland AE. Methylation not a frequent "second hit" in tumors with germline *BRCA* mutations. Fam Cancer. 2009;8(4):339–46.
10. Collins N, Wooster R, Stratton MR. Absence of methylation of CpG dinucleotides within the promoter of the breast cancer susceptibility gene *BRCA2* in normal tissues and in breast and ovarian cancers. Br J Cancer. 1997;76(9):1150–6.
11. Rich TA, Woodson AH, Litton J, Arun B. Hereditary breast cancer syndromes and genetic testing. J Surgical Oncol. 2014.
12. Foulkes WD. Inherited susceptibility to common cancers. N Engl J Med. 2008;359(20): 2143–53.
13. Honrado E, Benitez J, Palacios J. The molecular pathology of hereditary breast cancer: genetic testing and therapeutic implications. Mod Pathol. 2005;18(10):1305–20.
14. Gerdes AM, Cruger DG, Thomassen M, Kruse TA. Evaluation of two different models to predict *BRCA1* and *BRCA2* mutations in a cohort of Danish hereditary breast and/or ovarian cancer families. Clin Genet. 2006;69(2):171–8.
15. Melchor L, Benitez J. The complex genetic landscape of familial breast cancer. Human Genet. 2013;132(8):845–63.
16. Thompson D, Easton D. The genetic epidemiology of breast cancer genes. J Mammary Gland Biol Neoplasia. 2004;9(3):221–36.
17. Diez O, Osorio A, Duran M, Martinez-Ferrandis JI, de la Hoya M, Salazar R, et al. Analysis of *BRCA1* and *BRCA2* genes in Spanish breast/ovarian cancer patients: a high proportion of mutations unique to Spain and evidence of founder effects. Hum Mutat. 2003;22(4):301–12.
18. Mersch J, Jackson MA, Park M, Nebgen D, Peterson SK, Singletary C, et al. Cancers associated with *BRCA1* and *BRCA2* mutations other than breast and ovarian. Cancer. 2015;121(2):269–75.
19. Mavaddat N, Barrowdale D, Andrulis IL, Domchek SM, Eccles D, Nevanlinna H, et al. Pathology of breast and ovarian cancers among *BRCA1* and *BRCA2* mutation carriers: results from the Consortium of Investigators of Modifiers of *BRCA1/2* (CIMBA). Cancer Epidemiol Biomarkers Prev. 2012;21(1):134–47.

20. King MC, Marks JH, Mandell JB. New York breast cancer study G. Breast and ovarian cancer risks due to inherited mutations in *BRCA1* and *BRCA2*. Science. 2003;302(5645): 643–6.
21. Antoniou A, Pharoah PD, Narod S, Risch HA, Eyfjord JE, Hopper JL, et al. Average risks of breast and ovarian cancer associated with *BRCA1* or *BRCA2* mutations detected in case Series unselected for family history: a combined analysis of 22 studies. Am J Human Genet. 2003;72(5):1117–30.
22. Chen S, Parmigiani G. Meta-analysis of *BRCA1* and *BRCA2* penetrance. J Clin Oncol. 2007;25(11):1329–33.
23. Rebbeck TR, Kauff ND, Domchek SM. Meta-analysis of risk reduction estimates associated with risk-reducing salpingo-oophorectomy in *BRCA1* or *BRCA2* mutation carriers. J Nat Cancer Inst. 2009;101(2):80–7.
24. Easton DF, Ford D, Bishop DT. Breast and ovarian cancer incidence in *BRCA1*-mutation carriers. Breast Cancer Linkage Consortium. Am JHuman Genet. 1995;56(1):265–71.
25. Metcalfe KA, Lynch HT, Ghadirian P, Tung N, Olivotto IA, Foulkes WD, et al. The risk of ovarian cancer after breast cancer in *BRCA1* and *BRCA2* carriers. Gynecol Oncol. 2005;96(1): 222–6.
26. Atchley DP, Albarracin CT, Lopez A, Valero V, Amos CI, Gonzalez-Angulo AM, et al. Clinical and pathologic characteristics of patients with BRCA-positive and BRCA-negative breast cancer. J Clin Oncol. 2008;26(26):4282–8.
27. Marcus JN, Watson P, Page DL, Narod SA, Lenoir GM, Tonin P, et al. Hereditary breast cancer: pathobiology, prognosis, and *BRCA1* and *BRCA2* gene linkage. Cancer. 1996;77(4): 697–709.
28. Eisinger F, Jacquemier J, Charpin C, Stoppa-Lyonnet D, Bressac-de Paillerets B, Peyrat JP, et al. Mutations at *BRCA1*: the medullary breast carcinoma revisited. Cancer Res. 1998;58 (8):1588–92.
29. Lakhani SR, Gusterson BA, Jacquemier J, Sloane JP, Anderson TJ, van de Vijver MJ, et al. The pathology of familial breast cancer: histological features of cancers in families not attributable to mutations in *BRCA1* or *BRCA2*. Clin Cancer Res. 2000;6(3):782–9.
30. Lakhani SR, Van De Vijver MJ, Jacquemier J, Anderson TJ, Osin PP, McGuffog L, et al. The pathology of familial breast cancer: predictive value of immunohistochemical markers estrogen receptor, progesterone receptor, HER-2, and *p53* in patients with mutations in *BRCA1* and *BRCA2*. J Clin Oncol. 2002;20(9):2310–8.
31. Arun B, Vogel KJ, Lopez A, Hernandez M, Atchley D, Broglio KR, et al. High prevalence of preinvasive lesions adjacent to *BRCA1/2*-associated breast cancers. Cancer Prev Res. 2009;2(2):122–7.
32. Kennedy RDQJ, Johnston PH, Harkin DP. *BRCA1*: mechanisms of inactivation and implications for management of patients. Lancet. 2002;360:1007–14.
33. Verhoog LCBC, Seynaeve C, Dahmen G, van Geel AN, Bartels CC, et al. Survival in hereditary breast cancer associated with germline mutations of *BRCA2*. J Clin Oncol. 1999;17:3396–402.
34. Rennert G, Bisland-Naggan S, Barnett-Griness O, Bar-Joseph N, Zhang S, Rennert HS, et al. Clinical outcomes of breast cancer in carriers of *BRCA1* and *BRCA2* mutations. N Engl J Med. 2007;357(2):115–23.
35. Brekelmans CT, Tilanus-Linthorst MM, Seynaeve C, vd Ouweland A, Menke-Pluymers MB, Bartels CC, et al. Tumour characteristics, survival and prognostic factors of hereditary breast cancer from *BRCA2*-, B. Eur J Cancer. 2007;43:867–76.
36. Kriege MSC, Meijers-Heijboer H, Collee JM, Menke-Pluymers MB, Bartels CC, et al. Distant disease-free interval, site of first relapse and post-relapse survival in B. Breast Cancer Res Treat. 2008;111:303–11.
37. Bonadona V D-MS, Voirin N, Sinilnikova OM, Mignotte H, Mathevet P, Bremond A, et al. Prognosis of early-onset breast cancer based on *BRCA1/2* mutation status in a French population-based cohort and review. Breast Cancer Res Treat. 2007;101(2):233–45.

38. Bordeleau L, Panchal S, Goodwin P. Prognosis of BRCA-associated breast cancer: a summary of evidence. Breast Cancer Res Treat. 119(1):13–24.
39. Robson MLD, Federici M, Satagopan J, Bogolminy F, Heerdt A, Borgen P, et al. Breast conservation therapy for invasive breast cancer in Ashkenazi women with *BRCA* gene founder mutations. J Natl Cancer Inst. 1999;91:2112–7.
40. Chappuis POKL, Bégin LR, Wong N, Brunet JS, Narod SA, et al. Germline *BRCA1/2* mutations and p27(Kip1) protein levels independently predict outcome after breast cancer. J Clin Oncol. 2000;18:4045–52.
41. Stoppa-Lyonnet D, Ansquer Y, Dreyfus H, Gautier C, Gauthier-Villars M, Bourstyn E, et al. Familial invasive breast cancer: worse outcome related to *BRCA1* mutations. J Clin Oncol. 2000;18:4053–9.
42. Moller P, Evans DG, Reis MM, Gregory H, Anderson E, Maehle L, et al. Surveillance for familial breast cancer: Differences in outcome according to *BRCA* mutation status. Int J Cancer. 2007;121(5):1017–20.
43. Lee EH, Park SK, Park B, Kim SW, Lee MH, Ahn SH, et al. Effect of *BRCA1/2* mutation on short-term and long-term breast cancer survival: a systematic review and meta-analysis. Breast Cancer Res Treat. 122(1):11–25.
44. Lee LJ, Alexander B, Schnitt SJ, Comander A, Gallagher B, Garber JE, et al. Clinical outcome of triple negative breast cancer in *BRCA1* mutation carriers and noncarriers. Cancer. 117(14):3093–100.
45. Bayraktar S, Gutierrez-Barrera AM, Liu D, Tasbas T, Akar U, Litton JK, et al. Outcome of triple-negative breast cancer in patients with or without deleterious *BRCA* mutations. Breast Cancer Res Treat. 2011;130(1):145–53.
46. Shah PD, Patil S, Dickler MN, Offit K, Hudis CA, Robson ME. Twenty-one-gene recurrence score assay in BRCA-associated versus sporadic breast cancers: differences based on germline mutation status. Cancer. 2016;122(8):1178–84.
47. Zhong Q, Peng HL, Zhao X, Zhang L, Hwang WT. Effects of *BRCA1/2* on Ovarian and breast cancer survival-response. Clin Cancer Res. 2015;21(16):3807.
48. Byrski T, Huzarski T, Dent R, Gronwald J, Zuziak D, Cybulski C, et al. Response to neoadjuvant therapy with cisplatin in *BRCA1*-positive breast cancer patients. Breast Cancer Res Treat. 2009;115(2):359–63.
49. Byrski T, Gronwald J, Huzarski T, Grzybowska E, Budryk M, Stawicka M, et al. Pathologic complete response rates in young women with *BRCA1*-positive breast cancers after neoadjuvant chemotherapy. J Clin Oncol. 2010;28(3):375–9.
50. Arun B, Bayraktar S, Liu DD, Gutierrez Barrera AM, Atchley D, Pusztai L, et al. Response to neoadjuvant systemic therapy for breast cancer in *BRCA* mutation carriers and noncarriers: a single-institution experience. J Clin Oncol. 2011;29(28):3739–46.
51. Silver DP, Richardson AL, Eklund AC, Wang ZC, Szallasi Z, Li Q, et al. Efficacy of neoadjuvant Cisplatin in triple-negative breast cancer. J Clin Oncol. 2010;28(7):1145–53.
52. Wen J, Li R, Lu Y, Shupnik MA. Decreased *BRCA1* confers tamoxifen resistance in breast cancer cells by altering estrogen receptor-coregulator interactions. Oncogene. 2009;28(4): 575–86.
53. Fong PC, Boss DS, Yap TA, Tutt A, Wu P, Mergui-Roelvink M, et al. Inhibition of poly (ADP-ribose) polymerase in tumors from *BRCA* mutation carriers. N Engl J Med. 2009; 361(2):123–34.
54. Iglehart JD, Silver DP. Synthetic lethality—a new direction in cancer-drug development. N Engl J Med. 2009;361(2):189–91.
55. Carey LA, Sharpless NE. PARP and cancer—if it's broke, don't fix it. N Engl J Med. 2011;364(3):277–9.
56. Kennedy RD, Quinn JE, Mullan PB, Johnston PG, Harkin DP. The role of *BRCA1* in the cellular response to chemotherapy. J Nat Cancer Inst. 2004;96(22):1659–68.

57. Tassone P, Tagliaferri P, Perricelli A, Blotta S, Quaresima B, Martelli ML, et al. *BRCA1* expression modulates chemosensitivity of *BRCA1*-defective HCC1937 human breast cancer cells. Br J Cancer. 2003;88(8):1285–91.
58. Chabalier C, Lamare C, Racca C, Privat M, Valette A, Larminat F. *BRCA1* downregulation leads to premature inactivation of spindle checkpoint and confers paclitaxel resistance. Cell Cycle. 2006;5(9):1001–7.
59. Byrski T, Huzarski T, Dent R, Marczyk E, Jasiowka M, Gronwald J, et al. Pathologic complete response to neoadjuvant cisplatin in *BRCA1*-positive breast cancer patients. Breast Cancer Res Treat. 2014;147(2):401–5.
60. Byrski T, Dent R, Blecharz P, Foszczynska-Kloda M, Gronwald J, Huzarski T, et al. Results of a phase II open-label, non-randomized trial of cisplatin chemotherapy in patients with *BRCA1*-positive metastatic breast cancer. Breast Cancer Res. 2012;14(4):R110.
61. Tutt A, Ellis P, Kilburn L, Gilett C, Pinder S, Abraham J, et al. The TNT trial: a randomized phase III trial of carboplatin compared with docetaxel for patients with metastatic or recurrent locally advanced triple negative or *BRCA* 1/2 breast cancer. Tutt A E. SABC. 2014;S3–01.
62. Leitch AM, Ellis MJ, Boughey JC, Unzeitig G, Royce M, Boughey JC, et al. Neoadjuvant weekly paclitaxel followed by anthracycline-based chemotherapy. San Antonio Breast Cancer Symp. 2015.
63. Kriege M, Jager A, Hooning MJ, Huijskens E, Blom J, van Deurzen CH, et al. The efficacy of taxane chemotherapy for metastatic breast cancer in *BRCA1* and *BRCA2* mutation carriers. Cancer. 118(4):899–907.
64. von Minckwitz G, Hahnen E, Fasching PA, et al. Pathological complete response (pCR) rates after carboplatin-containing neoadjuvant chemotherapy in patients with germline *BRCA* (gBRCA) mutation and triple-negative breast cancer (TNBC): Results from GeparSixto. 2014 ASCO Annual Meeting, J Clin Oncol. 2014;32(suppl; abstr 1005):5s.
65. Sharma P, Lopez-Tarruella S, Garcia-Saenz JA, Ward C, Connor C, Gomez HL, et al. Efficacy of neoadjuvant carboplatin plus docetaxel in triple negative breast cancer: Combined analysis of two cohorts. Clin Cancer Res. 2017;23(3):649–57.
66. McAndrew NP, Maxwell KN, Stopfer J, et al. Oncotype DX scores in *BRCA1* and *BRCA2* associated breast cancer. J Clin Oncol. 2015;33(suppl; abstr 541).
67. Byrski T, Blum JL, Goncalves A, Campone M, et al. Activity of trabectedin in germline *BRCA1/2*-mutated metastatic breast cancer: results of an international first-in-class phase II study. Ann Oncol. 2014;25(6):1152–8.
68. Tedesco KL. Final results of a phase II trial of trabectedin (T) in triple-negative H-p, and *BRCA1/2* germ-line-mutated metastatic breast cancer (MBC) patients (pts). ASCO Annu Meet. 2011, suppl abstract 1125.
69. Delaloge S, Wolp-Diniz R, Byrski T, et al. Activity of trabectedin in germline *BRCA1/2*-mutated metastatic breast cancer: results of an international first-in-class phase II study. Ann Oncol. 2014;25(6):1152–8.
70. Ame JC, Spenlehauer C, de Murcia G. The PARP superfamily. BioEssays. 2004;26(8): 882–93.
71. Dantzer F, de La Rubia G, Menissier-De Murcia J, Hostomsky Z, de Murcia G, Schreiber V. Base excision repair is impaired in mammalian cells lacking Poly(ADP-ribose) polymerase-1. Biochemistry. 2000;39(25):7559–69.
72. McCabe N, Turner NC, Lord CJ, Kluzek K, Bialkowska A, Swift S, et al. Deficiency in the repair of DNA damage by homologous recombination and sensitivity to poly(ADP-ribose) polymerase inhibition. Cancer Res. 2006;66(16):8109–15.
73. Ashworth A. A synthetic lethal therapeutic approach: poly(ADP) ribose polymerase inhibitors for the treatment of cancers deficient in DNA double-strand break repair. J Clin Oncol. 2008;26(22):3785–90.
74. Farmer H, McCabe N, Lord CJ, Tutt AN, Johnson DA, Richardson TB, et al. Targeting the DNA repair defect in *BRCA* mutant cells as a therapeutic strategy. Nature. 2005;434(7035): 917–21.

75. Bryant HE, Schultz N, Thomas HD, Parker KM, Flower D, Lopez E, et al. Specific killing of *BRCA2*-deficient tumours with inhibitors of poly(ADP-ribose) polymerase. Nature. 2005;434(7035):913–7.
76. Tutt A, Robson M, Garber JE, Domchek SM, Audeh MW, Weitzel JN, et al. Oral poly (ADP-ribose) polymerase inhibitor olaparib in patients with *BRCA1* or *BRCA2* mutations and advanced breast cancer: a proof-of-concept trial. Lancet. 2010;376(9737):235–44.
77. Audeh MW, Carmichael J, Penson RT, Friedlander M, Powell B, Bell-McGuinn KM, et al. Oral poly(ADP-ribose) polymerase inhibitor olaparib in patients with *BRCA1* or *BRCA2* mutations and recurrent ovarian cancer: a proof-of-concept trial. Lancet. 2010;376(9737): 245–51.
78. Kaufman B, Shapira-Frommer R, Schmutzler RK, Audeh MW, Friedlander M, Balmana J, et al. Olaparib monotherapy in patients with advanced cancer and a germline *BRCA1/2* mutation. J Clin Oncol. 2015;33(3):244–50.
79. Puhalla S, Beumer JH, Pahuja S, et al. Final results of a phase 1 study of single-agent veliparib (V) in patients (pts) with either *BRCA1/2*-mutated cancer (BRCA+), platinum-refractory ovarian, or basal-like breast cancer (BRCA-wt). J Clin Oncol. 2014; 32(suppl; abstr 2570):5s.
80. Somlo G, Frankel PH, Luu TH, et al.BRCA-associated metastatic breast cancer (MBC): California Cancer Consortium trial PHII-96. 2014 ASCO Annual Meeting. J Clin Oncol. 2014;32(suppl; abstr 1021):5s.
81. Drew Y, Ledermann JA, Jones A, Hall G, Jayson GC, Highley M, et al. Phase II trial of the poly(ADP-ribose) polymerase (PARP) inhibitor AG-014699 in BRCA-1 and -2 mutated, advanced ovarian and/or locally advanced or metastatic breast cancer. J Clin Oncol. 2011;29:3104.
82. Sk Sandhu, Schelman WR, Wilding G, Moreno V, Baird RD, Miranda S, et al. The poly (ADP-ribose) polymerase inhibitor niraparib (MK4827) in *BRCA* mutation carriers and patients with sporadic cancer: a phase 1 dose-escalation trial. Lancet Oncol. 2013;14(9): 882–92.
83. De Bono JS, Mina LA, Gonzalez M, et al.First-in-human trial of novel oral PARP inhibitor BMN 673 in patients with solid tumors. 2013ASCO Annual Meeting JCO, J Clin Oncol. 2013;31(suppl; abstr 2580).
84. Turner NC, Balmaña J, Fasching PA, et al. A phase 2 study (2-stage, 2-cohort) of the oral PARP inhibitor talazoparib (BMN 673) in patients with germline *BRCA* mutation and locally advanced and/or metastatic breast cancer (ABRAZO). 2015 ASCO Annual Meeting, J Clin Oncol. 2015;33(suppl; abstr TPS1108).
85. Litton JK, Blum JL, Im Y-H, et al. A phase 3 o-l, randomized, parallel, 2-arm international study of the oral PARP inhibitor talazoparib (BMN 673) in *BRCA* mutation subjects with locally advanced and/or metastatic breast cancer (EMBRACA). 2015 ASCO Annual Meeting, J Clin Oncol. 2015;33(suppl; abstr TPS1107).
86. Piha-Paul SA, Goldstein JB, Hess KR, et al. Phase II study of the PARP inhibitor talazoparib (BMN-673) in advanced cancer patients with somatic alterations in *BRCA1/2* mdiPoPl, a homologous recombination defect, mutations/deletions in other *BRCA* pathway genes and germline mutation S in *BRCA1/2* (not breast or ovarian cancer). 2015 ASCO Annual Meeting, J Clin Oncol. 2015;33(suppl; abstr TPS2617).
87. Gronwald J, Tung N, Foulkes WD, Offit K, Gershoni R, Daly M, et al. Tamoxifen and contralateral breast cancer in *BRCA1* and *BRCA2* carriers: an update. Int J Cancer. 2006; 118(9):2281–4.
88. Narod SA, Brunet JS, Ghadirian P, Robson M, Heimdal K, Neuhausen SL, et al. Tamoxifen and risk of contralateral breast cancer in *BRCA1* and *BRCA2* mutation carriers: a case-control study. Hereditary breast cancer clinical study group. Lancet. 2000;356(9245):1876–81.
89. Pierce LJ, Levin AM, Rebbeck TR, Ben-David MA, Friedman E, Solin LJ, et al. Ten-year multi-institutional results of breast-conserving surgery and radiotherapy in *BRCA1/2*-associated stage I/II breast cancer. J Clin Oncol. 2006;24(16):2437–43.

90. Malone KE, Begg CB, Haile RW, Borg A, Concannon P, Tellhed L, et al. Population-based study of the risk of second primary contralateral breast cancer associated with carrying a mutation in *BRCA1* or *BRCA2*. J Clin Oncol. 2010;28(14):2404–10.
91. Metcalfe K, Lynch HT, Ghadirian P, Tung N, Olivotto I, Warner E, et al. Contralateral breast cancer in *BRCA1* and *BRCA2* mutation carriers. J Clin Oncol. 2004;22(12):2328–35.
92. Reding KW, Bernstein JL, Langholz BM, Bernstein L, Haile RW, Begg CB, et al. Adjuvant systemic therapy for breast cancer in *BRCA1/BRCA2* mutation carriers in a population-based study of risk of contralateral breast cancer. Breast Cancer Res Treat. 2010;123(2):491–8.
93. Robson M, Svahn T, McCormick B, Borgen P, Hudis CA, Norton L, et al. Appropriateness of breast-conserving treatment of breast carcinoma in women with germline mutations in *BRCA1* or *BRCA2*: a clinic-based series. Cancer. 2005;103(1):44–51.
94. Robson ME, Chappuis PO, Satagopan J, Wong N, Boyd J, Goffin JR, et al. A combined analysis of outcome following breast cancer: differences in survival based on *BRCA1/BRCA2* mutation status and administration of adjuvant treatment. Breast Cancer Res. 2004; 6(1):R8–17.
95. Phillips K-A, Milne RL, Rookus MA, et al. Tamoxifen and risk of contralateral breast cancer for *BRCA1* and *BRCA2* mutation carriers. J Clin Oncol. 2013;31:3091–9.
96. Wesolowski R SA, Tao J, Moore HC, editor. Differential outcomes in patients treated with endocrine therapy for early or locally advanced breast cancer based on *BRCA* mutation status. 2009 ASCO Annual Meeting. J Clin Oncol. 2009;27(suppl; abstr e22065).
97. Metcalfe K, Gershman S, Lynch HT, Ghadirian P, Tung N, Kim-Sing C, et al. Predictors of contralateral breast cancer in *BRCA1* and *BRCA2* mutation carriers. Br J Cancer. 2011; 104(9):1384–92.
98. http://www.cancer.org/healthy/findcancerearly/cancerscreeningguidelines/CAf,%20american-cancer-society-guidelines-for-theearly-,%20detection-of-cancer. Accessed Dec 6.
99. Lee CH, Dershaw DD, Kopans D, Evans P, Monsees B, Monticciolo D, et al. Breast cancer screening with imaging: recommendations from the society of breast imaging and the ACR on the use of mammography, breast MRI, breast ultrasound, and other technologies for the detection of clinically occult breast cancer. J Am Coll Radiol JACR. 2010;7(1):18–27.
100. Le-Petross HT, Whitman GJ, Atchley DP, Yuan Y, Gutierrez-Barrera A, Hortobagyi GN, et al. Effectiveness of alternating mammography and magnetic resonance imaging for screening women with deleterious *BRCA* mutations at high risk of breast cancer. Cancer. 2011;117(17):3900–7.
101. NCCN GFH-RABaO, Available CPGiOv, http://www.nccn.org/professionals/physician_gls/pdf/f,%20genetics_screening.pdf. Accessed Dec 9.
102. Brentnall TA. Cancer surveillance of patients from familial pancreatic cancer kindreds. Med Clin North Am. 2000;84(3):707–18.
103. Canto MI, Goggins M, Yeo CJ, Griffin C, Axilbund JE, Brune K, et al. Screening for pancreatic neoplasia in high-risk individuals: an EUS-based approach. Clin Gastroenterol Hepatol. 2004;2(7):606–21.
104. Cancer: UPSTFSfC, AHRQ UPSTFRS, http://www.ahrq.gov/P-E-Af,%20clinic/uspstf08/colocancer/colors.htm. Accessed Dec 2.

Radiation Oncology Considerations in the Management of Mutation Carriers with Breast Cancer

9

Meena S. Moran

Introduction

The optimal management of these patients with hereditary breast carcinoma (hBC) continue to raise discussions and concerns, with only a small proportion of breast cancer cases diagnosed annually attributed to a documented autosomal dominant transmission of deleterious *BRCA1* and *BRCA2* genes. Compared with the mid-1990s when these 2 genes were first discovered, the increasing awareness of breast cancer, genetic testing, and more widespread screening and improvements in imaging modalities and prophylaxis have resulted in a larger proportion of hereditary BCs presenting with early stage disease. Breast conservation therapy (BCT), defined as breast-conserving surgery (quadrantectomy, partial mastectomy or lumpectomy) followed by whole breast radiation therapy (WBRT), is now a standard and proven alternative to mastectomy for early stage breast cancer and is supported by long-term follow-up from Level I data which have demonstrated the equivalence of BCS to mastectomy for early stage breast cancer [1–6]. Additionally, the meta-analysis pooling data from these trials has unequivocally demonstrated a small, statistically significant benefit in overall survival with the use of whole breast radiation therapy after lumpectomy [6]. Unfortunately, the era in which these trials were conducted significantly predates genetic testing for hereditary breast cancers and our knowledge of *BRCA* mutations, and thus, outcomes based on mutation status from these trials are lacking. Furthermore, given the

M.S. Moran (✉)
Therapeutic Radiology, Yale University School of Medicine,
15 York Street, New Haven, CT 06505, USA
e-mail: meena.moran@yale.edu

M.S. Moran
Yale New Haven Hospital, 15 York Street, New Haven, CT 06505, USA

A.B. Chagpar (ed.), *Managing BRCA Mutation Carriers*,
DOI 10.1007/978-3-319-59198-8_9

significant differences between the clinical manifestations of *BRCA1* (and, to a lesser extent, *BRCA2*)-associated breast cancers compared with sporadic breast cancer, it is difficult to extrapolate the data from the BCT trials to hereditary BC. Several clinical-pathologic features found be to highly prognostic in the general sporadic BC population, such as high-grade disease or triple-negative subtype, have also been found to be more frequently associated with *BRCA1*, which may suggest worse long-term outcomes for hereditary BC. Therefore, it is reasonable to consider that these differences in genotypes may also have implications for management of women with *BRCA*-associated BC.

This chapter reviews the clinical and radiobiologic treatment concepts of radiation therapy, reviews local-regional management principles with breast conservation therapy versus mastectomy in hereditary breast cancers, and provides an overview of the theories and implications of radiation therapy delivery in *BRCA* mutation carriers.

Radiobiology and Radiation Concerns Specific to *BRCA* Carriers

Radiation Biology of X-rays

The X-rays used in both diagnostic and therapeutic radiotherapy consist of ionizing radiation, derived from the ejection of electrons of a molecule, which then produce direct and indirect effects on DNA of targeted cells. The aim of therapeutic radiation is to induce DNA damage in the form of double-stand breaks, which ultimately leads to cell death (of tumor) and repair of damage (normal tissue). These therapeutic benefits of radiation therapy in eradicating tumor cells relative to surrounding normal tissue are principally based on the several important factors: total radiation dose delivered to the tumor, the interval of the fractionation (i.e., frequency: daily, once a week, twice a day), fractionation dose (i.e., 200, 300, 500 cGy), the tolerance of the surrounding normal tissue to radiation, and the enhanced ability of normal tissue to repair itself relative to tumor cells.

Radiation damage can be classified by its ability to cause potentially repairable effects to either the tumor or the surrounding normal tissue. Operationally, radiation damage can be divided into three categories: (a) lethal damage, which is irreparable and irreversible damage which leads to cell death (b) potentially lethal damage (PLD), which can be reversible or modified by post-irradiation environmental conditions, and (c) sub-lethal damage, which is reversible within hours of the radiation damage unless *additional* sub-lethal damage (an added fraction of radiation) is added. Separate targets within any individual (normal or cancer) cell that sustain sub-lethal damage can combine the effects of the damage to form lethal damage [7]. Alternatively, the repair of sub-lethal damage mandates that a complex cascade of events function appropriately to produce enzymes/proteins that detect DNA damage and repair.

Relationship Between Radiation and *BRCA*

Although the proteins produced by *BRCA1* and *BRCA2* are distinct and structurally unrelated, they are both known to be key components of large multi-protein complexes that have been identified to be involved in the repair of double-strand breaks by homologous recombination, and DNA repair through excision of nucleotides associated with transcription [8, 9]. The normal repair of double-stranded DNA breaks by homologous recombination occurs by use of the unmodified complementary strand of the DNA complementary strand of the DNA, called sister chromatid or homologous chromosome, as a template, to recover the original information. For sporadic breast cancers treated with radiation, normal tissue surrounding the tumor cells repair double-strand breaks by 3 distinct mechanisms: non-homologous end joining, microhomology-mediated end joining and homologous recombination. Conversely, the breast cancers cells, which inherently are mutated to a degree, have difficulty with double-strand breaks or other radiation-induced DNA damage, often rendering them irreparable because neither strand can then serve as a template for repair. This then results in cell death or in rarer instances, resulting in unregulated cell division, which can lead to the formation of a secondary malignant tumor.

Though the exact process of DNA repair and the activation of proteins controlling the cell cycle with radiotherapy remain to be further elucidated, it is known that proteins such as *p53* and *p21*, which function to stop the cell cycle and direct the cell to apoptosis (programmed cell death), are involved [10]. Various *BRCA* proteins bind to *RAD51*, a 339-amino acid essential recombinant protein known to be central in homologous recombination repair. This, along with other pathways, results in the failure of cells to arrest and/or repair DNA damage [11]. *RAD51* is thought to be involved in stages of the S and G2 phases of the cell cycle. When the *BRCA-RAD51* pathway is defective, the DNA repair process is re-directed through an alternative and often faulted pathway. Though this general association between the *BRCA* genes and *RAD51* has been elucidated, there are many unidentified pathways, associations, and roles of various proteins that remain to be deciphered [8, 11]. For example, *BRCA1* has been found to be involved in various cellular functions, including crucial roles in cell cycle checkpoint control and transcriptional regulation, X-chromosome inactivation, and mammary gland development [12]. Therefore, inactivation of genes encoding cell cycle checkpoint proteins (e.g., *p53*) results in uncontrolled, continuous divisions of the cell, and malfunctioning of these complex pathways results in the observed high incidence of cancer development that is experienced by *BRCA1* and *BRCA2* patients (i.e., breast/ovarian cancer).

Furthermore, these same pathways may have potential implications for increased toxicity or sensitivity to radiation in *BRCA* carriers. Though somewhat conflicting, the published laboratory-based data suggest that *BRCA1/2*-mutated mice or cells show elevated radio sensitivity among those with a defect in the *BRCA* genes [13, 14]. For example, murine embryos with *BRCA1* mutations have been shown to be exquisitely sensitive to radiation [15], and this radio sensitivity is most pronounced after irradiation to assays at low-dose rates [16]. Nevertheless, the clinical data, particularly with respect to therapeutic radiation in mutation carriers, remains limited and conflicting.

Our baseline understanding of the functions of *BRCA* genes has led to concerns that patients with a *BRCA1/2* mutation may be more sensitive to the deleterious effects of ionizing radiation due to an impaired capacity to repair double-strand DNA breaks. This may have consequences for the use of mammography in breast cancer screening, particularly for young *BRCA1* and *BRCA2* mutation carriers. Furthermore, for *BRCA* carriers afflicted with breast cancer, mutation status may have implications for treatment decisions such as breast conservation therapy versus mastectomy, the use of post-mastectomy radiation therapy, and potential added toxicity and risk of radiation-induced malignancies. These discussions are particularly relevant because both diagnostic and therapeutic ionizing radiations are well-documented risk factors for secondary malignancies for all patients [17].

The major concern for *BRCA* mutation carriers, along with other genetic syndromes that have defects in genes involved with DNA repair, is the potential radiation damage to adjacent normal tissue when targeting the tumor. Neighboring normal tissue, incidentally exposed to radiation, can also sustain sub-lethal damage, which is more susceptible to progress to lethal damage since the surrounding normal cells also harbor DNA repair mutations, raising concerns for the potential increase in radiation-induced acute or long-term toxicity effects such as severe fibrosis, telangiectasia, brachial plexopathy, lymphedema, and second malignancies. Another concern is the long-term effects of **low-dose** scatter radiation to the surrounding healthy tissue, which is inherent with any radiation delivery, and its augmented potential for carcinogenic effects in hereditary BC patients. Because multiple genetic hits are necessary for tumorigenesis, individuals that carry germ line mutations in DNA damage response genes are theoretically felt to be particularly prone to cancer development.

Due to the involvement of *BRCA1* and *BRCA2* in the repair functions described above which typically repair aberrations induced by radiation to normal tissue, *BRCA1* and *BRCA2* mutation carriers have been theoretically felt to be more sensitivity to radiation. A DNA damage-induced *BRCA1* protein complex has described as a part of the mRNA-splicing machinery which, in response to DNA damage, regulates the pre-mRNA splicing of a number of genes involved in DNA damage signaling and repair. Mutations in *BRCA1* and a number of genes encoding proteins found within this complex have been reported to increased sensitivity to DNA damage. Hence, these findings suggest an increased breast cancer risk from low-dose radiation for women with a familial or genetic risk compared with the general population [18].

Pre-clinical Data on Radio Sensitivity

The question of genetic deficiencies affecting the ability of normal tissue to repair DNA damage from radiation therapy have been studied, assessing chromosomal radio sensitivity in patients with DNA damage processing genes such as those of hereditary BC. For example, patients with hereditary disorders predisposing to cancer development, such as hereditary retinoblastoma, ataxia-telangiectasia, and Nijmegen breakage syndrome, have been demonstrated to have enhanced radio sensitivity [19]. If breast cancer precursor cells such as breast lobules have sustained a cancer-predisposing mutation at a locus such as *p53*, then entire breast lobule carries this mutation, explaining the elevated breast cancer risk observed in germ line *p53* mutation syndromes such as Li-Fraumeni syndromes. Similarly, high dose ionizing radiation delivered during childhood or early puberty predisposes women to adult-onset breast cancer [20].

Embryonic cells in mice with mutated *BRCA1* and *BRCA2* have been shown to display hypersensitivity to ionizing radiation [15, 21]. For *BRCA2*, radiation sensitivity was measured in blastocysts as a function of inner cell mass outgrowth and trophoblast cell number after 400 cGy of γ irradiation. While wild-type and heterozygous *BRCA2* embryos had a minimal reduction in inner cell mass outgrowth, there was complete ablation of the inner cell mass in homozygous mutated *BRCA2* embryos in addition to reduction in the number of trophoblast cells in homozygous mutant embryos compared with wild-type heterozygous and control embryos [15]. Similarly, studies of mutated mice embryos with homozygous *BRCA1* mutations have also demonstrated hypersensitivity to γ irradiation [21].

In addition, there are translational studies that show a higher sensitivity of *BRCA* mutation carriers to radiation. In one study, lymphocyte cultures of *BRCA* carriers exposed to 1 Gy of radiation had significantly higher mean chromatid breaks per cell than non-carriers and a higher maximum number of breaks compared with matched controls [22]. These findings are consistent with other similar studies suggesting a higher radiation sensitivity in *BRCA* carriers.

The sensitivity of cells to different types of DNA damage has also been explored in *BRCA* carriers. Peripheral lymphocytic cells containing a heterozygous mutation in *BRCA1* have been found to be more sensitive toward combinations of radiation and chemotherapy agents such as bleomycin, cisplatin, cyclophosphamide, and bis-chloroethylnitrosourea [23]. These findings suggest that *BRCA* carriers may be at higher risk for the induction of mutations and secondary cancers with standard therapies.

Nevertheless, the sensitivity of *BRCA1* and *2* mutation carriers to radiotherapy investigated by prelaboratory data remain conflicting. In contrast to the data discussed above, there are published data in which a relationship between mutation status and radio sensitivity was not able to be established [24, 25].

Therapeutic Management of *BRCA* Mutation Carriers with Radiation Therapy in the Breast Conservation Setting Breast Conservation

Breast-conserving therapy (BCT) consists of surgical removal of the primary tumor followed by radiation therapy to the intact breast and is utilized as a standard treatment approach for localized ductal carcinoma in situ (DCIS) and early stage breast cancer. With the incidence of breast cancer detection migrating to earlier stages of disease, in addition to the more widespread use of neoadjuvant chemotherapy to downsize tumors which otherwise would previously have required mastectomy, there are an increasing numbers of patients who are eligible for breast conservation therapy. In both the invasive and DCIS setting, multiple randomized prospective studies have demonstrated equivalent long-term survival outcomes for BCT compared to mastectomy and the benefits of whole breast radiation therapy after conservative surgery in reducing local-regional relapse compared with breast-conserving surgery alone [1, 3–6, 26, 27].

While defining the appropriate selection of patients for a breast-conserving approach is sometimes debated, the established criteria are based on the ability to (1) achieve an acceptable cosmetic result (i.e., tumor size with respect to breast size), (2) obtain negative margins at surgery, and (3) safely deliver radiation therapy and minimize potential for side effects. In terms of radiation, the technique utilized in each of the trials establishing breast conservation therapy utilized whole breast radiation therapy delivered using external beam techniques. Target volumes to encompass the entire breast volume typically extend from the clavicular head superiorly to approximately 2 cm below the inframammary fold inferiorly, and from the midline medially and approximately 2 cm beyond palpable breast tissue laterally (usually at mid-axillary line) (Fig. 9.1a–c). The current technology mandates an in-depth knowledge of the anatomy and treatment techniques to improve homogeneity in the treatment field while minimizing the dose to organs at risk. An anatomic atlas for volume contouring both tumor and normal tissues have been established [28]. The vast majority of patients treated in these trials were treated with conventionally fractionated, whole breast radiation delivered which entails 180–200 cGy per day to doses of 4500–5040 cGy delivered over a 5–6 week period. In most contemporary practices, the practice of boosting is also employed, where an additional radiation dose of 1000–1600 cGy is delivered to the lumpectomy bed to dose escalate the region at highest risk of recurrence, ultimately requiring patients to receive daily treatments from 5 to 7 weeks of treatment.

Results of Breast Conservation in *BRCA* Carriers Versus Sporadic Cancers

Early studies of familial breast cancer used a positive family history as a surrogate for genetic predisposition; therefore, many of these publications included patients

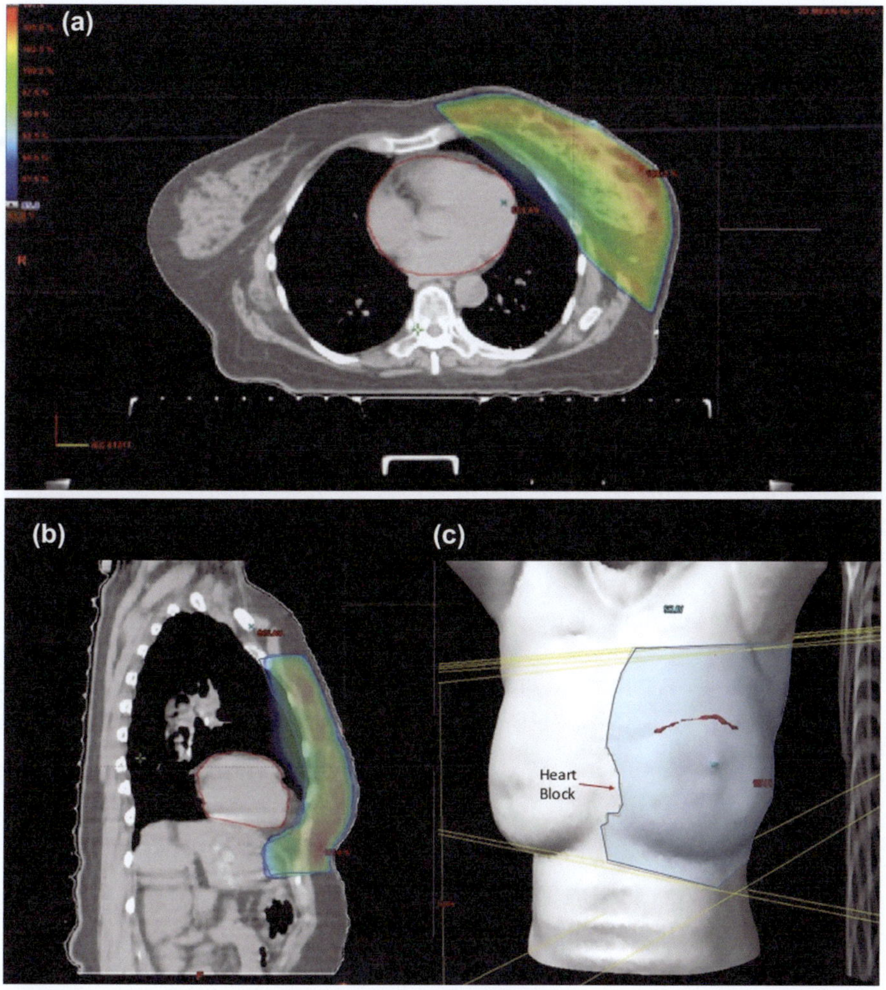

Fig. 9.1 CT treatment planning scan of a patient receiving radiation therapy to the *left breast* after breast-conserving surgery and sentinel node biopsy. Various cardiac avoidance techniques can be utilized to minimize heart and lung exposure. **a** Axial section showing dose distribution to the whole breast. **b** Sagittal section demonstrating the dose to the breast with avoidance of the heart. **c** The digital reconstruction illustrates the medial and lateral tangential beams as they intersect with on the skin. In this case, a small cardiac block is used to significantly decrease heart/lung dose

who did not necessarily harbor *BRCA1/2* gene germ line mutations. Hence, conflicting results reported across studies may be related to the inaccuracy of risk assessment when genetic predisposition was based on a positive family history rather than a true genetic mutation. For example, one investigation of patients with ≥ 3 first-degree relatives with breast or ovarian cancer or whose families had *BRCA1/2* mutations found that local recurrence rates after breast-conserving surgery were initially similar to patients with sporadic breast cancer; however, with longer follow-up, higher rates of recurrence were found in the hereditary group compared with age-matched patients with sporadic disease [29]. The majority of these earlier studies did not demonstrate genetic predisposition as an independent predictor of local recurrence after breast-conserving surgery and whole breast radiation therapy in patients with loosely defined criteria of risk [30–33].

Once the identification and cloning of *BRCA1* and *BRCA2* genes became readily available for patient testing, genetically predisposed patients began being sent for genetic screening and testing for the presence of *BRCA1/BRCA2* germ line mutations. One of the earliest reports of breast conservation therapy in genetically tested *BRCA* carriers was a case controlled study from Yale in which the frequency of *BRCA1/BRCA2* mutations was studied in a series of breast cancer patients who had experienced an in-breast recurrence. Of the 52 patients identified with a documented in-breast failure, 8 (15%) were considered to have had a deleterious mutation, though one of these mutations was subsequently classified as no longer being deleterious, thus reducing the frequency of *BRCA* carriers to 13% of the cohort [34]. Interestingly, the median time to IBTR in the *BRCA1/BRCA2* carriers was 7.8 years compared to 4.7 years for patients without a germ line mutation, suggesting that recurrences in mutation carriers were more likely to be new primaries with a longer time for development of the second breast cancer than a true cancer recurrence.

There is a body of published literature studying the rates of in breast recurrence for *BRCA1/BRCA2* compared to sporadic breast cancers. In a series from Memorial Sloan-Kettering Cancer Center, archival tissue samples were retrospectively collected from 305 women of Ashkenazi Jewish descent treated with BCT, of which 28 harbored genetic mutations. On multivariate analysis, only age remained a significant predictor for IBTR with a relative risk of 2.5; *BRCA* mutation status did not significantly predict for increased risk of IBTR [35].

Another series that was unable to show *BRCA1/2* status to be an independent predictor of ipsilateral breast tumor recurrence was a cohort of *BRCA1/2* carriers compared with women with sporadic breast cancer treated at Institute Curie. At a median follow-up of 8.8 years, the crude rates of local breast tumor recurrence in 27 *BRCA1* and *BRCA2* patients was 24% compared with 19% in 261 women with sporadic breast cancer ($p = 0.47$). These data were updated at a 13.4-year median follow-up and similarly reported no difference in the rates of ipsilateral recurrence between the 2 cohorts with extended follow-up (36% *BRCA* vs. 33% sporadic, $p = 0.42$) [36].

A multi-institutional series by Pierce, et al. published in 2000 reported on 71 women with a *BRCA1/2* mutation and stage I/II breast cancer treated with BCT that were matched 1:3 with 213 women with sporadic breast cancer from centers across North America. The objective of the study was to assess radiation-associated complications and patterns of recurrence between the 2 cohorts. While *BRCA*-associated tumors were associated with higher histologic and nuclear grades and were more likely to be estrogen/progesterone receptor negative than the sporadic cohort, none of the outcome endpoints (5-year actuarial overall survival, relapse-free survival, or in-breast local control) differed significantly between the 2 cohorts. There were no significant differences in acute or chronic morbidity in skin, subcutaneous tissue, lung, or bone based on the Radiation Therapy Oncology Group/European Organization for Research and Treatment of Cancer (RTOG/EORTC) toxicity scoring scale. Based on their findings that tumor control rates, survival, and radiation sensitivity in breast tissue heterozygous for a *BRCA1/2* germ line mutation were similar to matched controls at 5 years, the authors concluded that, though longer follow-up was needed, administering radiation to germ line *BRCA1/2* mutation carriers is safe [37].

This multi-institutional cohort was subsequently expanded to be of the largest series to analyze the role of breast conservation in *BRCA* carriers with data from 11 institutions. Again, the outcomes of 160 *BRCA1/BRCA2* patients with Stage I or II breast cancer treated with breast-conserving surgery and whole breast radiotherapy were matched with 445 patients similarly treated with sporadic disease. There was no significant difference in IBTR overall between carriers and controls; 10- and 15-year estimates were 12 and 24% for carriers and 9 and 17% for controls, respectively (hazard ratio [HR], 1.37; $P = 0.19$). Of note, the incidence of in-breast recurrences did not differ between the 2 cohorts for carriers who had undergone oophorectomy ($p = 0.37$). However, when multivariate analyses removed the patients who had undergone oophorectomy, *BRCA1/2* mutation status was found to be an independent predictor of IBTR by nearly twofold (HR, 1.99; $P = 0.04$); the incidence of IBTR in *BRCA* carriers who had undergone oophorectomy was not significantly different from that in sporadic controls ($P = 0.37$) [38].

An important series from Yale establishing the likelihood of late recurrences in *BRCA* patient evaluated the outcome of BCT in a series of 127 patients diagnosed with breast cancer by age 42 who underwent testing for germ line *BRCA1/BRCA2* mutations. Of these, 22 were found to have deleterious mutations. At a median follow-up of 12.7 years, the rate of in-breast recurrences was found to be similar in the two groups until approximately 6 years, after which significantly more recurrences (49%) were observed in the *BRCA* group compared with 21% in the sporadic group ($P = 0.001$). The multivariate analysis adjusting for the effects of age revealed that the mutation status remained a significant predictor for IBTR. In comparing the histology and sites of the recurrences to the original cancers, 9/11 cancers (82%) were felt to be new primaries as opposed to true clonogenic relapses. Of note, none of the patients in the *BRCA* cohort had received adjuvant hormonal therapy or prophylactic oophorectomy, in contrast to the aforementioned studies in which *BRCA*-associated breast cancers had received these interventions [38]. These high rates of in-breast relapses in the *BRCA* cohort led the authors to conclude that patients with germ line *BRCA1/2* mutations remain at significantly higher risk of second primary tumors in the residual tissue after breast conservation treatment [39]. These findings are consistent with the results of multi-institutional analysis described above that found increased rates of ipsilateral and contralateral breast cancers in the absence of oophorectomy and tamoxifen [38]. These findings are also consistent with another series published in 2009 that reported on 54 *BRCA1/2* patients which also demonstrated a statistically significant higher risk of local events after breast-conserving surgery and radiation. At 5 years, they reported a local relapse rates of 15% in *BRCA1/2* carriers compared with 4% in sporadic controls ($p = 0.03$) [40].

Though some of the results across studies appear conflicting in their conclusions, this may be due to their retrospective nature and differences in factors such as follow-up time, sample size, and ascertainment bias or confounding by factors that affect local recurrence such as margin status, radiation dose, and use of oophorectomy and hormonal therapy. These limitations should be considered when interpreting these results. Ultimately, evaluation of the aggregate data surprisingly suggests consistent findings across the studies: outcomes with breast conservation therapy are similar in terms of **early** IBTR between *BRCA*-related and sporadic breast cancers; with longer follow-up; however, *BRCA*-associate breast cancers appear to have a higher risk of in-breast recurrences (or new primaries) with the use of breast-conserving therapy. This was recently confirmed by a recent meta-analysis which concluded that breast conservation therapy does not increase the risk of in-breast relapses in *BRCA* mutation carriers compared with non-carriers, that radiation therapy was at least as effective, though *BRCA* carriers were at increased risk for new primaries given the continued risk in the residual mutated breast tissue. Ultimately, there was no overall survival difference between carriers and non-carriers after breast conservation therapy [41].

Breast Conservation Versus Mastectomy in *BRCA*-Associated Breast Cancers

Assessing the equivalence of breast conservation therapy versus mastectomy in *BRCA* carriers will likely never be determined in a prospective fashion. The only study to assess this specific question was a multi-institutional series in which *BRCA* patients across 9 institutions were analyzed by treatment type to determine outcomes of breast conservation compared with mastectomy. In this series, 655 *BRCA1* or *BRCA2* patients were treated with breast conservation ($n = 302$) or mastectomy ($n = 353$) were analyzed by treatment type and outcomes. They reported in-breast failures to be significantly more likely after breast conservation compared with mastectomy (15 year: 23.5% vs. 5.5%, $p < 0.0001$). Concordant with other previously discussed data, the failures after breast conservation appeared to be new primary cancers, and not failures in controlling the primary tumor [42].

Radiation Toxicity Considerations for *BRCA* Mutation Carriers

The issue of radio sensitivity in *BRCA*-mutated tumors stems from theoretical and preclinical data that suggest the inherent inability of *BRCA* carriers to repair treatment-induced damage, which would potentially provide a benefit in tumor-cell killing, but simultaneously would make surrounding normal tissue susceptible to radiation-related complications. Theoretic and laboratory-based concerns of increased sensitivity of *BRCA* heterozygote normal tissue resulting in increased complications with radiation exposure have been explored (see discussion above, radiobiologic considerations). The earliest clinical studies of radiotoxicity predate routine genetic testing in patients perceived as being high risk for genetic susceptibility and were limited by their use of a loosely defined family history as a surrogate for mutation status [43].

Risk of Acute/Long-Term Radiation-Related Toxicity in *BRCA* Carriers

In order to evaluate the risk of radiation-related toxicity to the breast and surrounding heart, lung, or soft tissue in *BRCA*-associated cancers versus controls in a precise fashion, a variety of factors need to be controlled for (apart from the radiation dose delivered), which include, but are not limited to, patients' body habitus/breast cup size, inhomogeneity across radiation treatment plans, and differences in ethnicity which may result in inherent differences in skin reactions. Thus, determining whether toxicity of radiation therapy is increased in *BRCA* mutation carriers relative to sporadic controls, controlling for confounding variables, is challenging.

In 1998, Gaffney, et al. evaluated *acute* radiotherapy reactions in 30 *BRCA1* and 20 *BRCA2* carriers, without a control for comparison, and reported the most severe radiation reaction to be moist desquamation which was described as self-limited in 29% (6/21) of irradiated patients [44].

Subsequently, larger clinical studies where radiation response has been measured in *BRCA* carriers have not found any differences in acute and chronic toxicity. The multi-institutional series by Pierce, et al. evaluated differences in acute or chronic morbidity in their cohorts of patients and reported on skin, subcutaneous tissue, lung, or bone based on the RTOG/EORTC toxicity scoring scale. They did not find any significant differences between *BRCA* carriers and sporadic controls [37]. Another series from the UK assessed acute and late toxicity in *BRCA* mutation-associated breast cancers compared with sporadic controls. They reported no differences in breast erythema, moist desquamation, or fatigue as acute side effects (all $p > 0.05$) and no differences in rib fractures, lung fibrosis, necrosis of soft tissue/bone, and pericarditis as late effects ($p > 0.05$). In addition, LENT-SOMA scores and clinical photography failed to detect any differences in patient-reported pain, edema, fibrosis, telangiectasia, or ulcerations. Further, no differences were found in terms of atrophy (assessed by physical examination), use of medications for pain, atrophy or ulcers, or need for lymphedema management [45]. Thus, it is generally felt that though there are theoretic and in vitro-based radiobiologic concerns regarding increased sensitivity to radiation resulting in increased acute and long-term toxicity for *BRCA* carriers, the existing limited clinical data does not support these concerns.

Secondary Radiation-Induced Malignancies and/or Increase in Contralateral Breast Cancer Risk

Age is the one the most important risk factors for determining likelihood of developing a secondary malignancy. The younger the patient is at the time of radiation treatment, the higher the risk is of a future second cancer. It is well documented that the overall risk of contralateral breast cancers in *BRCA* germ line carriers at a baseline is high. When considering factors that can modify the risk of contralateral breast cancers, radiation needs to be discussed in the context of breast conservation therapy; specifically, whether the use of radiation for *BRCA*-associated breast cancers puts patients at a higher risk of additional second malignancies including contralateral events and in field recurrences. It is well documented from studies risk of radiation-induced malignancies in atomic bomb survivors and patients exposed to X-rays in the diagnostic setting that a minimal latency period of 10–12 years is required between radiation exposure and secondary cancer onset. Thus, when interpreting studies on radiation-induced breast cancer risk or second malignancy risk, duration of follow-up remains a critical factor. It is also important to recognize that a threshold dose for secondary malignancy development has not been identified, and the dose-effect relationship from these studies appears to be somewhat linear. This suggests that any exposure to ionizing radiation can be carcinogenic depending on radiation variables (higher total dose, larger individual exposure dose, cumulative dose, larger field size etc.) [46]. In addition, there are numerous other additional factors (other than direct radiation-related cell killing and fractionation) that can affect the magnitude or the shape of the dose-response curve after therapeutic radiation exposure. These include host variables such as mutations in genes affecting repair and hypoxia.

BRCA1 or *BRCA2* carriers with breast cancer have been shown to be at increased risk of contralateral breast cancer compared with non-carriers after receiving radiotherapy [47, 48]. This increased risk is particularly apparent in genetically predisposed young women (<40 years). In the Yale series, contralateral events were significantly higher in *BRCA1/2* carriers after whole breast radiation compared with the sporadic controls (42% vs 9%, $p = 0.001$) and remained significant on multivariate regression analysis adjusted for age [39]. Another series similarly found higher rates of second malignancies in *BRCA* deleterious mutation carriers compared with controls. At 10 years of follow-up, secondary malignancies were seen in 14% in sporadic patients versus 39% *BRCA* patients [49]. In the multi-institutional series by Pierce, et al., contralateral breast cancers were also significantly increased in carriers versus controls, with 10-year estimates of 26% versus 3% and 15 year estimates of 39% versus 7% (HR: 10.43, $p < 0.001$) [38].

Other studies have failed to show an increase in contralateral breast cancer risk [38, 50, 51]. For example, a nested case-control analysis from the WECARE study, a population-based study of a large genetically tested cohort of 603 metachronous contralateral breast cancers matched with 1199 patients with unilateral breast cancer (of which 158 were *BRCA1/2* mutation carriers), analyzed risk of contralateral events after radiation therapy. They reported no increased risk of contralateral breast cancer events after radiotherapy for primary breast cancer, adjusting for age of diagnosis, age at menarche and menopause, number of full-term pregnancies, family history, and receipt of adjuvant systemic treatment, histology, and stage of the first primary [51]. Furthermore, the WECARE study found no clear evidence of increased contralateral breast cancer risk for patients treated with breast radiotherapy among carriers of *BRCA1*/*BRCA2* mutations [52].

Lastly, another cohort study of *BRCA1/2*-breast cancer patients found no association between radiotherapy for primary breast cancer and risk of contralateral breast cancer, and no increased risk of additional (second) primary breast cancers or contralateral cancers in the *BRCA1/2* breast cancer patients who were irradiated before the age of 40 years [48].

Existing published data to support a higher risk of scatter ionizing *therapeutic* radiation increasing the carcinogenic effect in *BRCA1/2*-associated breast cancers (relative to sporadic breast cancers) is weak. However, this does not likely apply to young *BRCA1/2* mutation carriers since there are relatively good data to support the development of secondary cancers from diagnostic (low-dose) radiation in young mutation carriers. Hence, clinicians rightfully remain concerned regarding a carcinogenic effect of therapeutic ionizing radiation in young *BRCA1/2* mutation carriers. The major factors associated with decreased risk of contralateral breast cancers in *BRCA+* patients are oophorectomy, increased age at diagnosis, and use of adjuvant tamoxifen

In summary, though both therapeutic and diagnostic radiation exposure is known to increase the risk of secondary malignancies, particularly in the young age group, clinical data analyzing increased risk of *BRCA1* and *BRCA2* mutation carriers relative to sporadic breast cancer patients is limited due to the inherent difficulties (such as selection bias) in conducting well designed studies to accurately assess the magnitude of increased risk.

Faster Radiation Delivery Methods

Although conventionally fractionated radiotherapy course treats the entire breast, typically followed by a boost to the tumor bed over six to seven weeks, and the efficacy and cosmetic outcomes of such regimens are well-established, two strategies are being investigated to address the lengthy treatment time associated with adjuvant radiotherapy for early stage breast cancer. The techniques and applicability for *BRCA*-associated breast cancers is discussed below.

Hypo-fractionated Whole Breast Radiation Therapy

Recent efforts to shorten the course of conventionally fractionated whole breast radiation have established the use of hypo-fractionated whole breast radiation, whereby with larger daily fractions (i.e., 267 cGy) are delivered to a *biologically equivalent dose* to the **total dose** of standard fractionated (i.e., 39 Gy in 13 fractions, or 42.9 Gy given in 13 fractions) to the whole breast volume. Several prospective randomized trials have demonstrated long-term efficacy that is non-inferior to standard fractionation of 5–7 weeks with similar toxicity profiles [53–57]. Despite the excellent local control and cosmetic results demonstrated in these studies with the use of hypo-fractionation, there remains controversy as to which patients are eligible for these shorter fractionation schemas based on patients that were included in the original trials [58]. Although an American Society of Radiation Oncology (ASTRO) 2010 consensus guideline was developed with recommendations for appropriate patient selection for hypo-fractionation, [59] patient eligibility is being re-evaluated and will likely evolve to include a broader group of patients in the near future. Nevertheless, the vast majority of patients in the randomized trials were >50 years of age, and the existing data and guidelines do not support the use of hypo-fractionated whole breast radiation in younger patients because the higher risk of local recurrence associated with younger age [54, 56, 57]. Furthermore, because the preponderance of existing data supporting hypo-fractionated WBRT is derived from sporadic breast cancers that were ER-positive tumors in patients >50 years of age, the outcomes for hypo-fractionated WBRT have not been adequately assessed for *BRCA*-associated breast cancers. Given that different breast cancer subtypes may have different sensitivities to changes in radiation fractionation schedules (which forms the basis of hypo-fractionation schemas), assessment with radiobiological modeling and clinical trials needs to be done before routine use of this treatment in patients with *BRCA* mutations. Until then, hypo-fractionated WBRT should be used with caution in *BRCA*-associated cancers [60]. Regardless, the association of young age and triple-negative subtype with genetic breast cancers typically precludes the clinical applicability of hypo-fractionated WBRT for *BRCA* carriers.

Accelerated Partial Breast Irradiation

Another treatment modality under active investigation to shorten the lengthy treatment time associated with whole breast radiation is delivery with accelerated partial breast radiotherapy (APBI) techniques. The rationale for treatment of a select portion of the breast around the lumpectomy cavity (as opposed to whole breast) comes from data collected in both retrospective and prospective studies demonstrating that the majority of local recurrences occur in close proximity to the tumor bed. Local recurrences in other quadrants remote from the tumor bed are rare, representing only 3–4% of recurrences [61]. Based on these concepts, APBI techniques were designed to treat the highest risk portion of the breast, encompassing the lumpectomy cavity plus a margin of approximately 1–2 cm, typically delivered twice a day over the course of one week. APBI can be delivered using methods that utilize external beam photons (3D conformal EBRT), brachytherapy (such as interstitial or balloon-based brachytherapy), or techniques that deliver photons or electrons to the tumor bed intra-operatively (IORT: intra-operative radiation therapy). The vast majority of the existing data is retrospective and prospective randomized trials with long-term follow-up are lacking. Current guidelines support the use of APBI in highly selected low-risk patients using specific modalities, [62–65] but patient selection criteria vary across organizations and will remain controversial until data from additional randomized trials become available. Similar to hypo-fractioned whole breast radiation, caution should be utilized using APBI in *BRCA*-associated breast cancers until additional data become available. The ASTRO consensus guideline developed in 2009 with recommendations for appropriate patient selection for APBI specifically state that *BRCA* mutation carriers fall under the "unsuitable" category [62].

Post-mastectomy Radiation Therapy (PMRT)

The use of radiation therapy is an integral component in the multidisciplinary management of locally advanced breast cancer in the post-mastectomy setting. Post-mastectomy radiation has been used for decades to reduce the risk of local-regional recurrence in higher risk patients after complete removal of the breast tissue and regional nodes, and is supported by many randomized trials with long-term follow-up [66–68]. The techniques for radiation treatment to the chest wall have evolved significantly since their initial use in the 1940s, with progressive use of computed tomography (CT) simulation, modern-day linear accelerators, computerized treatment planning modalities, and onboard imaging techniques [69]. These improvements in technology have significantly decreased toxicity to heart and lung that was experienced with the older techniques, resulting in improvements to the therapeutic ratio for radiation therapy in the post-mastectomy setting. With meticulous treatment planning and improvements in techniques to decrease the exposure to surrounding normal tissue, the benefits in decreasing local-regional relapse have been shown to ultimately result in improvements in disease-free survival and overall survival due to significant decreases in cardiac and lung mortality associated with older techniques [70]. Recently, the recommendations for post-mastectomy radiation therapy have been revised to strongly consider radiation for patients with one or more positive nodes [71]. The most recent update of post-mastectomy radiation therapy guideline from the American Society of Clinical Oncology (ASCO) has updated to recommend radiating patients with T1-2 breast cancer with one to three positive axillary nodes after mastectomy, with the caveat that some subsets of these patients are likely to have such a low risk of local-regional failure, such that the absolute benefit of PMRT may be outweighed by its potential toxicities; risk-to-benefit ratios therefore need to be considered [72]. For *BRCA* mutation carriers undergoing mastectomy with high-risk features, consideration of post-mastectomy radiation should include a thorough discussion of the risks versus benefits of radiation and meticulous attention to minimizing dose to normal tissue structures.

Summary

BRCA mutation status has significant implications for local-regional management of breast cancer. In view of the strong association of *BRCA1*-associated breast cancer with TNBC, present guidelines for genetic screening have incorporated TNBC so that all patients younger than 60 years with this form of cancer, irrespective of family history, should be considered for genetic testing [73]. Patients should be informed that while *BRCA* carriers may have an overall increased lifetime risk of both ipsilateral and contralateral local recurrences in comparison with patients with sporadic breast cancers, this does not appear to affect cause-specific and overall survival outcomes. Thus, the presence of a *BRCA* mutation is not a contraindication for breast conservation in otherwise appropriately selected candidates. Importantly, these data do not suggest increased toxicity in germ line *BRCA1* and *2* carriers treated with radiotherapy compared with sporadic breast cancers, and therefore, radiotherapy should not be withheld when indicated for patient care.

The final recommendations for *BRCA* carriers with regard to radiation are still in evolution. Though any ionizing radiation is considered to be a significant risk factor for the development of primary breast cancer, there are no hard data suggesting increased acute or long-term toxicity. The risk of secondary malignancies, seen particularly in younger patients, is likely related to an underlying risk in patients harboring residual *BRCA*-mutated cells, rather than specific to the use of adjuvant radiotherapy in *BRCA1*/*2* mutation carriers.

References

1. Litière S, Werutsky G, Fentiman IS, Rutgers E, Christiaens M-R, Van Limbergen E, et al. Breast conserving therapy versus mastectomy for stage I-II breast cancer: 20 year follow-up of the EORTC 10801 phase 3 randomised trial. Lancet Oncol. 2012;13(4):412–9.
2. Poggi MM, Danforth DN, Sciuto LC, Smith SL, Steinberg SM, Liewehr DJ, et al. Eighteen-year results in the treatment of early breast carcinoma with mastectomy versus breast conservation therapy: the National Cancer Institute randomized trial. Cancer. 2003;98 (4):697–702.
3. Simone N, Dan T, Shih J, Smith S, Sciuto L, Lita E, et al. Twenty-five year results of the national cancer institute randomized breast conservation trial. Breast Cancer Res Treat. 2012;132(1):197–203.
4. Veronesi U, Cascinelli N, Mariani L, Greco M, Saccozzi R, Luini A, et al. Twenty-year follow-up of a randomized study comparing breast-conserving surgery with radical mastectomy for early breast cancer. N Engl J Med. 2002;347(16):1227–32.
5. Wickberg A, Holmberg L, Adami HO, Magnuson A, Villman K, Liljegren G. Sector resection with or without postoperative radiotherapy for stage I breast cancer: 20-year results of a randomized trial. J Clin Oncol. 2014;32(8):791–7.
6. Early Breast Cancer Trialists' Collaborative Group (EBCTCG), Darby S, McGale P, Correa C, Taylor C, Arriagada R, Clarke M, et al. Effect of radiotherapy after breast-conserving surgery on 10-year recurrence and 15-year breast cancer death: meta-analysis of individual patient data for 10 801 women in 17 randomised trials. Lancet. 2011;378(9804):1707–16.

7. Hall EJ, Giaccia AJ. Radiobiology for the radiologist (edition 7E, 2012). 6th ed. Philadelphia: Lippincott Williams & Wilkins; 2012. Vol. ix, 546 pgs.
8. Patel KJ, Yu VP, Lee H, Corcoran A, Thistlethwaite FC, Evans MJ, et al. Involvement of *BRCA2* in DNA repair. Mol Cell. 1998;1(3):347–57.
9. Moynahan M, Chiu J, Koller B, Jasint J. *BRCA1* controls homology-directed DNA repair. Mol Cell. 1999;4(4):511–8.
10. Kavanagh JN, Redmond KM, Schettino G, Prise KM. DNA double strand break repair: a radiation perspective. Antioxid Redox Signal. 2013;18(18):2458–72.
11. Jasin M. Homologous repair of DNA damage and tumorigenesis: the *BRCA* connection. Oncogene. 2002;21(58):8981–93.
12. Scully R, Livingston DM. In search of the tumour-suppressor functions of *BRCA1* and *BRCA2*. Nature. 2000;408:429–32.
13. Rothfuss A, Schutz P, Bochum S, Volm T, Eberhardt E, Kreienberg R, et al. Induced micronucleus frequencies in peripheral lymphocytes as a screening test for carriers of a *BRCA1* mutation in breast cancer families. Cancer Res. 2000;60(2):390–4.
14. Tutt A, Ashworth A. The relationship between the roles of *BRCA* genes in DNA repair and cancer predisposition. Trends Mol Med. 2002;8(12):571–6.
15. Sharan SK, Morimatsu M, Albrecht U, Lim DS, Regel E, Dinh C, et al. Embryonic lethality and radiation hypersensitivity mediated by *Rad51* in mice lacking *BRCA2*. Nature. 1997;386(6627):804–10.
16. Baeyens A, Thierens H, Claes K, Poppe B, Messiaen L, De Ridder L, et al. Chromosomal radiosensitivity in breast cancer patients with a known or putative genetic predisposition. Br J Cancer. 2002;87(12):1379–85.
17. van Leeuwen FE, Klokman WJ, Veer MB, Hagenbeek A, Krol AD, Vetter UA, et al. Long-term risk of second malignancy in survivors of Hodgkin's disease treated during adolescence or young adulthood. J Clin Oncol. 2000;18(3):487–97.
18. Jansen-van der Weide MC, Greuter MJW, Jansen L, Oosterwijk JC, Pijnappel RM, de Bock GH. Exposure to low-dose radiation and the risk of breast cancer among women with a familial or genetic predisposition: a meta-analysis. Eur Radiol. 2010;20(11):2547–56.
19. Sanford KK, Parshad R, Gantt R, Tarone RE, Jones GM, Price FM. Factors affecting and significance of G2 chromatin radiosensitivity in predisposition to cancer. Int J Radiat Biol. 1989;55:963–81.
20. Scully R, Livingston DM. In search of the tumour-suppressor functions of *BRCA1* and *BRCA2*. Nature. 2000;408(6811):429–32.
21. Shen SX, Weaver Z, Xu X, Li C, Weinstein M, Chen L, et al. A targeted disruption of the murine *BRCA1* gene causes g-irradiation hypersensitivity and genetic instability. Oncogene. 1998;17:3115–24.
22. Ernestos B, Nikolaos P, Koulis G, Eleni R, Konstantinos B, Alexandra G, et al. Increased chromosomal radiosensitivity in women carrying *BRCA1/BRCA2* mutations assessed with the G2 assay. Int J Radiat Oncol Biol Phys. 2010;76(4):1199–205.
23. Trenz K, Lugowski S, Jahrsdorfer U, Jainta S, Vogel W, Speit G. Enhanced sensitivity of peripheral blood lymphocytes from women carrying a *BRCA1* mutation towards the mutagenic effects of various cytostatics. Mutation Res. 2003;544(2–3):279–88.
24. Baeyens A, Thierens H, Claes K, Poppe B, De Ridder L, Vral A. Chromosomal radiosensitivity in *BRCA1* and *BRCA2* mutation carriers. IntJ Rad Biol. 2004;80(10):745–56.
25. Nieuwenhuis B, Van Assen-Bolt AJ, Van Waarde-Verhagen MA, Sijmons RH, Van der Hout AH, Bauch T, et al. *BRCA1* and *BRCA2* heterozygosity and repair of X-ray-induced DNA damage. Int J Radiat Biol. 2002;78(4):285–95.
26. Donker M, Litière S, Werutsky G, Julien J-P, Fentiman IS, Agresti R, et al. Breast-conserving treatment with or without radiotherapy in ductal carcinoma in situ: 15-year recurrence rates and outcome after a recurrence, from the EORTC 10853 randomized phase III trial. J Clin Oncol. 2013;31(32):4054–9 Epub 2013 Sep 16.

27. Early Breast Cancer Trialists' Collaborative Group (EBCTCG), Correa C, McGale P, Taylor C, Wang Y, Clarke M, Davies C, et al. Overview of the randomized trials of radiotherapy in ductal carcinoma in situ of the breast. J Natl Cancer Inst Monogr. 2010; 2010(41):162–77.
28. Li XA, Tai A, Arthur DW, Buchholz TA, Macdonald S, Marks LB, et al. Variability of target and normal structure delineation for breast cancer radiotherapy: an RTOG multi-institutional and multiobserver study. Int J Radiat Oncol Biol Phys. 2009;73(3):944–51.
29. Seynaeve C, Verhoog LC, Van De Bosch LM, Van Geel AN, Menke-Pluymers M, Meijers-Heijboer EJ, et al. Ipsilateral breast tumour recurrence in hereditary breast cancer following breast-conserving therapy. Eur J Cancer. 2004;40(8):1150–8.
30. Chabner E, Nixon A, Gelman R, Hetelekidis S, Recht A, Bornstein B, et al. Family history and treatment outcome in young women after breast-conserving surgery and radiation therapy for early-stage breast cancer. J Clin Oncol. 1998;16(6):2045–51.
31. Freedman LM, Buchholz TA, Thames HD, Strom EA, McNeese MD, Hortobagyi GN, et al. Local-regional control in breast cancer patients with a possible genetic predisposition. Int J Radiat Oncol Biol Phys. 2000;48(4):951–7.
32. Haas JA, Schultz DJ, Peterson ME, Solin LJ. An analysis of age and family history on outcome after breast-conservation treatment: the University of Pennsylvania experience. Cancer J Sci Am. 1998;4(5):308–15.
33. Harrold EV, Turner BC, Matloff ET, Pathare P, Beinfield M, McKhann C, et al. Local recurrence in the conservatively treated breast cancer patient: a correlation with age and family history. Cancer J Sci Am. 1998;4(5):302–7.
34. Turner BC, Harrold E, Matloff E, Smith T, Gumbs AA, Beinfield M, et al. *BRCA1/BRCA2* germline mutations in locally recurrent breast cancer patients after lumpectomy and radiation therapy: implications for breast-conserving management in patients with *BRCA1/BRCA2* mutations. J Clin Oncol. 1999;17(10):3017–24.
35. Robson M, Levin D, Federici M, Satagopan J, Bogolminy F, Heerdt A, et al. Breast conservation therapy for invasive breast cancer in Ashkenazi women with *BRCA* gene founder mutations. J Natl Cancer Inst. 1999;91(24):2112–7.
36. Kirova YM, Savignoni A, Sigal-Zafrani B, de La Rochefordiere A, Salmon RJ, This P, et al. Is the breast-conserving treatment with radiotherapy appropriate in *BRCA1/2* mutation carriers? Long-term results and review of the literature. Breast Cancer Res Treat. 2010; 120(1):119–26.
37. Pierce LJ, Strawderman M, Narod SA, Oliviotto I, Eisen A, Dawson L, et al. Effect of radiotherapy after breast-conserving treatment in women with breast cancer and germline *BRCA1/2* mutations. J Clin Oncol. 2000;18(19):3360–9.
38. Pierce LJ, Levin AM, Rebbeck TR, Ben-David MA, Friedman E, Solin LJ, et al. Ten-year multi-institutional results of breast-conserving surgery and radiotherapy in *BRCA1/2*-associated stage I/II breast cancer. J Clin Oncol. 2006;24(16):2437–43.
39. Haffty BG, Harrold E, Khan AJ, Pathare P, Smith TE, Turner BC, et al. Outcome of conservatively managed early-onset breast cancer by *BRCA1/2* status. Lancet. 2002;359 (9316):1471–7.
40. Garcia-Etienne CA, Barile M, Gentilini OD, Botteri E, Rotmensz N, Sagona A, et al. Breast-conserving surgery in *BRCA1/2* mutation carriers: are we approaching an answer? Ann Surg Oncol. 2009;16(12):3380–7.
41. Valachis A, Nearchou AD, Lind P. Surgical management of breast cancer in *BRCA*-mutation carriers: a systematic review and meta-analysis. Breast Cancer Res Treat. 2014;144(3): 443–55.
42. Pierce LJ, Phillips KA, Griffith KA, Buys S, Gaffney DK, Moran MS, et al. Local therapy in *BRCA1* and *BRCA2* mutation carriers with operable breast cancer: comparison of breast conservation and mastectomy. Breast Cancer Res Treat. 2010;121(2):389–98.

43. Chabner E, Nixon A, Gelman R, Hetelekidis S, Recht A, Bornstein B, et al. Family history and treatment outcome in young women after breast conserving surgery and radiation therapy for early stage breast cancer. J Clin Oncol. 1998;16:2045–51.
44. Gaffney DK, Brohet RM, Lewis CM, Holden JA, Buys SS, Neuhausen SL, et al. Response to radiation therapy and prognosis in breast cancer patients with *BRCA1* and *BRCA2* mutations. Radiother Oncol. 1998;47(2):129–36.
45. Shanley S, McReynolds K, Ardern-Jones A, Ahern R, Fernando I, Yarnold J, et al. Late toxicity is not increased in *BRCA1/BRCA2* mutation carriers undergoing breast radiotherapy in the United Kingdom. Clin Cancer Res. 2006;12(23):7025–32.
46. de Gonzalez AB, Gilbert E, Curtis R, Inskip P, Kleinerman R, Morton L, et al. Second solid cancers after radiotherapy: a systematic review of the epidemiological studies of the radiation dose-response relationship. Int J Radiation Oncol BiolPhysics. 2013;86(2):10.
47. Broeks A, Braaf LM, Huseinovic A, Nooijen A, Urbanus J, Hogervorst FB, et al. Identification of women with an increased risk of developing radiation-induced breast cancer: a case only study. Breast Cancer Res. 2007;9(2):R26.
48. Drooger J, Akdeniz D, Pignol JP, Koppert LB, McCool D, Seynaeve CM, et al. Adjuvant radiotherapy for primary breast cancer in *BRCA1* and *BRCA2* mutation carriers and risk of contralateral breast cancer with special attention to patients irradiated at younger age. Breast Cancer Res Treat. 2015;154(1):171–80.
49. Haffty BG, Choi DH, Goyal S, Silber A, Ranieri K, Matloff E, et al. Breast cancer in young women (YBC): prevalence of *BRCA1/2* mutations and risk of secondary malignancies across diverse racial groups. Ann Oncol. 2009;20(10):1653–9.
50. Metcalfe K, Lynch HT, Ghadirian P, Tung N, Kim-Sing C, Olopade OI, et al. Risk of ipsilateral breast cancer in *BRCA1* and *BRCA2* mutation carriers. Breast Cancer Res Treat. 2011;127(1):287–96.
51. Bernstein JL, Thomas DC, Shore RE, Robson M, Boice JD, Stovall M, et al. Contralateral breast cancer after radiotherapy among *BRCA1* and *BRCA2* mutation carriers: a WECARE study report. Eur J Cancer. 2013;49(14):2979–85 (Oxford, England: 1990).
52. Bernstein JL, Thomas DC, Shore RE, Robson M, Boice JD Jr, Stovall M, et al. Contralateral breast cancer after radiotherapy among *BRCA1* and *BRCA2* mutation carriers: a WECARE study report. Eur J Cancer. 2013;49(14):2979–85.
53. Haviland JS, Owen JR, Dewar JA, Agrawal RK, Barrett J, Barrett-Lee PJ, et al. The UK Standardisation of breast radiotherapy (START) trials of radiotherapy hypofractionation for treatment of early breast cancer: 10-year follow-up results of two randomised controlled trials. Lancet Oncol. 2013;14(11):1086–94.
54. Whelan TJ, Pignol JP, Levine MN, Julian JA, MacKenzie R, Parpia S, et al. Long-term results of hypofractionated radiation therapy for breast cancer. N Engl J Med. 2010;362(6):513–20.
55. FAST Trialists group, Agrawal RK, Alhasso A, Barrett-Lee PJ, Bliss JM, Bliss P, Bloomfield D, et al. First results of the randomised UK FAST trial of radiotherapy hypofractionation for treatment of early breast cancer (CRUKE/04/015). Radiother Oncol. 2011;100(1):93–100.
56. Bentzen SM, Agrawal RK, Aird EG, Barrett JM, Barrett-Lee PJ, Bentzen SM, et al. The UK standardisation of breast radiotherapy (START) Trial B of radiotherapy hypofractionation for treatment of early breast cancer: a randomised trial. Lancet. 2008;371(9618):1098–107.
57. Bentzen SM, Agrawal RK, Aird EG, Barrett JM, Barrett-Lee PJ, Bliss JM, et al. The UK standardisation of breast radiotherapy (START) Trial A of radiotherapy hypofractionation for treatment of early breast cancer: a randomised trial. Lancet Oncol. 2008;9(4):331–41.
58. Rowe B, Moran MS. Accelerated partial breast irradiation and hypofractionated whole breast radiation. US Oncol Hematol. 2011;7(1):31–7.
59. Smith BD, Bentzen SM, Correa CR, Hahn CA, Hardenbergh PH, Ibbott GS, et al. Fractionation for whole breast irradiation: an American Society for radiation oncology (ASTRO) evidence-based guideline. Int J Radiation Oncol Biol Physics. 2011;81(1):59–68 Epub 2010 Jul 16.

60. Moran MS. Radiation therapy in the locoregional treatment of triple-negative breast cancer. Lancet Oncol. 2015;16(3):e113–22.
61. Sanders ME, Scroggins T, Ampil FL, Li BD. Accelerated partial breast irradiation in early-stage breast cancer. J Clin Oncol. 2007;25(8):996–1002.
62. Smith BD, Arthur DW, Buchholz TA, Haffty BG, Hahn CA, Hardenbergh PH, et al. Accelerated partial breast irradiation consensus statement from the American Society for Radiation Oncology (ASTRO). J Am Coll Surgeons. 2009;209(2):269–77.
63. Surgeons TASoB. Consensus statement for accelerated partial breast irradiation online. 2008.
64. Shah C, Vicini F, Wazer DE, Arthur D, Patel RR. The American Brachytherapy Society consensus statement for accelerated partial breast irradiation. Brachytherapy. 2013;12(4): 267–77.
65. ASBS APBI Guideline. American society of breast surgeons' consensus statement for accelerated partial breast irradiation. https://www.breastsurgeonsorg/statements/PDF_Statements/APBIpdf. 2011.
66. Ragaz J, Jackson SM, Le N, Plenderleith IH, Spinelli JJ, Basco VE, et al. Adjuvant radiotherapy and chemotherapy in node-positive premenopausal women with breast cancer. N Engl J Med. 1997;337(14):956–62.
67. Overgaard M, Hansen PS, Overgaard J, Rose C, Andersson M, Bach F, et al. Postoperative radiotherapy in high-risk premenopausal women with breast cancer who receive adjuvant chemotherapy. Danish breast cancer cooperative group 82b trial. N Engl J Med. 1997;337 (14):949–55.
68. Overgaard M. Overview of randomized trials in high risk breast cancer patients treated with adjuvant systemic therapy with or without postmastectomy irradiation. Semin Radiat Oncol. 1999;9(3):292–9.
69. Moran MS, Haffty BG. Radiation techniques and toxicities for locally advanced breast cancer. Semin Radiat Oncol. 2009;19(4):244–55.
70. EBCTCG (Early Breast Cancer Trialists' Collaborative Group), McGale P, Taylor C, Correa C, Cutter D, Duane F, Ewertz M, et al. Effect of radiotherapy after mastectomy and axillary surgery on 10-year recurrence and 20-year breast cancer mortality: meta-analysis of individual patient data for 8135 women in 22 randomised trials. Lancet. 2014 Jun 21;383 (9935):2127–35. Epub 2014 Mar 19.
71. Gradishar WJ, Anderson BO, Balassanian R, Blair SL, Burstein HJ, Cyr A, et al. Invasive breast cancer version 1.2016, NCCN clinical practice guidelines in oncology. J Natl Compr Cancer Netw. 2016;14(3):324–54.
72. Recht A, Comen EA, Fine RE, Fleming GF, Hardenbergh PH, Ho AY, et al. Postmastectomy radiotherapy: an American Society of Clinical Oncology, American society for radiation oncology, and society of surgical oncology focused guideline update. J Clin Oncol. 2016 Nov–Dec;6(6):e219–e234. Epub 2016 Sep 19.
73. Adopted on March 1, by the American Society Of Clinical Oncology. American society of clinical oncology policy statement update: genetic testing for cancer susceptibility. J Clinl Oncol. 2003;21(12):2397–406.

Fertility Preservation in *BRCA* Carriers: Special Considerations

10

Iris G. Insogna, Elizabeth S. Ginsburg and Ann H. Partridge

Introduction

Breast cancer is a significant international public health problem and is the leading cause of cancer death in women worldwide [1]. In high-income countries, women under the age of 45 are more likely to die of breast cancer than of any other disease or injury [2]. Women with hereditary breast and ovarian cancer syndromes, including carriers of deleterious mutations in the *BRCA1* or *2* genes, are not only at an approximately fivefold increased risk of developing breast cancer, but also develop malignancy approximately 10 years earlier than those with non-hereditary disease, often during their reproductive years [3–5]. In such patients, often referred to as *BRCA*-positive/+ or carriers, the potential for early onset disease combined with the potential adverse impact on fertility of cancer prevention strategies and

I.G. Insogna (✉)
Department of Obstetrics and Gynecology, Brigham and Women's Hospital and Harvard Medical School, 75 Francis Street, Boston, MA 02115, USA
e-mail: iinsogna@partners.org

E.S. Ginsburg
Division of Reproductive Endocrinology and Infertility, Department of Obstetrics and Gynecology, Brigham and Women's Hospital and Harvard Medical School, 75 Francis Street, Boston, MA 02115, USA
e-mail: eginsburg@partners.org

A.H. Partridge
Department of Medical Oncology, Dana-Farber Cancer Institute, 450 Brookline Avenue, Yawkey 1250, Brookline, MA 02215, USA
e-mail: ahpartridge@partners.org

A.H. Partridge
Department of Medicine, Brigham and Women's Hospital and Harvard Medical School, Boston, USA

A.B. Chagpar (ed.), *Managing BRCA Mutation Carriers*,
DOI 10.1007/978-3-319-59198-8_10

treatment if disease develops necessitates proactive, careful counseling about reproductive planning. This should be considered for both previvors (patients who are *BRCA*+ but have not developed disease) and *BRCA*+ survivors (patients who have developed disease in the setting of a hereditary predisposition). While male *BRCA* carriers are at increased risk of developing malignancies, most present later in life, not during reproductive years, and there are not standard risk-reducing strategies employed that are associated with infertility for male previvors. Nevertheless, potential concerns regarding reproduction should be addressed for young male patients as well.

In this manuscript, we review the unique considerations surrounding reproduction, fertility, and fertility preservation for carriers of *BRCA* gene mutations. We address the relevant fertility concerns for the previvor and survivor populations, including strategies for patients with a new cancer diagnosis as well as consideration of preimplantation genetic diagnosis (PGD), even in the absence of infertility, to avoid passing on a deleterious germ line mutation to progeny. Available options are discussed, including gamete (egg or sperm) and embryo cryopreservation, as well as the more controversial use of LHRH agonists for ovarian suppression to preserve fertility. We also highlight future directions such as ovarian tissue cryopreservation. The available data regarding the safety of current assisted reproductive technologies (ART) for *BRCA* carriers will be reviewed. Finally, we describe some of the salient ethical concerns in this area.

Whether *BRCA* carriers are previvors, newly diagnosed with cancer, or longer term survivors, reproductive goals may weigh heavily on their minds, particularly regarding how treatment may impact their fecundity. Fertility preservation discussions prior to initiating treatment, and preservation of the ability to have biological children, have been shown to increase quality of life and improve psychological outcomes for cancer patients [6, 7]. In particular, for young women with breast cancer, international guidelines recommend early referral to reproductive endocrinology and infertility specialists (REIs) to discuss reproductive goals [2, 8]. However, there are a number of potential barriers and concerns surrounding fertility preservation and subsequent pregnancy in *BRCA* carriers with or without a history of cancer including utility, efficacy, timing, and safety of any intervention [9]. *BRCA* carriers require special consideration as detailed below.

Reproductive Considerations for Previvors (*BRCA* Carriers Who Have not Developed Cancer)

In order to reduce the lifetime risk of cancer in previvors, it is recommended that patients consider prophylactic mastectomy and bilateral salpingo-oophorectomy, surgery that results in sterilization. Carriers of the *BRCA1* mutation have a 15–45% lifetime risk of developing ovarian cancer and $\leq 85\%$ risk of developing breast cancer. A *BRCA2* mutation confers a lifetime risk of ovarian cancer of 10–20% and a breast cancer risk of 40% [3]. By undergoing prophylactic bilateral mastectomy, breast cancer risk and mortality can be reduced by 90–95% [10]. Bilateral salpingo-oophorectomy reduces the risk of ovarian cancer by approximately 85–90% and reduces the risk of breast cancer by 40–70% [11, 12]. Based on international guidelines, patients should undergo bilateral salpingo-oophorectomy once childbearing is completed, or by the age of 35–40, given the increased risk of ovarian cancer with age [2, 13].

Previvors may desire to complete childbearing prior to undergoing prophylactic surgery. These patients, in the absence of other fertility concerns, may be able to conceive naturally. The optimal timing of reproduction is a very personal decision. However, even in patients who are not infertile, IVF for the purposes of PGD is still a consideration for *BRCA* carriers.

Preimplantation Genetic Diagnosis

PGD is useful in hereditary disorders when intended parents wish to avoid passing a specific gene on to their offspring. *BRCA* genes demonstrate an autosomal dominant pattern of inheritance, conferring a 50% chance that a child will be a carrier. Following IVF, PGD allows parents to know in advance the carrier status of each embryo. By transferring only non-carrier embryos, intended parents can prevent the propagation of the *BRCA* gene in their family [14].

For PGD, the woman must undergo ovarian stimulation with subsequent oocyte harvesting, and sperm must be available for IVF. Once embryos are fertilized via IVF, they are allowed to develop in the embryology laboratory. At day five and/or six of development, the external cells of the blastocyst (trophectoderm cells, destined to become the placenta) can be biopsied and genetically tested for the presence of specific gene, such as *BRCA1* or *2* mutations. PGD may be an attractive option for pre-surgical *BRCA* patients, who can later have a cryopreserved embryo thawed and transferred. The ethical implications of PGD for *BRCA* carriers, and the disposition of non-carrier and indeterminate embryos after PGD testing, are discussed later. For some women, particularly those needing urgent neoadjuvant chemotherapy, *BRCA* status may be unknown at the time of embryo banking. Additional consideration should be taken when discussing the timing of embryo banking, and the option of PGD, under these circumstances.

Preimplantation genetic screening can also provide information about other chromosomal abnormalities, such as aneuploidy, prior to the transfer or storage of embryos. This may help eliminate the need for invasive testing in the first trimester of pregnancy, such as chorionic villus sampling or amniocentesis, after which some women opt to terminate. If under time constraints, this consideration may be particularly relevant.

IVF with or without PGD and cryopreservation of embryos allows patients to plan for the future, aligning with the modern trend toward delaying childbearing until later in life [15]. Since prophylactic salpingo-oophorectomy is recommended by age 40, some patients may wish to undergo surgery with the plan to delay pregnancy and family building. For patients who do not wish to carry a pregnancy, or are of an age where becoming pregnant may be difficult or dangerous, the option of using a gestational carrier may also be considered.

Reproductive Considerations for *BRCA* Carriers with a Cancer Diagnosis, Survivors

Options for reproduction and family building may be different for carriers of the *BRCA* gene with a new diagnosis of cancer, when time constraints may be more pressing. Prior to initiating treatment, these patients should be counseled on the potential risks of therapy to fertility, and on the current options for fertility preservation, namely gamete or embryo cryopreservation [16]. Alternatively, ovarian suppression, with the use of luteinizing hormone-releasing hormone agonists during chemotherapy, may offer some ovarian protection, an option discussed in more detail below [17]. In patients with ER-positive disease in particular, there may be concern surrounding the safety of fertility preservation strategies with regard to disease outcomes given the associated maintenance or temporary elevation of reproductive hormones, and this should be considered on a case-by-case basis. In *BRCA1*-associated breast cancers, 10–24% are ER-positive, compared to 65–79% ER positivity in *BRCA2*-associated breast cancers [18]. Survivors of cancer who may have completed chemotherapy and/or radiation but still have their ovaries in situ may consider ART after treatment. However, patients may have subsequent diminished ovarian reserve, chemotherapy-induced amenorrhea, or premature menopause due to treatments received and may require fertility assistance [15]. Barton and colleagues compared female cancer survivors with all other patients undergoing IVF/ICSI and demonstrated that survivors were low responders and had poorer outcomes overall. Survivors had significantly fewer oocytes retrieved and embryos available for transfer. Pregnancy and live birth rates were lower among survivors as well with odds ratios of 0.30 (95% CI 0.13–0.68) and 0.27 (95% CI 0.10–0.69), respectively [19]. Patients should be counseled that following treatment with chemotherapy or radiation, ART may be challenging; some may ultimately need a donor egg in order to achieve pregnancy.

Female Fertility Preservation: Existing Options

The most established method of fertility preservation is embryo cryopreservation [16]. Following a protocol of controlled ovarian stimulation (COS) to promote the development of the greatest numbers of follicles, oocytes are harvested, fertilized, then the resultant embryos cryopreserved and stored until ready for use.

Oocyte cryopreservation is also now a standard option, which involves similar COS, oocyte retrieval followed by cryopreservation of mature oocytes in their unfertilized state. Fertilization, pregnancy, and live birth rates reported with thawed oocytes appear to be equivalent to fresh oocytes, making oocyte cryopreservation an accepted, non-experimental practice [20]. Specific data for breast cancer patients regarding success with cryopreserved oocytes versus embryos are not yet available, though it is presumed these are roughly equivalent. As of 2012, of the 387 US IVF clinics registered as Society for Assisted Reproductive Technology members, 200 (51.7%) offered oocyte cryopreservation [21].

For patients that are either single or not in a relationship with a potential co-parent, gamete cryopreservation may be a good option. Alternatively, embryo cryopreservation is appropriate if a couple decides they would like to parent a biological child in the future. A single female can also opt to freeze embryos using her oocytes and a donor sperm. However, in cases of divorce, courts may not necessarily grant ownership and control of the embryos to the cancer survivor. Therefore, it is important to inform women that oocyte cryopreservation, or cryopreservation of eggs fertilized with anonymous donor sperm, is the only way to guarantee that they will have complete control of embryos in the future. The option of embryo or gamete freezing allows for substantial flexibility on a case-by-case basis (Table 10.1).

Table 10.1 Options for female *BRCA* carriers

	Oocyte cryopreservation	Embryo cryopreservation	Ovarian suppression	Ovarian tissue cryopreservation
Advantages	Established practice	Established practice	Allows for a delay in surgical removal of ovaries	Tissue is spared exposure to chemotherapy or radiation
	Does not require a partner	Option for PGD	Potentially reduces tissue exposure to chemotherapy or radiation	
Disadvantages	Potential delay to treatment	Potential delay to treatment	New practice with unclear role for *BRCA* carriers	Experimental
	Exposure to COS	Exposure to COS	Limited access	Potential to reintroduce (pre)malignant cells
	Limited access	Limited access		Limited access

Ovarian Suppression During Chemotherapy

Recently, regimens that suppress and thereby theoretically protect ovarian function during chemotherapy have emerged as potential options for reproductive age women with cancer. Chemotherapy can have a negative impact on ovarian reserve, and alkylating agents commonly used to treat breast cancer can be particularly gonadotoxic [22]. The extent of potential damage to ovarian reserve is related to the choice of chemotherapy regimen, dosage, and temporal exposure, as well as patient age [22].

For women who maintain their ovaries during therapy, luteinizing hormone-releasing hormone analogues (LHRHa), when given concomitantly with chemotherapy, may offer some ovarian protection against such gonadal cytotoxicity [23]. Expert opinion and meta-analyses of available data indicate that LHRH agonists do seem to help preserve ovarian function during chemotherapy exposure [17]. In a recent randomized control trial of 281 women, patients who received LHRHa were significantly more likely to resume menses after treatment compared to the control group (73% vs. 64%), and the LHRHa patients reported more pregnancies compared to the control group (8 vs. 3, $p < 0.05$) [17]. In this trial, there was no difference in disease-free survival between the two groups [17, 23].

Importantly, two large randomized controlled trials, the Tamoxifen and Exemestane Trial (TEXT) and the Suppression of Ovarian Function Trial (SOFT), evaluated potential benefits of ovarian suppression regimens during chemotherapy for premenopausal breast cancer with hormone-receptor-positive breast cancer. Combined analysis of these data demonstrates a reduced risk of breast cancer recurrence for women treated with ovarian suppression plus the aromatase inhibitor exemestane, compared to ovarian suppression plus tamoxifen [17, 24]. Thus, use of ovarian suppression is increasingly being used routinely for breast cancer treatment in the setting of early-stage hormone-receptor-positive breast cancer in pre-menopausal women. Further, there is no apparent harm with regard to disease outcome in this setting with the use of ovarian suppression through chemotherapy, whether for treatment or for fertility preservation. Nevertheless, the potential role for ovarian suppression for fertility preservation in women with early breast cancer remains controversial, whether they are mutation carriers or not. In treating *BRCA* carriers, this should be considered on a case-by-case basis.

Future Directions

Ovarian tissue cryopreservation is considered an experimental technique for fertility preservation. It involves surgical biopsy or stripping of the ovarian cortex, or removing the entire ovary, then cryopreserving the specimen for future use. The ovarian tissue can later be thawed and reimplanted in the patient once therapy is completed, either in or outside the pelvis [25]. As of January 2016, more than 60 live births have resulted from ovarian tissue cryopreservation, with all deliveries resulting from orthotopic (in the pelvis) tissue replacement [26]. A recent study from Israel of 20 patients, the majority of which had hematologic malignancies, reported 53% of patients successfully conceived following ovarian tissue transplantation. Sixteen pregnancies were achieved (six spontaneous and 10 following IVF), with 10 subsequent live births, and two ongoing pregnancies [26].

Ovarian tissue cryopreservation avoids exposing the ovaries to the toxic effects of chemotherapy and/or radiation. It also eliminates the need to delay treatment because the surgery can be done at any time in the cycle and does not require any ovarian stimulation. Similar to oocyte cryopreservation, ovarian tissue cryopreservation does not require a partner.

Ovarian tissue cryopreservation is performed with the intention to reimplant the tissue in the future. This may not be the safest option for *BRCA* carriers, given their increased risk of ovarian malignancy. It is theoretically possible that the ovarian tissue may be reimplanted for a brief time, stimulated to retrieve oocytes, and then removed soon thereafter, to minimize exposure to potentially pre-malignant or malignant cells. However, this remains experimental, without data related to outcomes in *BRCA* patients. In addition, it is likely contraindicated in patients with metastatic disease, as breast cancer may metastasize to ovaries. Only select centers are offering ovarian tissue cryopreservation, and it is not widely available [4, 25].

Female Fertility Preservation: Potential Risks and Challenges

BRCA carriers with a new diagnosis of cancer may have concerns that cryopreservation of oocytes or embryos may lead to a dangerous delay in treatment [27]. In order to maximize the yield of oocytes harvested for cryopreservation, most providers recommend controlled ovarian stimulation (COS), a process that requires an average of two weeks [4]. In order to reduce delay in cancer treatment, some providers will begin COS as soon as possible, rather than waiting for the early follicular phase, as is traditional practice. This random-start protocol using gonadotropin antagonists has been shown to have similar outcomes compared to follicular phase-start protocols, with no difference in the number of total and mature oocytes retrieved, oocyte maturity rate, and fertilization rates [28]. However, data including live birth rates are scant. In an emergency situation, random-start protocols are most appropriate.

Most women, however, may be able to undergo at least one IVF cycle without significantly delaying the initiation of their therapy. In 2007, a study by Madrigano et al. highlighted the importance of early referral to an REI in order to avoid delays while still maximizing potential fertility preservation. For 23 patients with breast cancer, the mean time from fertility evaluation to egg retrieval was 33.3 (10–65) days [29]. On average, patients underwent ovarian stimulation for 11.5 (9–20) days prior to egg retrieval [29]. The average time from definitive surgery to initiation of chemotherapy was 46.8 days in the women who underwent fertility preservation

[29]. Time from diagnosis to initiation of chemotherapy was not statistically different between women who opted for fertility preservation compared to those who did not (71 days vs. 67 days). For women undergoing egg retrieval, time from diagnosis to definitive surgery was longer by 15 days, but this was not statistically significant. Standard of care treatment for breast cancer was not delayed in any clinically significant way by undergoing fertility preservation treatment [29, 30].

A second concern exists regarding risks of exposure to the hormonal protocol used to induce ovarian stimulation. Typical regimens for stimulation protocols use high-dose gonadotropins that result in supraphysiologic systemic levels of estradiol, as much as 10–20 times normal [15, 31]. This allows for the recruitment of multiple follicles, maximizing the number of oocytes retrieved. Patients also receive GnRH agonists or antagonists to try to prevent premature luteinization and ovulation [31]. Exposure to such high levels of estradiol may be of particular concern for women that have estrogen-receptor-positive breast cancer, keeping in mind that 65–79% of breast cancers will be estrogen-receptor-positive in *BRCA2* carriers, compared to 10–24% of breast cancers in *BRCA1* carriers [18].

Alternative regimens, utilizing the selective estrogen-receptor modulator tamoxifen or the aromatase inhibitor letrozole during COS, have been well described and may help limit exposure to high levels of estradiol [4, 31, 32]. Both tamoxifen and letrozole can be used as ovulation induction agents, and the use of either agent is associated with improvement in IVF cycle outcomes [29, 33].

One study by Oktay et al. reported that the use of low-dose FSH with letrozole or tamoxifen results in improved embryo yield compared to the use of tamoxifen alone, but use of letrozole may be preferential as it produces lower levels of estradiol [33]. The same study indicated that tamoxifen alone may briefly result in increased levels of estradiol during stimulation. There are, however, no studies comparing live birth rates with different ovulation induction regimens in breast cancer patients. Again, this may be of particular relevance to patients with ER-positive breast cancers.

Importantly, limited data show that the use of letrozole or tamoxifen as part of the COS protocol does not appear to negatively impact breast cancer outcomes, or increase the risk of recurrence [31]. Data also show that fertility treatments in general do not increase risk of epithelial ovarian cancer for patients specifically with *BRCA* mutations [34].

A natural cycle is also a possibility, in which COS is not performed. The follicles produced by a woman in her natural cycle are aspirated, with as many eggs harvested as possible, though typically the yield is much lower than with ovarian stimulation and would be expected to result in much lower chances of future pregnancy [33]. This treatment has largely been abandoned by most large fertility preservation centers.

Ovarian Reserve and *BRCA* Carriers

Some data suggest that *BRCA* carriers may have inherent diminished ovarian reserve compared to non-carriers and may have fewer oocytes retrieved after COS [35]. There is evidence that *BRCA1* carriers have poorer serum markers of ovarian reserve, specifically a fourfold greater chance of having anti-Mullerian hormone (AMH)<1 ng/mL compared with non-*BRCA1* carriers [36]. However, having poorer ovarian reserve based on serum markers may not translate into meaningful clinical outcomes.

A recent study out of Israel reviewed IVF data for *BRCA* carriers and non-carriers, specifically evaluating response to COS. Carriers and non-carriers had comparable oocyte yield (13.75 vs. 14.75) and low response rates (8.06% vs. 6.45%) [37]. There were no differences in ovarian response, fertilization rates (70.6% vs. 59.66%), or resultant embryos (8.4 vs. 7.19) [37].

BRCA carriers and non-carriers appear to be at similar risk for chemotherapy-induced amenorrhea [38]. Though *BRCA* carriers generally experience menopause 1–2 years earlier than non-carriers, this does not appear to have any meaningful impact on fertility outcomes, such as age at first parity or need for fertility treatments [39]. Compared to non-*BRCA* breast cancer patients, *BRCA* carriers do not appear to be more susceptible to gonadotoxic side effects of chemotherapy and do not appear to have worse fertility outcomes.

Recent data reported by Sabatini et al. over a 17-year experience demonstrate that cancer patients in general have fertility outcomes that are equivalent to women without a history of cancer. The study reported no difference in outcomes for frozen embryo transfers between cancer patients and patients with tubal factor infertility. Both groups had equivalent number of oocytes retrieved and embryos frozen. Similarly, there were no differences between cumulative pregnancy rate per transfer for cancer patients (37%) compared to controls (43%), and cumulative live birth rate per transfer (30% vs. 32%, respectively) [40].

There has been concern regarding the safety of pregnancy for patients with a history of breast cancer, particularly with estrogen-receptor-positive cancers. Pregnancy, however, appears to be safe for patients who have undergone curative treatment. Regardless of receptor status, women who achieve pregnancy after therapy have no difference in survival or risk or recurrence compared to women who do not get pregnant, recognizing that data are limited to retrospective analyses or registry studies [41].

Male Fertility Preservation: Existing Options

Discussions regarding fertility preservation are relevant to male *BRCA* carriers as well. Although male carriers of *BRCA1* have a lifetime breast cancer risk of less than 2%, they may be twice as likely to develop prostate cancer before age 65, for example [42].

Table 10.2 Options for male *BRCA* carriers

	Sperm cryopreservation	Embryo cryopreservation
Advantages	Established practice	Established practice
	Does not require a partner	Option for PGD
	No delay in treatment	
Disadvantages	Limited access	Limited access

Though most of the cancers seen in male *BRCA* patients present later in life, they can affect a man of reproductive age. If so, the patient should be given the option to bank sperm prior to initiating therapy [16]. Semen collection and freezing is the standard of care for men of reproductive age with a new diagnosis of cancer and can usually be arranged in a matter of days [8]. These patients should also be offered the opportunity to pursue PGD if they wish to eliminate the *BRCA* gene in their offspring (Table 10.2).

Socioeconomic, Cultural, and Ethical Considerations

It is an inherent right to parent and to have a family. The Universal Declaration of Human Rights, proclaimed by the United Nations General Assembly in 1948, states men and women have a right to found a family. *BRCA* carriers may require additional efforts to protect their reproductive rights, given the heritability of the mutation, and that standard treatment for these patients often involves procedures that affect their fertility.

In young women with breast cancer, concerns about fertility are associated with significant psychosocial stress [43]. *BRCA* carriers need early discussion regarding reproductive goals and early referral to appropriate providers, namely REI physicians, to review the available options [2, 8]. Unfortunately, consultation with an REI is not yet the standard of care, and access to such subspecialists may be limited. One study found that in women with localized breast cancer diagnosed at 40 years of age or younger, only 68% had discussed fertility options and only 10% underwent fertility preservation prior to initiating treatment [9]. A 2011 survey of over 1000 women diagnosed with cancer at ages 15–40 reported only 61% of women were counseled about fertility risks prior to initiating treatment, and only 4% pursued fertility preservation [44].

Though it does not require stimulation protocols or invasive procedures, referral rates for sperm banking can be low as well. In one recent study, 29% of male cancer patients received fertility counseling and only 11% attempted sperm banking [45].

In the USA alone, access varies tremendously based on location. The highest percentages of IVF clinics are located in northeastern and western states [46]. In areas with fewer hospitals and fewer subspecialty practices, it may not be possible to refer patients to an REI physician. Access to genetic testing to even confirm *BRCA* carrier status may be difficult as well though most insurance policies, including Medicare, will cover the cost of genetic screening in appropriate patients [13].

Interestingly, even among women who undergo fertility preservation prior to therapy, few seem to pursue attempts at pregnancy after treatment [30]. This may in part be due to patient and/or provider concerns about pregnancy safety. It should be reinforced that pregnancy does not appear to increase risk of recurrence or mortality [41].

Cultural differences may also limit patients' access to care. In Italy, for example, embryo cryopreservation was forbidden in 2004 [47]. Federal law mandated that all created embryos were used for transfer and outlawed cryopreserving embryos for future use. Italy allows only gamete banking, but even this limitation can significantly impair options surrounding fertility preservation.

Socioeconomic and racial impacts on fertility options should not be underestimated either. Survey data suggest that women without bachelor's degrees, for example, are less likely to be counseled on reproductive risks of cancer treatment, with an odds ratio of 0.7. The same study found trends toward disparities in access to reproductive services for women over 35, Latina and African American women, and parous women [44].

Financial barriers can also pose a significant problem for many patients. There is no insurance mandate to cover fertility treatments for patients with hereditary cancer syndromes, so for many patients, the cost of IVF cycles and the cost of preserving gametes or embryos are paid out-of-pocket. This can be prohibitively expensive. In 2010, the average cost for a female cancer patient to undergo fertility preservation with oocyte or embryo cryopreservation was $8655. It cost approximately $1495 (17%) more for embryo cryopreservation than oocyte cryopreservation [48]. More recent data indicate that in 2016, it costs between $12,000–15,000 for egg freezing and $15,000–$18,000 for embryo freezing. Storage fees average $900 per year [49]. One retrospective review found that breast cancer patients who were wealthier and older were more likely to pursue fertility preservation treatment [50].

However, some large fertility centers are able to offer discount services for patients with cancer in recognition that these patients do not have time to save for this expense. There are 15 states (Arkansas, California, Connecticut, Hawaii, Illinois, Louisiana, Maryland, Massachusetts, Montana, New Jersey, New York, Ohio, Rhode Island, Texas, and West Virginia) that offer coverage for some infertility diagnosis and treatment, but the extent of coverage varies on a state-by-state basis [51].

The opportunity for PGD, given the heritability of the *BRCA* genes, introduces unique ethical concerns that merit consideration as well. The option to undergo PGD is increasingly desired in patients with hereditary cancer syndromes. A meta-analysis of 13 studies found that, of the 370 respondents affected by a hereditary cancer syndrome, 28% felt their syndrome impacted family planning, 72% felt that PGD should be offered, and 43% would consider using PGD [52]. In a survey of 22 couples affected by hereditary breast and ovarian cancer, half chose to undergo PGD because they "believed it was their moral duty to protect their future child(ren) from suffering" [14]. For patients who have had cancer as a direct result of a *BRCA* mutation, PGD may be a particularly important option for their psychological well-being and family planning goals.

PGD poses additional potential ethical quandaries as it is not universally agreed upon that *BRCA* carrier status is an appropriate indication for PGD. There is no mandated insurance coverage for PGD, and coverage for this indication again varies widely state-by-state. Since *BRCA* carrier status predisposes to a potentially fatal adult-onset disease, many IVF centers will allow PGD. However, some may have concerns that it inappropriately eliminates potential offspring that have a significant chance of being healthy and disease-free [53].

Identifying carrier and non-carrier embryos, as well as those with indeterminate status through the use of PGD, has led to ethical debate regarding the fate of such embryos [53]. The potential outcomes of PGD should be explicitly discussed with patients prior to initiating testing. Patients opting for PGD should understand that the goal of PGD is to identify and transfer a non-carrier embryo. If this does not align with the patient's goal, the patient should not elect for PGD.

Finally, emphasis should be placed on safeguarding the future of any gametes or embryos produced via ART. Following any fertility preservation technique, it is advisable to create legal documents guiding the disposition of any embryos or gametes that are stored for the future, particularly outlining ownership [54]. This can help avoid debate regarding the posthumous use of stored gametes or embryos.

Conclusion

BRCA carriers require special consideration regarding reproduction. Prophylactic and therapeutic management of these patients may compromise their ability to achieve their reproductive goals. A multidisciplinary team approach to this sensitive issue is warranted including oncology, reproductive endocrinology as well as psychosocial supportive providers to assist patients and their loved ones in making the best decisions for themselves in their medical and social situations.

References

1. Ferlay J, Shin HR, Bray F, Forman D, Mathers C, Parkin DM. Estimates of worldwide burden of cancer in 2008: GLOBOCAN 2008. Int J Cancer (Internet). 2010;127(12):2893–917. Available from: http://doi.wiley.com/10.1002/ijc.25516 (Cited 30 July 2016).
2. Paluch-Shimon S, Pagani O, Partridge AH, Bar-Meir E, Fallowfield L, Fenlon D, et al. Second international consensus guidelines for breast cancer in young women (BCY2). Breast. 2016;26:87–99.
3. Ledermann JA, Raja FA, Fotopoulou C, Gonzalez-Martin A, Colombo N, Sessa C, et al. Newly diagnosed and relapsed epithelial ovarian carcinoma: ESMO clinical practice guidelines for diagnosis, treatment and follow-up. Ann Oncol (Internet). 2013;24 Suppl 6(suppl 6):vi24–32. Available from: http://www.ncbi.nlm.nih.gov/pubmed/24078660 (cited 31 July 2016).
4. Lambertini M, Ginsburg ES, Partridge AH. Update on fertility preservation in young women undergoing breast cancer and ovarian cancer therapy. Curr Opin Obstet Gynecol (Internet).

2015;27(1):98–107. Available from: http://content.wkhealth.com/linkback/openurl?sid=WKPTLP:landingpage&an=00001703-201502000-00017 (Cited 30 Jul 2016).

5. Antoniou A, Pharoah PD, Narod S, Risch HA, Eyfjord JE, Hopper JL, et al. Average risks of breast and ovarian cancer associated with *BRCA1* or *BRCA2* mutations detected in case series unselected for family history: a combined analysis of 22 studies. Am J Hum Genet. 2003;72:1117–30.
6. Letourneau JM, Ebbel EE, Katz PP, Katz A, Ai WZ, Chien AJ, et al. Pretreatment fertility counseling and fertility preservation improve quality of life in reproductive age women with cancer. Cancer (Internet). 2012;118(6):1710–7. Available from: http://doi.wiley.com/10.1002/cncr.26459 (Cited 18 July 2016).
7. Schover LR, Rybicki LA, Martin BA, Bringelsen KA. Having children after cancer. Cancer (Internet). 1999;86(4):697–709. Available from: http://doi.wiley.com/10.1002/%28SICI%291097-0142%2819990815%2986%3A4%3C697%3A%3AAID-CNCR20%3E3.0.CO%3B2-J (Cited 18 July 2016).
8. Lee SJ, Schover LR, Partridge AH, Patrizio P, Wallace WH, Hagerty K, et al. American Society of Clinical Oncology recommendations on fertility preservation in cancer patients. J Clin Oncol [Internet). 2006;24(18):2917–31. Available from: http://www.ncbi.nlm.nih.gov/pubmed/16651642 (Cited 2 Aug 2016).
9. Ruddy KJ, Gelber SI, Tamimi RM, Ginsburg ES, Schapira L, Come SE, et al. Prospective Study of Fertility Concerns and Preservation Strategies in Young Women With Breast Cancer. J Clin Oncol (Internet). 2014;32(11):1151–6. Available from: http://jco.ascopubs.org/cgi/doi/10.1200/JCO.2013.52.8877 (Cited 31 July 2016).
10. Senkus E, Kyriakides S, Ohno S, Penault-Llorca F, Poortmans P, Rutgers E, et al. Primary breast cancer: ESMO clinical practice guidelines for diagnosis, treatment and follow-up. Ann Oncol (Internet). 2015;26 Suppl 5(suppl 5):v8–30. Available from: http://www.ncbi.nlm.nih.gov/pubmed/26314782 (Cited 31 July 2016).
11. Kauff ND, Satagopan JM, Robson ME, Scheuer L, Hensley M, Hudis CA, et al. Risk-reducing salpingo-oophorectomy in Women with a *BRCA1* or *BRCA2* Mutation. J Clin Oncol. 2008;26 (8):1331–37. http://dx.doi.org.ezp-prod1.hul.harvard.edu/101056/NEJMoa020119.
12. Haber D. prophylactic oophorectomy to reduce the risk of ovarian and breast cancer in carriers of *BRCA* mutations. N Engl J Med 2002;346:1660–62. http://dx.doi.org.ezp-prod1.hul.harvard.edu/101056/NEJMed020044.
13. ACOG. Pb103_Brca. Obstet Gynecol (Internet). 2009;114(106):192–202. Available from: http://www.ncbi.nlm.nih.gov/pubmed/21422879.
14. Derks-Smeets IAP, Gietel-Habets JJG, Tibben A, Tjan-Heijnen VCG, Meijer-Hoogeveen M, Geraedts JPM, et al. Decision-making on preimplantation genetic diagnosis and prenatal diagnosis: a challenge for couples with hereditary breast and ovarian cancer. Hum Reprod (Internet). 2014;29(5):1103–12. Available from: http://www.ncbi.nlm.nih.gov/pubmed/24603131 (Cited 31 July 2016).
15. Shapira M, Raanani H, Meirow D. IVF for fertility preservation in breast cancer patients—efficacy and safety issues. J Assist Reprod Genet (Internet). 2015;32(8):1171–8. Available from: http://link.springer.com/10.1007/s10815-015-0519-x (Cited 18 July 2016).
16. Loren AW, Mangu PB, Beck LN, Brennan L, Magdalinski AJ, Partridge AH, et al. Fertility preservation for patients with cancer: American Society of Clinical Oncology clinical practice guideline update. J Clin Oncol (Internet). 2013;31(19):2500–10. Available from: http://jco.ascopubs.org/cgi/doi/10.1200/JCO.2013.49.2678 (Cited 3 Aug 2016).
17. Lambertini M, Ceppi M, Poggio F, Peccatori FA, Azim HA, Ugolini D, et al. Ovarian suppression using luteinizing hormone-releasing hormone agonists during chemotherapy to preserve ovarian function and fertility of breast cancer patients: a meta-analysis of randomized studies. Ann Oncol (Internet). 2015;26(12):2408–19. Available from: http://www.ncbi.nlm.nih.gov/pubmed/26347105 (Cited 30 July 2016).
18. Wieand S, Hale K, Lee M, Walsh T, Owens K, Tait J, et al. Tamoxifen Breast Cancer Incidence. 2014;286(18):2251–6.

19. Barton SE, Missmer SA, Berry KF, Ginsburg ES. Female cancer survivors are low responders and have reduced success compared with other patients undergoing assisted reproductive technologies. Fertil Steril. 2012;97(2):381–6.
20. Pfeifer S, Gldberg J, McClure R, Lobo R, Thomas M, Widra E, et al. Practice Committees of American Society for reproductive medicine; society for assisted reproductive technology. mature oocyte cryopreservation: a guideline. Fertil Steril. 2013;99(1):37–43.
21. Avraham S, Machtinger R, Cahan T, Sokolov A, Racowsky C, Seidman DS. What is the quality of information on social oocyte cryopreservation provided by websites of society for assisted reproductive technology member fertility clinics? Fertil Steril. 2014;101(1):222–6.
22. Meirow D, Biederman H, Anderson RA, Wallace WHB. Toxicity of chemotherapy and radiation on female reproduction. Clin Obstet Gynecol (Internet). 2010;53(4):727–39. Available from: http://content.wkhealth.com/linkback/openurl?sid=WKPTLP:landingpage&an=00003081-201012000-00004 (Cited 18 Jul 2016).
23. Partridge AH. Prevention of premature menopause and preservation of fertility in young cancer survivors. JAMA (Internet). 2015;314(24):2625. Available from: http://jama.jamanetwork.com/article.aspx?doi=10.1001/jama.2015.17299 (Cited 29 July 2016).
24. Pagani O, Regan MM, Walley BA, Fleming GF, Colleoni M, Láng I, et al. Adjuvant exemestane with ovarian suppression in premenopausal breast cancer. N Engl J Med (Internet). 2014;371(2):107–18. Available from: http://www.nejm.org/doi/abs/10.1056/NEJMoa1404037 (Cited 3 Aug 2016).
25. Pfeifer S, Goldberg J, Lobo R, Pisarska M, Thomas M, Widra E, et al. Practice Committee of American Society for reproductive medicine. Ovarian tissue cryopreservation: a committee opinion. Fertil Steril. 2014;101(5):1237–43.
26. Meirow D, Ra H, Shapira M, Brenghausen M, Derech Chaim S, Aviel-Ronen S, et al. Transplantations of frozen-thawed ovarian tissue demonstrate high reproductive performance and the need to revise restrictive criteria. Clin trial regist number NCT02659592. Fertil Steril (Internet). 2016;106:467–74. Available from: http://fertstertforum.com/meirowd-ovarian-transplantation/ (Cited 3 Aug 2016).
27. Cakmak H, Rosen MP. Random-start ovarian stimulation in patients with cancer. Curr Opin Obstet Gynecol (Internet). 2015;27(3):215–21. Available from: http://content.wkhealth.com/linkback/openurl?sid=WKPTLP:landingpage&an=00001703-201506000-00011 (Cited 2 Aug 2016).
28. Cakmak H, Katz A, Cedars MI, Rosen MP. Effective method for emergency fertility preservation: random-start controlled ovarian stimulation. Fertil Steril. 2013;100(6):1673–80.
29. Madrigrano A, Westphal L, Wapnir I. Egg retrieval with cryopreservation does not delay breast cancer treatment. Am J Surg. 2007;194(4):477–81.
30. Westphal LM, Wapnir IL. Integration and safety of fertility preservation in a breast cancer program. Gynecol Oncol. 2012;124(3):474–6.
31. Azim AA, Costantini-Ferrando M, Oktay K. Safety of fertility preservation by Ovarian stimulation With Letrozole and gonadotropins in patients with breast cancer: A prospective controlled study. J Clin Oncol (Internet). 2008;26(16):2630–5. Available from: http://jco.ascopubs.org/cgi/doi/10.1200/JCO.2007.14.8700 (Cited 1 Aug 2016).
32. Meirow D, Raanani H, Maman E, Paluch-Shimon S, Shapira M, Cohen Y, et al. Tamoxifen co-administration during controlled ovarian hyperstimulation for in vitro fertilization in breast cancer patients increases the safety of fertility-preservation treatment strategies. Fertil Steril. 2014;102(2):488–495.e3.
33. Oktay K, Buyuk E, Libertella N, Akar M, Rosenwaks Z. Fertility preservation in breast cancer patients: a prospective controlled comparison of ovarian stimulation with tamoxifen and letrozole for embryo cryopreservation. J Clin Oncol (Internet). 2005;23(19):4347–53. Available from: http://www.jco.org/cgi/doi/10.1200/JCO.2005.05.037 (Cited 31 Jul 2016).
34. Perri T, Lifshitz D, Sadetzki S, Oberman B, Meirow D, Ben-Baruch G, et al. Fertility treatments and invasive epithelial ovarian cancer risk in Jewish Israeli *BRCA1* or *BRCA2* mutation carriers. Fertil Steril. 2015;103(5):1305–12.

35. Oktay K, Kim JY, Barad D, Babayev SN. Association of *BRCA1* mutations with occult primary ovarian insufficiency: a possible explanation for the link between infertility and breast/ovarian cancer risks. J Clin Oncol (Internet). 2010;28(2):240–4. Available from: http://jco.ascopubs.org/cgi/doi/10.1200/JCO.2009.24.2057 (Cited 18 July 2016).
36. Wang ET, Pisarska MD, Bresee C, Ida Chen Y-D, Lester J, Afshar Y, et al. BRCA1 germline mutations may be associated with reduced ovarian reserve. Fertil Steril. 2014;102(6):1723–8.
37. Shapira M, Raanani H, Feldman B, Srebnik N, Dereck-Haim S, Manela D, et al. BRCA mutation carriers show normal ovarian response in in vitro fertilization cycles. Fertil Steril. 2015;104(5):1162–7.
38. Valentini A, Finch A, Lubinski J, Byrski T, Ghadirian P, Kim-Sing C, et al. Chemotherapy-induced amenorrhea in patients with breast cancer with a *BRCA1* or *BRCA2* Mutation. J Clin Oncol (Internet). 2013;31(31):3914–9. Available from: http://jco.ascopubs.org/cgi/doi/10.1200/JCO.2012.47.7893 (Cited 31 July 2016).
39. Finch A, Valentini A, Greenblatt E, Lynch HT, Ghadirian P, Armel S, et al. Frequency of premature menopause in women who carry a *BRCA1* or *BRCA2* mutation. Fertil Steril (Internet). 2013;99:1724–8 (Cited 18 Jul 2016).
40. Cardozo ER, Thomson AP, Karmon AE, Dickinson KA, Wright DL, Sabatini ME. Ovarian stimulation and in-vitro fertilization outcomes of cancer patients undergoing fertility preservation compared to age matched controls: a 17-year experience. J Assist Reprod Genet. 2015;32(4):587–96.
41. Azim HA, Kroman N, Paesmans M, Gelber S, Rotmensz N, Ameye L, et al. Prognostic impact of pregnancy after breast cancer according to estrogen receptor status: a multicenter retrospective study. J Clin Oncol (Internet). 2013;31(1):73–9. Available from: http://jco.ascopubs.org/cgi/doi/10.1200/JCO.2012.44.2285 (Cited 18 Jul 2016).
42. Euhus DM, Robinson L. Genetic predisposition syndromes and their management. Surg Clin North Am. 2013;93(2):341–62.
43. Howard-Anderson J, Ganz PA, Bower JE, Stanton AL. Quality of life, fertility concerns, and behavioral health outcomes in younger breast cancer survivors: a systematic review. J Natl Cancer Inst (Internet). 2012;104(5):386–405. Available from: http://www.ncbi.nlm.nih.gov/pubmed/22271773 (Cited 31 July 2016).
44. Letourneau JM, Smith JF, Ebbel EE, Craig A, Katz PP, Cedars MI, et al. Racial, socioeconomic, and demographic disparities in access to fertility preservation in young women diagnosed with cancer. Cancer. 2012;118(18):4579–88.
45. Grover NS, Deal AM, Wood WA, Mersereau JE. Young men with cancer experience low referral rates for fertility counseling and sperm banking. J Oncol Pract (Internet). 2016;12(5):465–471. Available from: http://www.ncbi.nlm.nih.gov/pubmed/27118159.
46. Rudick B, Opper N, Paulson R, Bendikson K, Chung K. The status of oocyte cryopreservation in the United States. Fertil Steril. 2010;94:2642–6.
47. IVF Italy 2004.pdf.
48. Beck L, Sileo M, Copperman A. The average cost of fertility preservation for female cancer patients. Fertil Steril. 2010;94:S105.
49. Center MSKC. Fertility preservation: options for women who are starting cancer treatment.
50. Kim J, Oktay K, Gracia C, Lee S, Morse C, Mersereau JE. Which patients pursue fertility preservation treatments? A multicenter analysis of the predictors of fertility preservation in women with breast cancer. Fertil Steril. 2012;97(3):671–6.
51. American Society for Reproductive Medicine. ASRM: state infertility coverage at a glance (Internet). Available from: http://www.asrm.org/insurance.aspx.
52. Rich TA, Liu M, Etzel CJ, Bannon SA, Mork ME, Ready K, et al. Comparison of attitudes regarding preimplantation genetic diagnosis among patients with hereditary cancer syndromes. Fam Cancer (Internet). 2014;13(2):291–9. Available from: http://link.springer.com/10.1007/s10689-013-9685-0 (Cited 31 July 2016).

53. Insogna IG, Ginsburg E, Rich T, Liu M, Etzel C, Bannon S, et al. Transferring embryos with indeterminate PGD results: the ethical implications. Fertil Res Pract (Internet). 2016;2(1):2. Available from: http://fertilityresearchandpractice.biomedcentral.com/articles/10.1186/s40738-016-0014-9 (Cited 12 Sept 2016).
54. Ethics Committee of the American Society for Reproductive Medicine T. Fertility preservation and reproduction in patients facing gonadotoxic therapies: a committee opinion. Fertil Steril (Internet). 2013;100:1224–31. (Cited 12 Sept 2016).

11 Management of Male *BRCA* Mutation Carriers

Donna-Marie Manasseh, Ryland J. Gore and Patrick I. Borgen

Introduction

Male breast cancer is rare, making up less than 1% of all breast cancers, but certainly not irrelevant—particularly in the setting of *BRCA* mutation carriers. While studies in male breast cancer have been limited and therefore this is often seen as the same disease as in women, Table 11.1 illustrates important similarities and differences between the two phenomena. Recent molecular studies [1] have revealed fundamental biological differences between FBC and MBC, and it is hoped that these findings will guide treatment strategies toward a more tailored approach to the male form of the disease. Significantly more MBCs (proportionally) than FBCs arise with an underlying germline cancer predisposition and display a vastly different penetrance compared with females. Furthermore, the genophenotypical association of basal-epithelioid-like cancer with *BRCA1*, present in FBC, is not observed in MBC. Differences in somatic changes between male and female breast cancer have also been reported, particularly a higher number of *PIK3CA* mutations and a paucity of *TP53* mutations, which are the most common genetic aberration in FBC. In general, chromosomal-based changes, in particular regions of gains, are seen more frequently in male than female breast cancer and methylation (by far the most common mechanism of tumor suppressor gene silencing) is seen

D.-M. Manasseh · R.J. Gore · P.I. Borgen (✉)
Department of General Surgery, Maimonides Medical Center, 4802 10th Avenue, Brooklyn, NY 11219, USA
e-mail: pborgen@maimonidesmed.org

D.-M. Manasseh
e-mail: dmanasseh@maimonidesmed.org

R.J. Gore
e-mail: rgore@maimonidesmed.org

A.B. Chagpar (ed.), *Managing BRCA Mutation Carriers*,
DOI 10.1007/978-3-319-59198-8_11

Table 11.1 Comparison of male and female breast cancer

	Male breast cancer	Female breast cancer
Cell type of origin	Overwhelmingly ductal	Mixed ductal and lobular
Tumor location	Always subjacent to nipple areolar complex	Upper outer quadrant most common but can occur anywhere
Estrogen receptors	90% ER+	75% ER+
Impact of *BRCA2* mutation	6% rate of ductal breast cancer	65–85% of invasive breast cancer
Contralateral breast cancer risk (sporadic)	7% lifetime risk	0.7% per year cumulative risk plateau at 15%
Contralateral breast cancer risk (*BRCA* associated)	4–40% lifetime risk	65% lifetime risk
HER-2 amplification	Wide range of positive HER-2	25% HER-2 positive
Endocrine ablative strategy	Tamoxifen citrate	Tamoxifen in premenopausal; aromatase inhibitor in postmenopausal
Chromosomal base changes	Regions of gains	Regions of deletions
Surgical management	Total mastectomy	Breast conservation Mastectomy
Axillary staging	Sentinel node biopsy	Sentinel node biopsy
Adjuvant systemic treatment	±Radiation Mainstay of treatment is tamoxifen Chemotherapy poorly studied in males	±Radiation ±Chemotherapy ±Endocrine therapy
Postoperative surveillance	H&P q6 months for 5 years, annually thereafter Imaging and laboratory studies if suspicion for recurrence or metastatic disease	H&P q6 months for 5 years, annually thereafter Annual mammography Laboratory studies if suspicion for metastatic disease
Associated malignancies	Prostate in males; pancreas in both genders; ovarian in females	

less frequently. Clinically, several molecular subtypes with prognostic relevance have been described. These include tumors displaying chromosomal complex high and methylation high subgroups. Profiling signatures pertaining to epithelial mesenchymal transition and hormonal therapy insensitivity have also been reported suggesting a far more uncommon phenotype than the typical estrogen responsive MBC. As with FBC, attention to male-specific multicenter trials based on the individual characteristics is needed but these trials are impractical. There is a far greater chance of success with the establishment of national tumor repositories and prospective registries coupled with the type of prospective/retrospective studies that produced assays such as the 21-gene genomic assay in estrogen receptor-positive, HER-2 negative breast cancer. Continued emphasis on the development of reliable preclinical models to elucidate more clearly the pathogenesis of MBC should provide attractive targets for study and hold out hope of improving the often poor prognosis seen in this disease.

Incidence and Epidemiology

Breast cancer, although a common disease, is highly uncommon in men. The American Cancer Society estimates that, in 2016, approximately 2600 new cases of invasive male breast cancer will be diagnosed and 440 men will die from their disease [2]. The annual incidence of (MBC) is less than 1% of all breast cancers and less than 1% of all male cancers [3]. The incidence of MBC has been increasing over the past 25 years, although this likely reflects an aging population with a shift of the bell curve of age in the USA to the right. It is reasonable to assume that decreases in deaths due to cardiovascular disease may result in more men surviving only to develop uncommon conditions such as male breast cancer in the future.

The average male breast cancer patient is diagnosed at a slightly older age than female breast cancer patients, 67 years versus 61 years, respectively [4, 5]. Men tend to present with more advanced tumor characteristics ($T > 2.0$ cm, positive axillary nodes) as compared to women. However, they also tend to present more commonly with ER+ tumors. The frequency of HER-2 gene amplification appears to be very similar to the observed rate in female breast cancer [6].

Approximately 20% of all MBCs have a family history of breast cancer and 10% have a germline mutation in *BRCA1* and *BRCA2* [7], with *BRCA2* being the most commonly associated mutation.

Risk Factors

The etiology of male breast cancer is not well understood. However, a number of risk factors are believed to be associated with the development of male breast cancer. There are virtually no risk determinants associated with MBC. Most risk factors are either environmental or genetically based. Among environmental factors are disorders that are associated with increased estrogen levels, testicular disorders, as well as occupational and environmental exposures. Age and racial differences play a role in the incidence of MBC. Caucasian men have a lower incidence as compared to African-American and Afro-Caribbean men (1/100,000 vs. 1.8/100,000) [8] and better prognostic features [5]. The incidence of MBC increases as men age but levels off at approximately age 80 [9]. On the contrary, FBCs tend to have two peaks of age-related risks.

Gynecomastia, defined as increased in vestigial breast tissue in men, results from increased estrogen exposure and has been noted in 40–50% of MBC patients [10, 11]. However, it is unclear whether gynecomastia itself imparts an elevated risk, as gynecomastia is also fairly common in men without breast cancer [12]. The highly common nature of gynecomastia and the very rare incidence of MBC suggest very little increase in risk from benign gynecomastia.

Other etiologies of increased levels of circulating estrogen have been associated with increased risk for male breast cancer. Testicular disorders, such as orchiectomy, testicular injury, cryptorchidism, and orchitis (usually secondary to mumps), can all result in an imbalance in the androgen-to-estrogen ratio and as a result can increase the risk of breast cancer [13–15]. Obesity, alcohol, cirrhosis, prostate cancer treatment, and hormonal treatments for trans-sexuality with their resultant higher circulating estrogen levels have also been associated with MBC risk [16–19]. Environmental exposures, such as chest wall irradiation, similar to female breast cancers, have been shown to increase the risk of breast cancer [20].

Family history is a significant risk factor for MBC. Some series have noted a 15–20% risk of MBC in individuals with at least one first-degree relative with breast cancer [21–23]. Approximately, 4–40% of these cancers are associated with some type of inherited genetic mutation [24].

Genetics

Genetic mutations play a role in male breast cancer. Despite the fact that a variety of germline mutations have been noted to be associated with male breast cancer, *BRCA* mutations, particularly *BRCA2*, are implicated most often. Some series estimate that *BRCA2* accounts for 40–75% of mutation carrier-related male breast cancer. *BRCA1* mutations account for 10–16% of male breast cancers [25, 26]. In these individuals, the estimated lifetime risk of developing breast cancer ranges from 1 to 5% in men with *BRCA1* mutations and 5–10% with *BRCA2* mutations [27]. In both instances this represents a vastly larger proportional increase in risk of breast cancer in men than is seen in female *BRCA* heterozygotes. Other genetic lesions believed to be associated with MBC include *PTEN*, *CHEK2*, *CYP17*, Klinefelter's syndrome (XXY), single nucleotide polymorphisms (SNPs), and *PALB2* [28]. The influence of these genetic abnormalities can be subdivided into degrees of penetrance: high (e.g., *BRCA1* and *BRCA2*)-, moderate (e.g., *CHEK2*, *PALB2*)-, and low-penetrance genetic lesions (e.g., SNPs).

BRCA1 and BRCA2 Genes

Both *BRCA1* and *BRCA2* work to preserve chromosome structure, yet the precise nature of their contribution and the specifics of gene–gene interaction have proven difficult to define. Both proteins have been implicated in a multitude of different processes including DNA repair and recombination, cell cycle control, and transcription. The similarities between the observed phenotypes induced by mutations or disruptions in function of *BRCA1* or *BRCA2* and their apparent cohabitation in certain macromolecular complexes have prompted speculation that they work in unison in common cellular pathways.

One defective copy of *BRCA1* or *BRCA2* in the germline is enough to cause a predisposition to cancer. However, the rate of mutation of the second allele (loss of heterozygosity) is less clear and may not be an obligate precursor to tumorigenesis as described in Knudsen's Two-Hit Hypothesis. King et al. demonstrated that loss of heterozygosity in *BRCA*-associated breast cancers has not occurred in 10–35% of cases [29].

Cancer susceptibility gene mutations fall into two broad, general classes [30]. Genes whose mutation or altered expression relieves normal controls on cell division, death, or lifespan, promoting the outgrowth of cancer cells, have been termed "gatekeepers." Those whose disruption causes genome instability, increasing the frequency of alterations in gatekeeper genes, work instead as "caretakers."

BRCA1 and *BRCA2* have multiple functions. Their main role as tumor suppressor genes is to maintain DNA integrity. Simply stated, when single-stranded or double-stranded DNA is damaged, *BRCA1* and *BRCA2* protein products initiate DNA repair or apoptosis. Cells then accumulate genetic instability allowing for additional mutations to occur in genes that are associated with cell cycle checkpoint activation. In cells carrying a *BRCA* mutation, cell death is avoided permanently, unchecked proliferation occurs, and the result is tumor formation. *BRCA2* mutations in men tend to result in breast cancer at a younger age and have a poorer prognosis compared to men with non-genetic-associated breast cancers. It has also been noted that males with *BRCA2* mutation-related breast cancers tend to have a higher proliferative index and higher grade than their female counterparts [31] (Fig. 11.1).

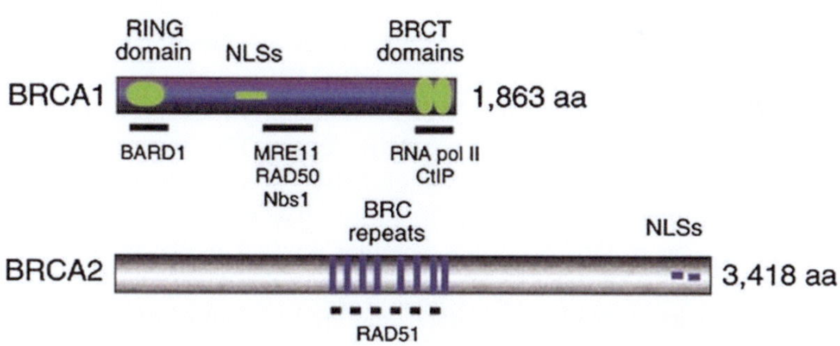

Fig. 11.1 Features of the human *BRCA* proteins *BRCA1* contains an N-terminal RING domain, nuclear localization signals (NLSs), and two C-terminal BRCT domains of ~110 residues (also found in several proteins with functions in DNA repair or cell cycle control). Interacting proteins discussed in the text are shown below approximate regions of binding. *BRCA2* contains eight repeats of the ~40 residue BRC motifs. Six of the eight motifs in human *BRCA2* can bind directly to *RAD51* when expressed in vitro [32]. Reprinted from Cell, Vol 108/No. 2, Venkitarraman AR, Cancer Susceptibility and the Functions of *BRCA1* and *BRCA2*, pgs 171–182, © 2002, with permission from Elsevier

BRCA1 and BRCA2 in Male Breast Cancer

A recent report from the Consortium of Investigators of Modifiers of *BRCA1/ BRCA2* [33] examined whether MBCs arising in the setting of *BRCA1* and/or *BRCA2* mutations display specific pathologic features and whether those features differ from *BRCA*-associated female breast.

The investigators studied the pathologic features of 419 *BRCA1/ BRCA2* MBCs and, using logistic regression analysis, contrasted those with data from 9675 *BRCA1/ BRCA2* FBCs and with population-based data from 6351 MBCs in the Surveillance, Epidemiology, and End Results (SEER) database. Among *BRCA2* MBCs, grade significantly decreased with increasing age at diagnosis ($P = 0.005$). Compared with *BRCA2* female breast cancers, *BRCA2* MBCs were of significantly higher stage (P for trend = $2 \times 10(-5)$) and higher grade (P for trend = 0.005) and were more likely to be estrogen receptor-positive [odds ratio (OR) 10.59; 95% confidence interval (CI) 5.15–21.80] and progesterone receptor-positive (OR 5.04; 95% CI 3.17–8.04). With the exception of grade, similar patterns of associations emerged when *BRCA1* MBCs were compared to matched FBCs. *BRCA2* MBCs also presented with higher grade than MBCs from the SEER database (P for trend = $4 \times 10(-12)$). The authors of this study concluded that MBCs associated with *BRCA* genes display distinct pathologic characteristics compared to matched female patients. A specific *BRCA2*-associated MBC phenotype characterized by greater biological aggressiveness (higher histologic grade) was reported.

Somatic Mutations in Male Breast Cancer

Somatic genetic alterations commonly seen in female breast cancer have been reported in MBC. A recent study sequenced 59 MBCs subtyped by immunohistochemistry (IHC) for all exons of 241 genes frequently mutated in female breast cancer and/or those that are related to DNA repair. The panorama of somatic mutations and copy number alterations of MBC were compared with that of carefully subtyped, matched female breast cancer. Twenty-nine percent of females and 71% of males were IHC classified as either luminal A-like or luminal B-like [34].

Male breast cancers displayed a wide and diverse range of somatic genetic alterations that, to some extent, recapitulate that of estrogen receptor-positive/HER2-negative female breast cancers, including recurrent mutations affecting *PIK3CA* (20%) and GATA3 (15%). Estrogen receptor-positive/HER2-negative male breast cancers, however, less frequently harbored 16q losses and *PIK3CA* and *TP53* mutations than estrogen receptor-positive/HER2-negative female breast cancers. In addition, male breast cancers were found to be significantly enriched for mutations affecting DNA repair-related genes. The authors concluded that at least a subset of MBCs are driven by a distinct repertoire of somatic changes [34].

The majority of male breast cancers (approximately 90%) are found to be invasive ductal carcinoma, followed by DCIS (10%) [35]. Evaluation of the SEER data registry reveals that 93.7% of male breast cancers are ductal or unclassified; 2.6% papillary; 1.8% mucinous; and only 1.5% lobular [4]. This is in contrast to female breast cancers in which 12% are lobular cancers. The majority of male breast cancers are ER+/PR+ with approximately 90% being ER+ and 80% PR+ [4]. Male breast cancers are more likely to be ER+/PR+ as compared to female breast cancers, and the rates of positivity increase with age. In contrast, overexpression of HER2 is somewhat less likely in male breast cancers [36, 37], ranging from 0 to 15% in some studies [38]. Therefore, the most common male breast cancer profile is ER/PR+, HER2 (−).

Diagnosis of Male Breast Cancer

The diagnostic workup of a suspicious mass in the male breast includes a combination of the clinical examination in addition to mammogram and ultrasound, with core needle biopsy for verification. Studies estimate that mammograms are 92% sensitive and 90% specific for male breast cancers [39]. Suspicious findings on a mammogram include the presence of a mass subjacent to the nipple, spiculations, and calcifications (Fig. 11.2). An ultrasound may be used as an adjunct and will demonstrate a hypoechoic mass with irregular borders. Cystic masses should also be viewed with suspicion, as neoplastic papillary lesions may resemble complex cystic lesions [40].

Any suspicious clinical or imaging examination should be confirmed by histopathology. This can be performed via fine needle aspiration (FNA) or by core needle biopsy. FNA may be used for initial evaluation, in most cases, if core needle biopsy is not readily available. However, when the FNA is inadequate or equivocal, core needle biopsy is indicated. As stated previously, the majority of male breast cancers are invasive ductal carcinomas.

The use of breast MRI in men is yet to be defined as most men present with a palpable mass and are subsequently treated with mastectomy, precluding the need for MRI in the majority of cases. MRI may be useful if there is a concern for chest wall invasion, unclear diagnostic features, or possibly in mutation carriers who are at a higher risk for developing breast cancer than the average male. There are limited data available to recommend breast MRI more than on a case-by-case basis.

In addition to breast imaging, the extent of disease workup should be considered on a case-by-case basis as most men present at later stages.

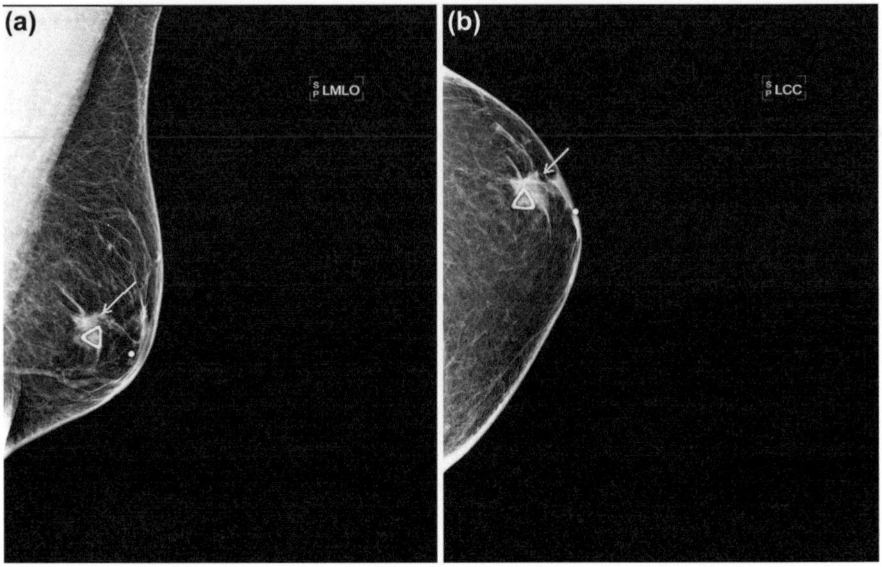

Fig. 11.2 *Left* mammogram of a male breast cancer patient who presented with a palpable mass (as indicated by *triangular* marker). This lesion is dense with spiculated edges

Treatment

Little data exist concerning precise efficacy of male breast cancer treatment due to the rare nature of the disease. There is even less known for male cancer susceptibility gene mutation carriers. Treatment strategies therefore follow general guidelines established for postmenopausal women. The majority of men with breast cancer are treated with mastectomy and axillary evaluation, either axillary dissection or sentinel lymph node biopsy.

Historically, radical mastectomy was recommended for the first half of the twentieth century, eventually supplanted by the modified radical mastectomy. However, most retrospective reviews of male breast cancer prognosis demonstrate survivals that are comparable to FBC [41]. Simple mastectomy is preferred over breast conservation due to the relatively small amount of tissue present, the fact that the tumors are virtually always immediately subjacent to the nipple areolar complex and the frequently seen advanced stage at presentation. Axillary evaluation is essential as it is a powerful prognostic indicator. Axillary node evaluation can be done with either axillary dissection in clinically positive axillary disease or via sentinel lymph node biopsy in clinically node negative disease as sentinel node biopsy has been demonstrated to be feasible and effective in male breast cancer [42, 43].

Radiation therapy for male breast cancer generally follows the same guidelines established for women with breast cancer. Radiation therapy has been found to be effective in decreasing local recurrence in men [44] although the potential effect on survival has been harder to establish. In a study from MD Anderson looking at 142 male patients, approximately 18% experience loco-regional failure, most commonly at the chest wall and supraclavicular area. Factors believed to result in failure include margin status, tumor size, and number of positive axillary nodes [45].

The data on systemic therapy for male mutation carriers and non-mutation carriers are limited due to the rarity of the disease. Therefore, the general approach has been to follow well-established guidelines used in the management of female breast cancers. Chemotherapy in both the adjuvant and neoadjuvant setting has been infrequently studied in men. Commonly used regimens include CMF [cyclophosphamide/methotrexate/5-FU] [46] and FAC [5-FU/doxorubicin/cyclophosphamide] [47].

The majority of male breast cancers are steroid binding hormone receptor responsive, and the mainstay of systemic therapy has been tamoxifen. Tamoxifen has been demonstrated in small studies to be of benefit in ER-positive male breast cancers. Its efficacy has been demonstrated to range anywhere from 25 to 80% [48–51] in advanced male breast cancer. In the adjuvant setting, it has improved 5-year overall survival from 44 to 61% and 5-year disease-free survival from 28 to 56% [52]. While the majority of male breast cancers are strongly hormone receptor-positive, up to 25% of men will discontinue tamoxifen [53]. Adverse side effects in male patients are similar to that of their female counterparts and include hot flashes, weight gain, and thromboembolic events, in addition to decreased libido, sexual dysfunction, and depression. It is important that providers are knowledgeable about treatment side effects and are able to manage patient expectations, in addition to offering strategies to manage side effects [54].

Aromatase inhibitors are used successfully in the treatment of female breast cancers. However, concerns about inadequate control of testicular endogenous estrogen production, has led some to postulate that the benefit of aromatase inhibitors in men, especially in the absence of concurrent gonadotropin-releasing hormone (GnRH) agonist use, [55] is limited. Therefore, aromatase inhibitors are typically used when tamoxifen is contraindicated (e.g., failure to tolerate tamoxifen secondary to side effect profile). Aromatase inhibitors may also be associated with a surge in endogenous testosterone production, which may be symptomatic [56].

In the metastatic disease setting, endocrine therapy strategies are first line followed by chemotherapy for certain clinical situations. Historically, surgical hormonal ablative therapy consisted of orchiectomy, hypophysectomy, or adrenalectomy. Although these strategies were somewhat effective, they were also quite morbid. These procedures have largely been supplanted by tamoxifen therapy [57]. If the disease is recalcitrant to treatment, several chemotherapy regimens have been demonstrated to provide adequate palliation. The role of trastuzumab in the HER2+ male breast cancer patient is unclear, but may be of benefit based on its success in female breast cancer [58].

Oncotype Dx has been demonstrated to be effective in female breast cancers in several studies. In male breast cancer, there is very limited information. In a study by Shak et al. [59], 347 male breast cancer patients were evaluated using the 21-gene Oncotype Dx RT-PCR assay. Recurrence scores were found to be similar in distribution to female breast cancers. The expression of ER was higher in males than females, as was Ki-67. The prognostic relevance for male breast cancer was not reported due to small sample size. More research is therefore needed to better delineate its role in male breast cancer. It is currently our practice to obtain the 21-gene assay in selected MBC patients with estrogen receptor-positive, HER-2-negative, and limited axillary node-positive patients to aid our decision analysis about treatment strategies.

Clinical Outcome

A number of matched pair analyses comparing MBC to the female form of the disease have demonstrated comparable survival statistics for both. However, the risk of second, non-breast primary cancers, however, seems to be higher in male than female breast cancer survivors. A recent national survey including more than 100,000 female and 578 sporadic male breast cancer patients investigated the risk of second primary malignancies [60]. Male breast cancer patients displayed significantly higher risk for thyroid (SIR: 13.2, 95% CI: 1.60–47.69), skin (SIR: 8.24, 95% CI: 3.02–17.94), and head and neck (SIR: 4.41, 95% CI: 2.35–7.54) cancers. Among breast cancer patients, risk factors significantly associated with second primary malignancies included male gender, older age, chemotherapy treatment, and comorbidity with liver cirrhosis.

Another study suggested that a male breast cancer patient's insurance status may impact his survival rate [61]. This study included 8828 male breast cancer patients diagnosed between 1998 and 2006 and followed to 2011 in the National Cancer Data Base. Cox regression was used to investigate the effect of payer's status and other factors on overall survival. Patients had 36.2, 42.7, 14.7, and 6.5% of stage I to IV cancer, respectively. Payer status was private 47.7%, Medicare 42.6%, Medicaid 3.24%, unknown 3.59%, and uninsured 2.95%. Median overall survival (MOS) for all patients was 10.6 years. In multivariate analysis, direct adjusted MOS was 12.46, 11.89, 9.99, 9.02, and 8.29 years for private, "unknown," Medicare, uninsured, and Medicaid payer's status, respectively. Patients with private and "unknown" payer's status showed a significantly better survival compared to uninsured patients, while Medicaid and Medicare patients did not. Age, race, stage, grade, income, comorbidity, distance travelled, and diagnosing/treating facility were also significant predictors of survival. Obvious biases with this analysis include a myriad of other controllable lifestyle factors that may be different between privately insured versus government-insured patients. Treatment delay and treatment in a dedicated cancer program did not have a significant influence on survival.

Surveillance

There is no role for screening for breast cancer in the general male population. However, it should be noted that male survivors of breast cancer have an increased risk of developing a contralateral breast cancer than the general population. The risk is even greater in male survivors less than 50 years of age. As such, it stands to reason that male survivors who are mutation carriers would benefit from routine breast imaging [55]. There are other significant risk factors for the development of male breast cancer. These patients SHOULD be screened for the disease.

For male mutation carriers, the current NCCN guidelines recommend self-examination and annual clinical breast examinations beginning at 35 years of age, as well as prostate cancer screening for carriers that specifically have a *BRCA* mutation [62]. Extrapolating from the NCCN guidelines that are currently in place for female mutation carriers with breast cancer who require surgical intervention, it seems appropriate that male breast cancer survivors should also undergo a history and physical examination every six months for the first five years after surgery, and then annually thereafter. If there is clinical suspicion for a recurrence or metastatic disease, the appropriate imaging and laboratory workup should be done at that time [63].

Lastly, special attention should be given to male mutation carriers, who unlike their non-mutation counterparts are at higher risk for the development of other primary cancers. Prostate, pancreatic, colorectal, gastric, and second breast primaries, as well as leukemia, have all been demonstrated to be of increased risk in mutation carriers [64–66]. It has been estimated that approximately 18% of male survivors will develop a second primary cancer over a median follow-up of 51 months. It is uncertain if adjuvant therapies of their breast cancer contribute to this risk. Therefore, surveillance, risk management strategies, and counseling need to be considered in these individuals. Such strategies include PSA at age 45 (prostate), spiral CT, EUS, ERCP (pancreatic), and colonoscopy (colon cancer) [67, 68].

Discussion

It has become customary to extrapolate from the results of treatment trials for female breast cancer and apply them to males with the disease. However, a closer examination of available data reveals that aspects of the etiology, pathogenesis, and treatment of MBC do not fit the simplistic model that men usually have endocrine-sensitive tumors which behave like those in postmenopausal women. Most males with breast cancer have none of the commonly recognized risk factors, indicating the gaps in our knowledge of the epidemiology of this very rare disease. In comparison with FBC, there is a larger proportion of *BRCA2* tumors (occurring in 10% of MBC), and under-representation of *BRCA1* tumors (found in only 1%), suggesting significant differences in the genetic etiology of MBC and FBC.

Genome-wide association studies [69] in FBC reported single nucleotide polymorphisms (SNPs) in no fewer than 12 novel independent loci that were consistently associated with disease but for MBC only 2 SNPs had a significantly increased risk. Molecular profiles of matched cancers in males and females showed a gender-associated modulation of major processes including energy metabolism, regulation of translation, matrix remodeling, and immune recruitment. Immunohistochemistry for kinase inhibitor proteins (KIPs) p27Kip1, and p21Waf1 indicates a significant difference in the immunostaining of tumors from male patients compared with females. These important biological differences point the way to the development of new therapies for MBC based on differences rather than similarities with female breast cancer.

There is growing evidence that male breast cancer patients are under-referred for genetic counseling and/or testing. This may be due to misconceptions about the autosomal location of *BRCA* tumor suppressor genes and a lack of awareness of their mode of transmission. A recent report from the Department of Veterans Affairs (VA) in 2016 sought to determine whether male and female Veterans diagnosed with breast cancer received *BRCA* testing as recommended by the National Comprehensive Cancer Network (NCCN) guidelines on Genetic/Familial High-Risk Assessment in Breast and Ovarian Cancer (v. 1.2010–1.2012). Of the 462 Veterans who met NCCN testing criteria, 126 (27%) received guideline-concordant care, either a referral for counseling or actual testing. No *BRCA* testing was recommended in 49 (50%) Veterans Administration Medical Centers that provide cancer treatment. Surprisingly, patients with second primary breast cancer were less likely to be referred/tested (OR 0.39; CI 0.17, 0.89; $p = 0.025$). For patients under age 51, a yearly increase in age decreased likelihood of referral or testing (OR 0.85; CI 0.76, 0.94; $p < 0.001$). There were no differences in testing by race. In conclusion, there was significant underutilization and lack of access to *BRCA* testing for Veterans diagnosed with breast cancer. Our research suggests the need for clinical decision support tools to facilitate delivery of guideline-concordant cancer care and improve Veteran access to *BRCA* testing [70].

It is also likely that there are gender differences in decision making with regard to *BRCA* gene testing. A recent study employing in-depth interviews of MBC survivors with *BRCA2* mutations revealed that 70.3% ($n = 45$) considered “Family Risk” as the primary reason for getting *BRCA* tested; 21.9% ($n = 14$) considered “Medical Considerations,” and 7.8% ($n = 5$) considered “Social Support” as their primary reason. Participants who were 50 years old or younger or who did not have children were more likely to consider medical reasons as the primary reason to get tested for *BRCA* gene status. In terms of self-concept, younger men reported feeling more stigmatized than their older counterparts; married men reported feeling a greater loss of control with regard to their *BRCA*-positive mutation status than did single men; and professional, gainfully employed men felt more vulnerable to the negative influences of the disease than those who were unemployed or who had already retired. Regression analysis results indicated that negative self-concept was strongly related to sampled males’ *BRCA* status 6 months after testing [71].

How have *BRCA* mutations favoring cancer loss of life years been able to survive selection pressure, in the case of the founder effect mutations, for thousands of years? Kwiatkowski et al. have proposed that deleterious mutations must provide other advantages that compensate for the loss of life expectancy, the most obvious being an increase in fertility in mutation carriers. This hypothesis was tested within 2150 hereditary breast and ovarian cancer families encompassing 96,325 individuals. Variables that were collected included incidence of breast/ovarian cancer, age at diagnosis, male breast cancer, and other cancer locations. Detailed reproductive histories were collected and evaluated. The authors reported that men in *BRCA*-mutated families had lower first and mean age at paternity, and fewer remained childless. For women in *BRCA* families, the overall miscarriage rate was lower and, in a logistic regression analysis including clinical factors, the difference in miscarriage rates and the male mean age at paternity remained significant. Fertility advantages were confirmed in a subgroup of 746 *BRCA* mutation carriers and 483 non-carriers from *BRCA*-mutated families. In particular, female carriers were less often nulliparous (9.1% of carriers vs 16.0%, $p = 0.003$) and both female and male carriers were found to have more children than non-carriers (1.8 ± 1.4 SD vs 1.5 ± 1.3, $p = 0.002$ for women and 1.7 ± 1.3 versus 1.4 ± 1.3, $p = 0.024$ for men) [72].

Another emerging benefit from *BRCA* testing in men lies in the realm of prostate cancer. The evidence to support the association of *BRCA* mutations with prostate cancer risk is based on retrospective studies involving multi-institutional registries of families affected by these mutations, or, more commonly, large cohorts of prostate cancer patients. The two- to sevenfold relative risk range of increased risk of developing prostate cancer in *BRCA* heterozygotes appears to be both clinically and statistically significant. Several small retrospective studies suggest an association between *BRCA* mutations and a more aggressive form of prostate cancer. Compared with non-carriers, patients with *BRCA2* mutations experienced significantly shorter overall survival and prostate cancer-specific survival [73]. While studies in male breast cancer have been limited and therefore this is often seen as the same disease as in women, Table 11.1 illustrates important similarities and differences between the two phenomena.

Conclusion

The management of male *BRCA* gene mutation carriers is challenging. The rarity of the disease makes prospective studies, which are the basis for evidence-based treatment guidelines, nearly impossible to perform with sufficient numbers and statistical power. Furthermore, the limited awareness of male breast cancer in the public sector contributes to later stage at presentation as well as a host of misperceptions of the disease. The American College of Surgeons estimates that a general surgeon in private practice in the USA will treat one male breast cancer patient during his/her career.

Men may feel stigmatized by the diagnosis of a "female cancer." Many conceal the diagnosis from all but immediate family. The systemic hormonal therapies utilized in men have significant side effects which may not be well addressed. Men on endocrine therapy can also have emotional and physical sequelae that they may suffer with "in silence" due to the perceived stigma associated with their disease.

Although stage for stage MBC and FBC have similar survival rates [74] the genetics of MBC suggests significant differences between male mutation carrier-related breast cancers and the sporadic form of the disease. These differences will surely impact future therapeutic options for male breast cancer patients. However, further research is needed to better delineate the fundamental features of these differences and their potential role in the treatment of this disease. Mutation and non-mutation carriers alike benefit from early detection and early intervention. Increased awareness and monitoring in known mutation carriers in conjunction with psychosocial support is critical in the successful management of this rare disease.

References

1. Deb S, Lakhani SR, Ottini L, et al. The cancer genetics and pathology of male breast cancer. Histopathology. 2016;68(1):110–8.
2. Cancer Facts and Figures 2016. American Cancer Society.
3. Jemal A, Siegel R, Xu J, et al. Cancer statistic. CA Cancer J Clin. 2010;60:277–300.
4. Giordano SH, Cohen DS, Buzdar AU, et al. Breast carcinoma in men: a population-based study. Cancer. 2004;101:51–7.
5. Korde LA, Zujewski JA, Kamin L, et al. Multidisciplinary meeting on male breast cancer: summary and research recommendations. J Clin Oncol. 2010;28(12):2114–22.
6. Ferzoco RM, Ruddy KJ. The epidemiology of male breast cancer. Curr Oncol Rep. 2016;18:1–6.
7. Haraldsson K, Loman N, Zhang QX, et al. BRCA germ-line mutations are frequent in male breast cancer patients without a family history of the disease. Cancer Res. 1998;58:1367–71.
8. Anderson WF, Althius MD, Brinton LA, et al. Is male breast cancer similar or different than female breast cancer? Breast Cancer Res Treat. 2004;83:77–86.
9. Ewertz M, Holmberg L, Karjalainen S, et al. Incidence of male breast cancer in Scandinavia, 1943–1982. Int J Cancer. 1989;43:27–31.
10. Newman J. Breast cancer in men and mammography of the male breast. Radiol Technol. 1997;69:17–28.
11. Heller KS, Rosen PP, Schottenfeld D, et al. Male breast cancer: a clinicopathologic study of 97 cases. Ann Surg. 1978;188:60–5.
12. Braunstein GD. Gynecomastia. N Eng J Med. 1993;328:490–5.
13. Thomas DB, Jimenez LM, McTiernan A, et al. Breast cancer in men: risk factors with hormonal implications. Am J Epidemiol. 1992;135:734–48.
14. Mabuchi K, Bross DS, Kessler II. Risk factors for male breast cancer. J Natl Cancer Inst. 1985;74:371–5.
15. Casagrande JT, Hanisch R, Pike MC, et al. A case-control study of male breast cancer. Cancer Res. 1988;48:1326–30.
16. Ewertz M, Holmberg L, Tretli S, et al. Risk factors for male breast cancer—a case control study from Scandinavia. Acta Oncol. 2001;40:467–71.
17. Hsing AW, McLaughlin JK, Cocco, et al. Risk factors for male breast cancer (United States). Cancer Causes Control. 1998;9:269–75.

18. Kanhai RC, Hage JJ, van Diest PJ, et al. Short term and long term histologic effects of castration and estrogen treatment on breast tissue of 14 male to female transsexuals in comparison with two chemically castrated men. Am J Surg Pathol. 2000;24:74–80.
19. Thellenberg C, Malmer B, Tavelin B, et al. Second primary cancers in men with prostate cancer: an increased risk of male breast cancer. J Urol. 2003;169:1345–8.
20. Thomas DB, Rosenblatt K, Jimenez LM, et al. Ionizing radiation and breast cancer in men (United States). Cancer Causes Control. 1994;5:9–14.
21. Hill A, Yagmur Y, Tran KN, et al. Localized male breast carcinoma and family history. An analysis of 142 patients. Cancer. 1999;86:821–5.
22. Johnson KC, Pan S, Mao Y. Risk factors for male breast cancer in Canada, 1994–1998. Eur J Cancer Prev. 2002;11:253–63.
23. Rosenblatt KA, Thomas DB, McTiernan, et al. Breast cancer in men: aspects of familial aggregation. J Natl Cancer Inst. 1991;83:849–54.
24. Haraldsson K, Loman N, Zhang QX, et al. BRCA2 germ line mutations are frequent in male breast cancer patients without a family history of the disease. Cancer Res. 1998;58:1367–71.
25. Ford D, Easton DF, Stratton M, et al. Genetic heterogeneity and penetrance analysis of the *BRCA1* and *BRCA2* genes in breast cancer families. The Breast Cancer Linkage Consortium. Am J Hum Genet. 1998;62:676–89.
26. Frank TS, Deffenbaugh AM, Reid JE, et al. Clinical characteristics of individuals with germline mutations in *BRCA1* and *BRCA 2*: analysis of 10,000 individuals. J Clin Oncol. 2002;20:1480–90.
27. Tai YC, Domchek S, Parmigiani G, et al. Breast cancer risk among male *BRCA1* and *BRCA2* mutation carriers. J Natl Cancer Inst. 2007;99:1811–4.
28. Silvestri V, Zelli V, Valentini V, et al. Whole-exome sequencing and targeted gene sequencing provide insights into the role of *PALB2* as a male breast cancer susceptibility gene. Cancer. 2017;123(2):210–8.
29. King TA, Li W, Brogi E, et al. Heterogenic loss of the wild-type *BRCA* allele in human breast tumorigenesis. Ann Surg Oncol. 2007;14(9):2510–8.
30. Kinzler KW, Vogelstein B. Cancer-susceptibility genes. Gatekeepers and caretakers. Nature. 1997;386(6627):761, 763.
31. Ottini L, Silvestri V, Rizzolo P, et al. Clinical and pathologic characteristics of *BRCA*—positive and *BRCA*-negative male breast cancer patients: results from a collaborative multicenter study in Italy. Breast Cancer Res Treat. 2012;134:411–8.
32. Venkitarraman AR. Cancer Susceptibility and the functions of *BRCA1* and *BRCA2*. Cell. 2002;108(2):171–82.
33. Silvestri V, Barrowdale D, Mulligan AM, et al. Male breast cancer in *BRCA1* and *BRCA2* mutation carriers: pathology data from the Consortium of Investigators of Modifiers of *BRCA1/2*. Breast Cancer Res. 2016;18(1):15.
34. Piscuoglio S, Ng CK, Murray MP, et al. The genomic landscape of male breast cancers. Clin Cancer Res. 2016;22(16):4045–56.
35. Stalsberg H, Thomas DB, Rosenblatt KA, et al. Histologic types and hormone receptors in breast cancer in men: a population-based study in 282 United States men. Cancer Causes Control. 1993;4:143–51.
36. Bloom KJ, Govil H, Gattuso P, et al. Status of HER-2 in male and female breast carcinoma. Am J Surg. 2001;182:389–92.
37. Muir D, Kanthan R, Kanthan SC. Male versus female breast cancers. A population- based comparative immunohistochemical analysis. Arch Pathol Lab Med. 2003;127:36–41.
38. Curigliano G, Colleoni M, Renne G, et al. Recognizing features that are dissimilar in male and female breast cancer: Expression of p21Waf1 and p27Kip1 using an immunohistochemical assay. Ann Oncol. 2002;13:895–902.
39. Evans GF, Anthony T, Turnage RH, et al. The diagnostic accuracy of mammography in the evaluation of male breast disease. Am J Surg. 2001;181:96–100.

40. Yang WT, Whitman GJ, Yuen EH, et al. Sonographic features of primary breast cancer in men. AJR Am J Roentgenol. 2001;176:413–6.
41. Gough DB, Donohue JH, Evans MM, et al. A 50 year experience of male breast cancer: is outcome changing? Surg Oncol. 1993;2:325–33.
42. DeCicco C, Baio SM, Veronesi P, et al. Sentinel node biopsy in male breast cancer. Nucl Med Commun. 2004;25:139–43.
43. Port ER, Fey JV, Cody HS3rd et al. Sentinel lymph node biopsy in patients with male breast carcinoma. Cancer. 2001;91:319–323.
44. Donegan WL, Redlich PN, Lang PJ, et al. Carcinoma of the breast in males: a multiinstitutional survey. Cancer. 1998;83:498–509.
45. Perkins GH, Middleton LP, Garcia SM, et al. Male breast carcinoma: outcomes and predictors of local-regional failure in patients treated without radiation therapy. Breast Cancer Res Treat. 2002;76:121.
46. Bagley CS, Wesley MN, Young RC, et al. Adjuvant chemotherapy in males with cancer of the breast. An J Clin Oncol. 1987;10:55–60.
47. Giordano SH, Perkins GH, Broglio K, et al. Adjuvant systemic therapy for male breast carcinoma. Cancer. 2005;104:2359–64.
48. Jaiyesimi IA, Buzdar AU, Sahin AA, et al. Carcinoma of the male breast. Ann Intern Med. 1992;117(9):771–7.
49. Kantarjian H, Yap HY, Hortobagyi G, et al. Hormonal therapy for metastatic male breast cancer. Arch Intern Med. 1983;143(2):237–40.
50. Riberio G. Tamoxifen in the treatment of male breast carcinoma. Clin Radiol. 1983;34:625–8.
51. Riberio G, Swindell R. Adjuvant tamoxifen for male breast cancer. Br J Cancer. 1992;65:252–4.
52. Anelli TF, Anelli A, Tran KN, et al. Tamoxifen administration is associated with a high rate of treatment-limiting symptoms in male breast cancer patients. Cancer. 1994;74(1):74–7.
53. Kantarjian H, Yap HY, Hortobagyi G, et al. Hormonal therapy for metastatic male breast cancer. Arch Intern Med. 1983;143(2):237–40.
54. Pemmaraju N, Munsell MF, Hortobagyi GN, et al. Retrospective review of male breast cancer patients: analysis of tamoxifen-related side effects. Ann Oncol. 2012;23(6):1471–4.
55. Ferzoco RM, Ruddy KJ. Optimal delivery of male breast cancer follow-up care: improving outcomes. Breast Cancer Targets Ther. 2015;7:371–9.
56. Nahleh Z, Girnius S. Male breast cancer: a gender issue. Nat Clin Pract Oncol. 2015;3:428–37.
57. Czene K, Bergqvist J, Hall P, Bergh J. How to treat male breast cancer. Breast. 2007;16: S147–54.
58. Giordano SH. A review of the diagnosis and management of male breast cancer. Oncologist. 2005;10:471–9.
59. Shak S, Palmer G, Baehner FL, et al. Molecular characterization of male breast cancer by standardized quantitative RT-PCR analysis: first large genomic study of 347 male breast cancers compared to 82,434 female breast cancers. J Clin Oncol (Meet Abstr). 2009;27:549.
60. Hung MH, Liu CJ, Teng CJ, et al. Risk of second non-breast primary cancer in male and female breast cancer patients: a population-based cohort study. PLoS ONE. 2016;11(2): e01485972016.
61. Shi R, Taylor H, Liu L, et al. Private Payer's status improves male breast cancer survival. Breast J. 2016;22(1):101–4.
62. National Comprehensive Cancer Network. (2016). NCCN Guidelines: Genetic/Familial High Risk Assessment: Breast and Ovarian. https://www.nccn.org/professionals/physician_gls/pdf/genetics_screening.pdf.
63. National Comprehensive Cancer Network. (2016). NCCN Guidelines: Breast Cancer. https://www.nccn.org/professionals/physician_gls/pdf/breast.pdf.
64. Kirchhoff T, Kauff N, Kolachana P, et al. *BRCA* mutations and the risk of prostate and colorectal cancer in Ashkenazi Jews. Am J Human Genet. 2003;73(5):246.

65. Goggins M, Schutte M, Lu J, et al. Germline *BRCA2* gene mutations in patients with apparently sporadic pancreatic carcinomas. Cancer Res. 1996;56(23):5360–4.
66. Risch HA, McLaughlin JR, Cole DEC, et al. Prevalence and penetrance of germline *BRCA1* and *BRCA2* mutations in a population series of 649 women with ovarian cancer. Am J Human Genet. 2001;68(3):700–10.
67. Liede A, Karlan BY, Narod SA. Cancer risks for male carriers of germline mutations in *BRCA1* or *BRCA2*: a review of the literature. J Clin Oncol. 2004;22(4):735–42.
68. Mohamad HB, Apffelstaedt JP. Counseling for male *BRCA* mutation carriers: a review. Breast. 2008;17:441–50.
69. Fentiman IS. Male breast cancer is not congruent with the female disease. Crit Rev Oncol Hematol. 2016;101:119–24.
70. Chun DS, Berse B, Venne VL, et al. *BRCA* testing within the department of veterans affairs: concordance with clinical practice guidelines. Fam Cancer 2016 Epub ahead of print. (PMID: 27589855 Doi:10.1007/s10689-016-9921-5).
71. Hesse-Bilber S, An C. Within-gender differences in medical decision making among male carriers of the *BRCA* genetic mutation for hereditary breast cancer. Am J Mens Health. 2015 Oct 14. pii: 1557988315610806. [Epub ahead of print].
72. Kwiatkowski F, Arbre M, Bidet Y, et al. *BRCA* Mutation increase fertility in families at heterditary breast/ovarian cancer risk. PLos One 2015 June 5:10(6) doi:10.1371.
73. Alanee S, Glogowski E, Kasminta A, et al. Clinical features and management of *BRCA1* and *BRCA2* associated prostate cancer. Front Biosci. 2014;6:15–30.
74. Borgen PI, Senie RT, McKinnon WM, et al. Carcinoma of the male breast: analysis of prognosis compared with matched female patients. Ann Surg Oncol. 1997 Jul-Aug;4(5):385–388.

Psychosocial, Ethical, and Legal Implications for Mutation Carriers

12

Jacquelyn M. Powers, Jessica M. Long and Willonie Mendonca

Psychosocial Considerations: Genetic Counseling, Testing and Disclosure Process

There is a common misperception that genetic testing is simply a blood draw and thus a single brief interaction with a genetics healthcare provider. From initially learning about a *BRCA1* or *BRCA2* gene mutation in the family, to preparing to learn one's own *BRCA* mutation status and, finally, to the subsequent adaptation to results (whether positive or negative), genetic testing is better addressed as a multi-dimensional, multi-step process tailored to an individual's unique needs, concerns, and circumstances. The National Society of Genetic Counselors (NSGC) has defined genetic counseling as "the process of helping people understand and adapt to the medical, psychological, and familial implications of genetic contributions to disease" with a goal of promoting informed choices [1]. While hereditary cancer risk may be addressed or even mitigated by enhanced surveillance, chemoprevention, and risk-reducing surgeries, careful attention also must be paid to the psychosocial experience of genetic testing, family dynamics surrounding risk communication, and ultimately the impact of the psychosocial experience on

J.M. Powers (✉) · J.M. Long
Department of Medicine, Division of Hematology-Oncology and Basser Center for BRCA, University of Pennsylvania, Perelman Center for Advanced Medicine, 3 West, 3400 Civic Center Boulevard, Philadelphia, PA 19104, USA
e-mail: jacquelyn.powers@uphs.upenn.edu

J.M. Long
e-mail: jessica.long@uphs.upenn.edu

W. Mendonca
GeneDx Laboratories, 207 Perry Parkway, Gaithersburg, MD 20877, USA
e-mail: wmendonca@genedx.com

A.B. Chagpar (ed.), *Managing BRCA Mutation Carriers*,
DOI 10.1007/978-3-319-59198-8_12

decisions related to medical management. This chapter will review psychosocial considerations from the initial *BRCA* genetic counseling and testing process, the dissemination of *BRCA*-positive genetic test results throughout the family, adaptation to *BRCA*-positive results via utilization of risk reduction and enhanced screening methods, as well as legal implications and ethical challenges pertaining to a genetic diagnosis of inherited susceptibility to cancer.

Importance of Pre- and Post-test Genetic Counseling

NSGC Practice Guidelines emphasize healthcare professionals providing genetic counseling for hereditary breast and ovarian cancer (HBOC) should explain "why the test is being offered, how the results might affect the patient's risk for cancer, and what medical management options may be offered depending upon the results" [2]. The components of this discussion have increased in complexity since the advent of multi-gene panel testing and clinical exome or genomic sequencing [3, 4]. Nonetheless, the goals of cancer genetic counseling remain the same: to facilitate informed decision-making in regard to genetic testing and to promote understanding and subsequent adaptation to results, such as through the implementation of appropriate cancer screening and prevention strategies [1, 2, 5].

Various professional societies have published statements emphasizing need for appropriate pre- and post-test counseling when providing genetic risk assessment for hereditary breast and ovarian cancer [2, 6, 7]. Pre-test counseling and the process of informed consent enable patients to consider in advance the implications of the test result for themselves and their relatives, including medical options like cancer risk reduction and surveillance [6]. In addition, pre-test counseling may involve provision of psychosocial support and discussion of the patient's concerns regarding risks for genetic discrimination, potential impact on emotional coping, sensitivity and specificity of testing, cost of testing, and testing procedures [2, 8]. The American Society of Clinical Oncologists policy statement describes the disclosure of genetic testing results as "a valuable opportunity" for post-test counseling wherein providers "interpret test results, recommend appropriate follow-up, and emphasize the importance of continuing regular prevention activities" [6].

In terms of the impact of cancer genetic counseling services, Braithwaite, et al. conducted a meta-analysis in 2004 of controlled trials of cancer genetic counseling; their findings suggested the provision of genetic counseling services increased knowledge about cancer genetics among patients "without an adverse effect on cancer-specific worry, general anxiety, distress, and depression" [5]. More recently, a comprehensive review of the literature for the United States Preventative Services Task Force documented a lack of reported harms of risk assessment. In addition, sixteen studies were identified supporting decreased cancer worry, anxiety, and depression, as well as improved accuracy of risk perception, among patients undergoing cancer genetic counseling [9].

One of the challenges of provision of cancer genetic counseling services is the limited, albeit growing, field of providers with special expertise in cancer risk

Table 12.1 Resources for locating cancer genetic counseling services

Resources for locating cancer genetic counseling services	
National society of genetic counselors (NSGC)	http://www.nsgc.org/ 312-321-6834
NCI cancer genetics services directory	https://www.cancer.gov/about-cancer/causes-prevention/genetics/directory 1-800-4-CANCER
InformedDNA	http://www.informeddna.com/ 800-975-4819

assessment and genetic test results interpretation. The American College of Surgeons Commission on Cancer requires qualified genetics services providers be available on-site or by referral for program accreditation [10]. However, such providers traditionally have been distributed unequally across the USA, most often located in or near larger cities, due to frequent affiliation with academic medical centers. Some insurance companies, such as CIGNA, recently have begun to require genetic counseling by a certified genetic counselor [11]. In other circumstances, cancer genetic testing and counseling might be provided by oncologists, surgeons, gynecologists, and/or nurses with advanced training in genetics. Beyond the traditional face-to-face model, other methods of cancer genetics service delivery have been explored in attempts to expand the reach of services while retaining high quality of care. Alternative methods of service delivery used by genetic counselors have included telephone, videoconference, and group counseling. While promising outcomes, researchers continue to evaluate whether these are equally effective in terms of patient care and cost [12–14] (Table 12.1).

Learning *BRCA* Genetic Test Results

Receiving a *BRCA1*- or *BRCA2*-positive genetic test result is a life-altering experience. Whether affected or unaffected by cancer, the way in which an individual views their health status and mortality is forever modified. Several studies exploring the short-term effects ($\leq$ 1 year) of *BRCA1/2* testing have demonstrated that *BRCA* mutation carriers experience a transient increase in general distress after receiving test results [15]. Studies over the last decade are discordant as to whether long-term distress (>1 year) increases [16], remains stable [17], or decreases across time [18]. These discordances may be somewhat attributable to study design. More recently, there seems to be emerging agreement that while a positive *BRCA1/2* result remains salient among carriers years after testing, testing does not appear to impact long-term psychological dysfunction [19]. A prospective comparison (3–9 years after result disclosure) of 464 women with positive, negative, or uninformative genetic testing found that mutation carriers with and without cancer reported modest increases in distress, but no evidence of clinically significant dysfunction [19]. Another study of 237 participants evaluated long-term adjustment in *BRCA1/2* carriers by measuring self-esteem and mastery. Higher self-esteem and mastery were associated with less cancer-specific anxiety. Of note, participants who underwent genetic testing were a self-selected sample, who may have prepared themselves for the possibility of a positive genetic testing result. Mean self-esteem in this cohort of women was no different than general population norms and was not related to carriers' age, affected status, time since disclosure, or prophylactic surgery status [20].

While large sample studies such as these found no lasting psychological distress overall, select populations (such as younger women and those who report the death of a relative from breast/ovarian cancer) may be more vulnerable to poorer long-term adjustment [16, 21]. For premenopausal women with *BRCA* mutations, unique life challenges arise such as: desire for biological children versus the possibility of early menopause due to risk-reducing salpingo-oophorectomy, and transference of fear onto their children [22]. Also, for those who have a "lived cancer experience" via diagnosis in themselves or a family member, they may have an exaggerated perception of cancer risk. Hence, what may have been a predictive blood or saliva test, often is bundled with all of these complex issues.

For some patients, reported cancer-specific distress may be best explained by past and/or ongoing experiences of cancer in the family, rather than the acute distress of receiving their genetic result [20]. In addition, adjustment can be influenced by an individual's medical and social environments. Some individuals report a sense of disorientation given discrepant medical opinions across physicians about the best management approach while others feel supported [20]. Feelings of distress may not directly correlate to adjustment and coping to mutation positive status, but rather frustration experienced with navigating care [20]. Furthermore, individuals may receive a positive test result in the absence of illness. Such individuals continue to manage everyday social duties and responsibilities, while coping with new risk information and navigating next steps [20] Without a physical ailment or proof or disease, carriers may find that others (unaffected by HBOC) may not understand the gravity of this type of risk information and therefore may not be able to provide appropriate support. Given this, *BRCA*-positive individuals may appreciate and benefit from support resources, which enable them to connect to others with a shared experience. Some examples are listed below in Table 12.2.

Table 12.2 Resources for support, information, and advocacy

Resources for support, information, and advocacy	
Facing our risk of cancer empowered (FORCE)	www.facingourrisk.org 1-866-288-RISK (7475)
Bright pink	www.brightpink.org
Basser center for *BRCA*	basser.org
HIS breast cancer awareness	hisbreastcancer.org
Sharsheret	www.sharsheret.org 866-474-2774

Risk Communication and Dissemination of Information to Relatives

After a *BRCA* gene mutation is identified in a family, opportunities arise for: (1) highly informative predictive genetic testing for carrier status and cancer risk assessment, (2) cancer risk reduction measures, and/or (3) enhanced cancer surveillance in relatives with a *BRCA* mutation. However, the great value of this information is diminished if not shared with at-risk relatives in a timely and an accurate manner. While many individuals initially pursue genetic testing for the sake of their relatives, it is not uncommon for a person with a newly identified *BRCA* mutation to articulate concerns and apprehensions about communicating the genetic testing information to family members. This section shares options for healthcare providers to support or enhance family health communication and also discusses parental disclosure of *BRCA*-positive results to minor and young adult offspring.

Frequently, the responsibility of sharing a positive *BRCA1/2* genetic test result with relatives rests on the proband, the first person in the family to undergo genetic testing [23–25]. If the evaluation for HBOC syndrome was prompted by a recent cancer diagnosis in the proband, that individual may be overwhelmed by their own health crisis and challenged by their own healthcare needs, leaving less time and ability to focus on the needs of relatives. Thus, the distress caused by a recent cancer diagnosis can prohibit the timely disclosure of *BRCA* genetic testing information [23, 26].

Family relationships and interpersonal dynamics also influence whether, when, and with whom genetic testing results are shared. It has been recognized that women act as "kin-keepers," more often taking on roles of maintaining relationships and facilitating communication among relatives, and this extends to the gathering and sharing of health information, including *BRCA1/2* genetic test results [27, 28]. Family leaders, including parents, may play a crucial role in gathering relevant health information, since they often have better recollection and access to more distant family history and relatives [27]. Hearing about an identified *BRCA1/2* mutation from a family member with a related cancer may help lend weight to the importance of the information, as well as create opportunities for discussion as "teachable moments" may arise over the course of treatment and follow-up care [27]. Thus, when encouraging a person to inform their at-risk relatives, it may be helpful to encourage careful thought about which relatives often serve as family leaders, providing emotional support and fostering communication between relatives. It may be useful to a person with a newly identified *BRCA* mutation to consider whom they might enlist to help disseminate genetic risk information.

Despite best efforts by healthcare providers to emphasize the salience of a *BRCA*-positive genetic test result for all at-risk relatives, resistance may be encountered. Branches of the family may have lost contact over the years due to geographic distance or even emotional distance, in the case of family conflict. Various studies have shown more distant branches of a family, such as second- and third-degree relatives, are less likely to be informed about a genetic risk in the family, compared to first-degree relatives [29–31]. Individuals who find it difficult to communicate with relatives may negatively experience a healthcare provider's encouragement as pressure to act or behave in a responsible manner [28, 32]. "Passive non-disclosure" has been described as the situation in which patients at first seem willing to share genetic testing information with relatives, but ultimately do not take such action. It may therefore be necessary to follow-up with patients after the initial results appointment to verify whether disclosure occurred [33].

Family notification letters are a tool frequently employed by genetic counselors to aid individuals with known gene mutations to educate relatives regarding the availability and importance of informative predictive genetic testing [34]. The Basser center for *BRCA* has dedicated portions of the organization's website to *BRCA* support and family communication, which includes template *BRCA1* and *BRCA2* Family Letters [35]. Despite this, challenges remain in terms of dissemination of genetic test results to relatives, leading some clinician–researchers to explore possible interventions to facilitate family communication [36, 37].

Even in scenarios where disclosure of genetic test results occurs, relatives may not recall or adequately understand the information received. A recent survey of relatives whose family member pursued *BRCA1/2* genetic testing found 10.5% were unable to recall the result at all, and 17.9% incorrectly interpreted the result they recalled. It should be noted, however, that this study included families with *BRCA* uninformative negative and inconclusive results, which were significantly less likely to be accurately understood [23].

In terms of gender differences, men may be less likely to accurately remember the genetic test result and more likely to deny having heard it [23]. It has also been shown that men in families with HBOC syndrome are less likely to be informed about the diagnosis [23, 28, 38], perhaps due to its disproportionate impact on female cancer risk. Female first-degree relatives and adult children are more likely to be informed [23, 30].

For parents with young children, questions often arise regarding whether and when to disclose a familial *BRCA* mutation to offspring. Professional societies caution against *BRCA* genetic testing for individuals under 18 years old, given the need to preserve the child's autonomy and the lack of proven medical benefit of interventions to reduce risk prior to this age (discussed in greater depth elsewhere in this chapter) [39–41]. However, these recommendations do not address the timing of disclosure to offspring, and few parents report involving a healthcare provider in the decision to disclose [42].

Despite this, studies have shown that a majority of parents do discuss *BRCA* genetic test results with their children, some even sharing this information with children as young as age 7 [28, 42, 43]. Both the age of the child and the result itself (whether positive or negative) appear to influence when parents tell their offspring; for positive results, 44% disclosed within one month, but parents tended to wait longer when if their children were younger [42]. Among a surveyed cohort of older offspring, daughters of mutation carriers aged 18–24, more than half indicated being informed within one month from the time the mother received a *BRCA*-positive result [44]. Often, these disclosures occurred in relation to plans for the mother to undergo preventative surgery or in the setting of her diagnosis of cancer [44]. When parents delay sharing a *BRCA*-positive result with children, they often cite wanting time to adjust to their own test result and to plan how to share the information [42]. One helpful tool for parents deciding when and how to discuss genetic risk information was created via a joint effort between NSGC and the organization Facing Our Risk of Cancer Empowered (FORCE); this booklet, *Talking About BRCA in Your Family Tree*, is publically accessible online (http://www.facingourrisk.org/understanding-brca-and-hboc/publications/documents/booklet-talking-about-brca-family.pdf) and serves as a valuable resource for parents struggling with these issues [45].

Cultural Considerations

Those of Ashkenazi (Central/Eastern European) Jewish ancestry have an elevated chance (1 in 40) of carrying a mutation in either *BRCA1* or *BRCA2* due to a phenomenon known as "founder effect" [46]. Therefore, knowledge about Jewish ancestry can profoundly impact breast cancer risk assessment [47]. For Orthodox women in particular, a positive *BRCA1* or *BRCA2* result brings forth unique considerations. Obligations under the Jewish code of ethics, referred to as Jewish Law (Halacha), emphasize protection of one's own health, which may be in conflict with consensus medical recommendations provided to mutation carriers. From a Halachic perspective, risk-reducing surgery involves "mutilation of a healthy organ," specifically breasts and ovaries that impact a woman's sexuality and appearance [48]. Additionally, removal of ovaries may prevent a *BRCA*-positive woman from fulfilling the mitzvah to "be fruitful and multiply" [48]. Individuals subsequently may be hesitant to learn or to share their mutation status, so as not to threaten marriageability. These are just a few examples which must be addressed with both evidence-based medicine and cultural sensitivity. Typically, a woman (and spouse) may present to their rabbinic authority for guidance and assistance. Of late, a growing number of Jewish authorities have been considering the health benefits of surgical interventions and permitting based upon the Halachic obligation to prevent disease [48].

Beyond this, it is important to note that most *BRCA*-specific research has been conducted in predominantly Caucasian populations. There exists a current under-representation of minority group engagement and awareness of genetic testing and, consequently, research specific to the psychosocial implications of medical management in mutation carriers. While there is a desire among minority populations, to embrace genetic testing [49, 50], these patients remain overwhelmingly less likely to undergo risk assessment and genetic testing [51]. Although certain barriers exist in terms of insurance coverage, awareness of testing, and service availability [52], these three factors do not fully explain these disparities. With the advent of health information technology, some have suggested employing electronic medical record-based decision support as well as utilization of online social media to identify a greater catchment of individuals who may be eligible for testing [53].

Psychosocial Impact and Utilization of Medical Management Options in *BRCA1/2* Mutation Carriers

Decisions about medical management among women and men with *BRCA1* and *BRCA2* gene mutations are complex and driven by many factors that can be both cognitive and affective. Among high-risk populations such as *BRCA1* and *BRCA2* mutation carriers, worry and other emotional states have emerged as strong predictors of behavior [54–56]. Risk perception, baseline depression and distress, having young children, and, in particular, previous cancer-related experience strongly influence psychological status and appear to drive decision-making [57].

Women may reduce their risk of cancer-related morbidity and mortality through increased surveillance and risk-reducing strategies [58–62]. Men also may consider enhanced surveillance; although at present, existing recommendations for men are not as well-defined or evidence-based as they are for women. A number of psychological implications arise from lifelong risk management and vary based upon gender, life stage, and presence or lack of illness in the family. This section will explore some of these issues with an emphasis on risk-reducing surgeries among women.

High-Risk Breast Surveillance

For the majority of *BRCA* mutation carriers, current data supports enhanced breast surveillance (via annual mammography and breast MRI) as a reasonable alternative to bilateral risk-reducing mastectomy (BRRM). Women choosing enhanced surveillance appear to live as long as those choosing BRRM, though breast cancer may still develop and require treatment [63–66]. It may be helpful to recognize and discuss with *BRCA*-positive women the fact that decisions surrounding enhanced breast screening versus risk-reducing mastectomy are often intensely personal and influenced by various factors, such as cancer-related worry, risk tolerance, potential for surgical risks/complications, concerns regarding physical postoperative changes, body image, sexuality, age, desire to breastfeed, and personal history of previous cancer diagnosis. There also may be a period of deliberation after learning a genetic testing result, when a woman is gathering information to optimize informed medical decisions, clarifying her concerns and weighing her priorities.

Among women at general population risk for breast cancer, annual mammography does not appear to confer significant negative psychological consequences unless a false positive result, an acute life experience around time of screen, or a diagnosis of breast cancer occurs. Among women with false positive screens, any emotional disturbance is typically transient and returns to baseline after a benign evaluation, whether with repeat imaging or biopsy. However, for some women, this disturbance may persist six or more months after a benign evaluation, thus supporting there may be vulnerable subgroups of women who may benefit from close monitoring for adverse events after breast imaging recall events [67–71].

Mutation carriers opting for surveillance versus surgery are a self-selected group, generally with lower breast cancer-related anxiety. Despite a lower cancer-related anxiety, *BRCA1/2* mutation status remains a predictor of stress during times of screening recall events (either repeat imaging or breast biopsy). A study of 189 women undergoing MRI scans, including 34 *BRCA* mutation carriers, did not observe a clinically meaningful increase in stress across high-risk women overall. However, *BRCA* mutation carrier status, as well as having a previous history of breast cancer, predicted higher intrusion and avoidance scores, representing more disproportionate adverse thoughts and withdrawnness, respectively. Fortunately, this cognitive avoidance did not appear to impair screening adherence, and the majority of women later returned for additional follow-up [71]. In a separate cohort of 281 high-risk women undergoing breast MRI and mammography surveillance in the Netherlands, including 42 women with *BRCA* mutations, global anxiety scores and breast cancer distress scores were higher on the day of the screening appointment, but both scores decreased a month later [72].

During times of screening recall, adequate coping and support are noted as important variables in adjustment among *BRCA* carriers [57, 73]. Additional study is needed to determine whether a small subgroup of women, including a subgroup among *BRCA* mutation carriers, may be at greater risk for more severe and prolonged anxiety surrounding enhanced breast surveillance and breast biopsies. Based on anecdotal clinical experiences, these may be women better served by risk reduction measures.

Some *BRCA* carriers delay the decision for BRRM and choose enhanced surveillance in the interim. In a large single-center study, uptake of surgery in women with *BRCA* mutations was highest within 2 years of receiving a genetic test result; however, uptake continued, so that the predicted uptake by 7 years was 60% for *BRCA1* carriers and 43% for *BRCA2* carriers [74]. Women who delayed BRRM were typically younger with lower absolute breast cancer risks at time of genetic testing result, and/or were influenced to choose eventually surgery based upon cancer morbidity/mortality in their family. Overall, clinical experience suggests that many women desire a period of reflection, information-gathering, and consultation with experts before deciding to proceed with or decline risk-reducing surgery [74]. Given sufficient alternatives such as earlier and more frequent breast surveillance via mammography and breast MRI, the pursuit of BRRM remains a highly individualized decision that must be tailored to a woman's unique set of needs and concerns.

Chemoprevention Uptake

Decisions regarding chemoprevention can be particularly emotionally burdensome given trade-offs between the benefit of reducing breast cancer risk and the risks of taking the medication itself [75]. The uptake of chemoprevention to reduce breast cancer risk among *BRCA1* and *BRCA2* mutation carriers remains low [76, 77]. Among high-risk women, including those with a strong family history and/or a

BRCA mutation, most remain unaware or have limited understanding of chemoprevention medications and/or harbor misconceptions regarding its use [78]. Despite well-documented benefits of tamoxifen in reducing breast cancer risk considerably, the risks may cause worry, fear, and reluctance in using this option. Among 1134 unaffected *BRCA1* and *BRCA2* mutation carriers ascertained via a formal cancer risk evaluation program, only 12.4% considered chemoprevention (tamoxifen or raloxifene) [77]. Although 379 patients within this cohort were under the age of 35, and therefore not eligible for chemoprevention, 912 individuals were between the ages of 35 and 60. Of these, 811 did NOT proceed with tamoxifen or raloxifene [77]. Reasons for low uptake included worry about side effects, plans to undergo risk-reducing bilateral mastectomies, and issues regarding communication between the individual and their healthcare provider [77].

A growing body of research regarding "affective forecasting" suggests people routinely base life decisions on the expected impact these choices will confer on their emotional well-being [75]. When applied to discussions about decision-making for chemoprevention, a large cohort of high-risk women were randomized to either an online educational tool about chemoprevention use ($N = 661$) versus no intervention ($N = 322$). At 3-month follow-up, only 0.5% ($N = 2$) had initiated chemoprevention medication, though 44% were still weighing their options. Researchers identified a negative perception of chemoprevention, in that taking such a medication would increase their health-related stress. Karavites et al. theorized that affective forecasts are consistently biased by 3 cognitive pillars:

1. *Focalism*: The tendency for people to focus on the most stressful aspect of a situation. This could be a disproportionate focus on possible side effects of a medication.
2. *Coping fallacy*: The underestimation of resilience in coping with stressful scenario. This could be the underestimation that people may adjust to taking a medication more easily than imagined.
3. *Dysphoric Forecasting Bias*: Stress in the moment, like receiving new risk information, may lead people to overestimate future stress. This could be the physician discussion regarding a medication, which causes acute stress, but if the topic were revisited later, stress may have somewhat dissipated.

Additional research is necessary to inform whether individuals' expectations are relatively realistic or biased by the above assumptions, and if biased, for whom and under what circumstances [75].

In addition to patient willingness, many patients also do not recall their healthcare provider discussing the option of chemoprevention or felt information was insufficient to make an educated decision [76]. Among surveyed physicians, a considerable portion of providers are not convinced of the potency of chemoprevention in *BRCA1/2* carriers specifically [76, 79]. Potentially, additional education and discussion between patient and provider on the topic of chemoprevention might encourage more women to seriously consider and engage in this significant risk reduction option.

Risk-Reducing Bilateral and Contralateral Mastectomies

In an international study, the USA demonstrated the highest uptake of surgical prophylaxis with 36.3% of 703 *BRCA* mutation carriers undergoing BRRM and 71.1% undergoing risk-reducing salpingo-oophorectomy, or RRSO [77]. Unlike RRSO, where the mortality benefit is well-evidenced, the mortality benefit of BRRM is modest to marginal [62]. Although enhanced surveillance by mammography and breast MRI may be advantageous in breast cancer detection at earlier and ideally more treatable stages, it is important to note that treatment may still require chemotherapy. For many women with *BRCA* mutations, the decision to undergo BRRM is primarily motivated by the paramount desire to avoid a breast cancer diagnosis and subsequent treatment.

Higher perceived risk, higher baseline worry and anxiety, as well as personal experience with cancer family history (first- and second-degree relatives deceased from breast cancer, particularly a mother), positively predict BRRM [56]. Some women pursue BRRM soon after receipt of a *BRCA*-positive genetic test result. Others defer until completion of childbearing, formation of committed relationships, or time of first breast cancer diagnosis.

Overall, women at risk for hereditary breast cancer who pursue BRRM are satisfied with their decisions. The urge to reduce breast cancer risk is the predominant concern prior to surgery [80–83], and several studies conclude that BRRM confers psychological benefits in terms of reduced cancer-related anxiety [80, 84, 85]. Nevertheless, many women do share the adverse impact BRRM (even with reconstruction) has on body image [73, 80, 81, 86]. Recognition of these potential benefits and drawbacks encourages a woman to fully investigate her position on risk-reducing surgery. As providers, it is important to engage in this balanced discussion and to understand this conversation may not be singular, instead requiring multiple discussions over time.

Bilateral Risk-Reducing Mastectomies: Short-Term Psychological Impact

Within the year of surgery and reconstruction, studies support reduced cancer-related worry and distress, but less favorable body image and decreased sexual satisfaction [87]. Feelings of "not being happy with breast appearance" and/or "not feeling feminine" can persist, even if not explicitly expressed by the woman. Higher baseline cancer distress and lower baseline self-esteem predict greater dissatisfaction with body image [16, 73, 80–84, 86, 87]. Critics of short-term follow-up studies suggest that measuring body image within 1-year post-surgery may not be truly representative of long-term adjustment. Some women may require more time to complete reconstruction, overcome post-surgical complications, and/or adjust to loss of breasts or new body proportions.

Bilateral Risk-Reducing Mastectomies: Long-Term Psychological Impact

Follow-up at longer time points after BRRM reveal that the adverse outcomes evident immediately following surgery, including the emotional aspects, will peak and then decline over time. A long-term study of women up to 6–9 years after BRRM with breast reconstruction revealed significant increase in general body image issues at 6-month and 1-year, which eventually diminished over time. However, this decline never fully returned to baseline. This suggests that some feelings of decreased femininity and sexual attractiveness, despite improvement, still persist [88]. In particular, poor body image and general distress prior to surgery negatively affect a woman's sense of femininity and attractiveness after surgery. Women with more favorable impressions, who sought social support, and who exhibited strong coping skills had better long-term adjustment.

It is known from interview studies that most women experience feelings of loneliness and isolation post-surgery, which may be counterbalanced by the process of sharing the physical and emotional effects of surgery with others [89, 90]. Partners, in particular, are an important source of support for women. Partner's acceptance of a changed appearance of the woman's body and his/her reassurance of their desirability may help to maintain a sense of attractiveness. Furthermore, women opting for BRRM while not having a partner may be more reluctant to pursue future intimate relationships due to concern about sharing their post-surgical appearance [16]. Such anticipatory fear might affect their breast-specific body image.

Contralateral Risk-Reducing Mastectomy

Women who undergo contralateral risk-reducing mastectomy (CRRM) often share similar motives and predictors to those who pursue BRRM. However, by nature of having had a breast cancer diagnosis, this population will have other factors play into their decision-making. Predictors of CRRM include younger age at time of diagnosis, node negative disease, positive family history for breast or ovarian cancer, previous unilateral mastectomy versus breast conservation, as well as *BRCA* mutation status. Of note, many of these predictors also impact a woman's decision for CRRM, even if *BRCA1/2* mutation testing is negative (no mutation identified). Except for some suggestion of CRRM improving disease-free survival in younger women with hormone receptor negative disease [91], there is currently little to no data to support that CRRM improves breast-specific mortality [92–94].

In a large multinational study of over 900 women with *BRCA1/2* mutations, 253 (27.3%) underwent CRRM after a breast cancer diagnosis. Of these 253 women, over ninety percent completed their CRRM as a second surgery [82]. Over the last decade, there has been an increase in *BRCA* genetic testing prior to surgery. Rates of BRRM are consistently higher when *BRCA* mutation status is known before surgery; identification of a *BRCA1/2* mutation post-surgically often leads to a subsequent risk-reducing surgery [94–96]. Although a particularly overwhelming time, establishing *BRCA* mutation status at time of cancer diagnosis enables women to consider surgical decisions for risk reduction if *BRCA*-positive [94–96].

Among women unaffected by breast cancer and those affected with unilateral breast cancer, satisfaction with BRRM or CRRM, respectively, is high [82]. Longitudinal data indicates no clinically meaningful variations in anxiety and depression scores pre- and post-surgery. Additionally, significant reduction in perceived cancer risk occurs across both groups (unaffected and affected), as well as reduced cancer worry among unaffected women [82].

Risk-Reducing Bilateral Salpingo-Oophorectomy

Several studies have demonstrated the efficacy of risk-reducing salpingo-oophorectomy (RRSO) in *BRCA* mutation carriers. RRSO leads to a significant reduction in all-cause, breast cancer-specific and ovarian cancer-specific mortality [58, 59, 62]. Therefore, RRSO is clearly indicated for female *BRCA* mutation carriers and is recommended between the ages of 35 and 40 [97].

Predictors of RRSO

Women recognize RRSO as a life-saving intervention, and the majority of mutation carriers do eventually proceed with RRSO [98–101]. Women opting for RRSO indicate high ovarian cancer risk (actual and perceived), lack of effective screening, and fear of developing ovarian cancer as primary motivators. Additional predictors of RRSO include older age (>40 years old), completion of childbearing, as well as higher baseline anxiety and worry [101, 102].

A 5-year follow-up study in women with *BRCA1/2* mutations who underwent risk-reducing surgery (both RRSO and BRRM) revealed that those who elected RRSO reported diminished fear of cancer, and most women felt surgery was worthwhile despite adverse consequences of surgical menopause [16]. Most women demonstrated adequate coping mechanisms and did not differ significantly from non-carriers in terms of psychological distress, cancer worry, and help-seeking behavior over the 5-year period. Carriers reporting solid family communication and supportive partners experienced a greater ease with coping and adjustment. Although generally satisfied with their decision, mutation carriers report feeling less satisfied with their bodies and had more problems with sexual and endocrine functioning. The biggest predictors of long-term distress were baseline general cancer-related anxiety, personal experience of losing loved one to cancer, and having young children (<15 years). The latter was attributed to "fear of leaving young children behind and to difficulties informing children about their cancer risks" [16]. Similar conclusions were drawn from a large series of 846 women with an increased ovarian cancer risk (368 of whom were *BRCA* mutation carriers). In this study, 44% of patients underwent RRSO, including 264 (72%) of *BRCA* mutation carrier cohort. There was favorable sentiment toward RRSO in high-risk women, and despite significant reports of decreased sexual and endocrine functioning, 86% of women would do it again. Proceeding with RRSO was associated with fewer breast/ovarian cancer worries and more favorable cancer risk perception. Favorable cancer risk perception was best appreciated in those women who had undergone both RRSO and BRRM [102].

"What I Wished I Had Known Before RRSO"

In the context of RRSO, patients may feel shy or embarrassed to discuss certain sensitive topics with their providers regarding post-surgical effects, such as libido or vaginal dryness. Indeed, they may not know enough information to generate such questions. Some may feel "survivor's guilt" that prevents them from sharing concerns about surgical side effects, convincing themselves to instead feel "lucky" or "thankful" for having the opportunity to ovarian cancer [103].

Ninety-eight mutation carriers, ascertained through the Yale Cancer Genetic Counseling program, completed a post-RRSO questionnaire, "What I wish I'd known before surgery: *BRCA* carriers' perspectives after bilateral salpingo-oophorectomy." Patients shared their desires, in hindsight, for more candid information from their providers regarding libido, impact on body image, changes to sex life, and possible relationship changes with partners. Their impressions suggested that pre-surgical dialogue predominantly focused on the surgical procedure itself, as well as the cancer risk reduction provided [103]. Although the latter is a critical piece, the pre-surgical discussion may be incomplete for many women.

The following adapted table (Table 12.3) highlights patient report of symptoms after RRSO.

Table 12.3 Patient report of symptoms after RRSO

Symptom	% reporting "frequent" or "very frequent"
Change in interest in sex	50.0
Change in sex life	43.9
Vaginal dryness	52.1
Hot flashes	42.9
Sleep disturbances	46.9
Night sweats	33.7
Painful intercourse	31.0
Change in body image	31.6
Depression	21.4

Table adapted from Campfield Bonadies et al. [103]

Hormone Replacement Therapy (HRT) and RRSO

Hormone replacement therapy (HRT) may be offered to unaffected, premenopausal *BRCA1/2* mutation carrier women who undergo RRSO. This may be used to combat symptoms of surgical menopause and perhaps reduce long-term effects such as osteoporosis, among other issues. For women concerned about surgical menopause, it may not be easy to elect RRSO, particularly if the woman lacks personal experience of ovarian cancer among relatives or friends, and if the woman's perception of ovarian and other cancers is that they are generally survivable [104, 105]. Conversations regarding use of HRT in this context should be discussed with a knowledgeable medical professional.

Male *BRCA1* and *BRCA2* Mutation Carriers

Little research attention has been paid to male experiences in HBOC families [106]. The sparse literature available describes men as mainly seeking genetic testing out of an obligation to their children [107–109], and it is reported that women have a strong influence upon male decision-making regarding genetic testing [107, 108]. Although study samples have been small, a majority of men are aware at some level of the hereditary nature of cancer in their families and harbor significant concerns about their own cancer risk [110].

Prostate Cancer Surveillance

Consistent with women in HBOC families, men also lack confidence in general practitioners to navigate management issues specific to *BRCA* mutations. However, in contrast to women, men are less likely to be included in surveillance programs [106]. Some *BRCA*-positive men, especially those under 50 years of age, report difficulty setting up serial prostate screening given resistance from their medical providers. This is likely due to the controversial status of prostate cancer surveillance among men in the general population [111–113]. Although controversial, there is reported high screening compliance among average risk men offered prostate surveillance (PSA and DRE) [111], and, although sample size has been small, among *BRCA1/2* mutation carrier males as well [109]. One limitation of the latter, however, is that this sampling was a highly motivated self-selected group.

In the largest international prospective cohort (IMPACT Study) comparing male *BRCA1/2* mutation carriers versus controls (true negatives for familial *BRCA* mutation), targeted PSA screening resulted in high positive predictive value and the identification of high-grade tumors [114–116]. This emerging data has supported PSA screening in *BRCA1/2* mutation carriers, especially given the younger cohort age (40–69 years of age) and identification of cancers requiring therapeutic intervention [114]. IMPACT also administered Quality of Life self-report surveys as a sub-study, recently been submitted for publication (pending). While analysis of both carriers and controls yielded population norms for psychological distress, some men with *BRCA* mutations have higher prostate cancer-specific distress than men who tested negative for a known familial *BRCA* mutation (personal communication from Elizabeth Bancroft, study coordinator and research nurse for the IMPACT project).

Male Breast Cancer Surveillance

There is the general perception that breast cancer is not a man's disease. For male *BRCA* carriers, breast cancer is no longer an abstract thought, but a real possibility based upon the 1–7% lifetime risk of developing breast cancer [117]. In age-matched studies, men are typically diagnosed at a later stage versus women, based upon failure to recognize early symptoms or changes to the breast tissue [118, 119]. Clinical breast examination (CBE) remains a key factor in early diagnosis and *BRCA*-positive men [118, 119]. Without proper education, men may be unfamiliar with breast symptoms they should be mindful of, and like self-advocacy with PSA/DRE, men can experience resistance when inquiring with their medical providers.

Men may feel stigma when sharing feelings about breast cancer risk, or having a breast cancer diagnosis [120]. Although literature is scant, patient testimonies can be found with similar resonating themes [120–122]:

1. Shock
2. Stigma in dealing with predominantly a woman's disease
3. Reluctance to appear vulnerable
4. Shift in body image after a surgery or treatment
5. Limited conversation about the breast cancer experience
6. Fear of judgment or dismissal from peers
7. Limited resources and what is available is generally intended for women
8. Resources felt to be most helpful: photos of a post-mastectomy male and information regarding side effects of hormonal therapy.

Despite the paucity of medical guidelines regarding male breast cancer and the minimal data regarding psychological impact in this population, it is important to identify actual and psychological barriers to care and to offer men adequate guidance in their journeys. Additionally, like women with HBOC, a subset of men with *BRCA* mutations may have less favorable coping with high anxiety and distress. It is important to identify mechanisms to identify these at-risk individuals and promote well-being and healthy behaviors.

Unique Considerations in Young Adulthood

A growing body of literature supports the fact that young women, aged 6–19 years of age, with a positive breast cancer family history with or without a familial *BRCA* mutation, do not have poorer psychosocial adjustment than their peers [123–126]. However, they do exhibit higher perceived breast cancer risks as well as cancer-specific distress. Young women have been shown to be quite adaptive to early communication and disclosure of *BRCA* genetic test results; this open sharing may improve healthy behaviors [124, 126]. More highly anxious young women from high-risk families tend to have more highly anxious mothers, and/or poorer communication within their own families [124, 126].

Testing for *BRCA1* and *BRCA2* mutations is generally advised against in minors under the age of 18. Increasingly, however, young women (18–24 year olds) from hereditary breast/ovarian cancer families are pursuing genetic testing despite low absolute risks of breast and ovarian cancer and despite the fact that evidence-based management options are not generally available or recommended until age 25 [127]. The decision-making process in which these young individuals engage is distinct from the larger population of women in HBOC families [127, 128]. In semi-structured interviews of 32 young *BRCA1/2* female mutation carriers, many women had feelings of vulnerability manifested as a sense of urgency in risk management [127]. Young women may experience external pressure from medical providers as well as close family members. Patients note they receive "life counseling" (for example, to have children early in preparation for ovary removal) in a similar manner to more straightforward recommendations of beginning yearly breast MRI at age 25. In addition, a significant feature within this age-group is reliance on parents for continued emotional, pragmatic, and financial support. This reliance is generally quite helpful; however, some young individuals can feel pressure to undergo genetic testing and engage in certain risk-reducing behaviors prematurely. The latter can be brought about by a parent who has experienced loss through a personal diagnosis and/or cancer diagnosis of a loved one.

Young women (and men) may desire genetic testing to resolve uncertainty, feeling that it is better to know whether they do or do not carry a *BRCA* mutation. The latter could be to relieve the burden of "not knowing," as well as aiding in preparedness for the future. In addition, although individuals are tending to start childbearing later than their predecessors, the opportunities for pre-implantation genetic diagnosis (PGD) provide the possibility for a different element of intentional reproductive family planning [128].

Given incomplete development within this age range as well as lack of "medical actionability," readiness to make independent and enduring life decisions should be carefully assessed by both the individual and their medical provider. Genetic education and decision-making should be an ongoing and inclusive process met with sensitivity and empathy. Motivators for testing should be thoroughly explored and vetted, and the young individual should be able to clearly articulate their perceived benefits, as well as risks and limitations to genetic testing at this time point in their lives.

Legal Issues in *BRCA* Genetic Testing and HBOC

Anti-genetic Discrimination Legislation, Protections and Limitations

Genetic testing has become an increasingly powerful diagnostic and prognostic tool, yet healthcare providers and their patients remain wary of the potential of genetic testing to trigger discrimination [129]. There is limited awareness of the true scope of legal protections afforded by legislation, and fear of misuse of genetic information remains prevalent. There is little evidence to support the concept that genetic discrimination by health insurers or employers actually occurs, although this may be due to a lack of such events being reported or confusion about what defines genetic discrimination [130, 131]. Still, patients have avoided clinical genetic testing and genetic research participation due to fear of discrimination [132–134]. This motivated the genetics community (providers and advocates) to lobby on behalf of these concerns, which resulted in the passage of Genetics Information Non-discrimination Act (GINA) in 2008. Although certainly not exclusive to *BRCA1* and *BRCA2* mutations carriers, healthcare providers and patients are continuously challenged to understand their legal protections and the limitations of these.

Federal and state laws are like patchwork. It can be difficult to fully comprehend the "who, what, when, where, and why." The following will highlight legislative milestones in the quest for genetic protections through the lens of a genetic counselor. This does not substitute for legal advice, nor does it provide every clause or subtext. There are four main pieces of legislation, that when utilized in combination, provide a minimum expectation for patient protections against the misuse of genetic information:

1. Health Insurance Portability and Accountability Act (HIPAA, 1996, final Privacy Rule 2002) [135];
2. Genetic Information Non-discrimination Act (GINA, 2008) [136];
3. The American Disabilities Act (ADA, Amendment 2009) [137, 138];
4. The Affordable Care Act (ACA, 2010) [139].

Historically, GINA is recognized as the first federal law whose primary objective was protection from genetic discrimination. GINA was built upon the privacy provisions implemented by the Health Insurance Portability and Accountability Act, as well as employment provisions set forth by the American Disabilities Act. More recently in 2010, the Affordable Health Care Act (ACA) worked to close certain gaps for which GINA fell short, although there remain important shortcomings that must be highlighted and understood.

Genetic Information Non-discrimination Act (GINA) of 2008

After 13 years of advocacy by the genetics community and US lawmakers, the Genetic Information Non-discrimination Act (GINA) was signed into law in 2008 [140]. GINA was the first application of the HIPAA privacy rule to genetic information, and it has two titles, health insurance (Title I) and employment (Title II). Broadly speaking, GINA prohibits employers and health insurance companies from discriminating against an individual based on his or her genetic information, which includes family history as well as direct-to-consumer genetic testing [136]. Importantly, health insurance companies and employers are not allowed to collect genetic information in order to use it to raise premium rates, deny coverage, or make adverse employment determination [136]. GINA clarifies HIPAA in documenting that genetic information is health-related information; however, this does not include symptomatic individuals who "manifest disease" (i.e., a woman with breast cancer who later tests positive for a *BRCA1* gene mutation). Additionally, GINA employment provisions do not apply to employers with fewer than 15 employees, and both health and employment provisions do not apply to Federal Employee Health Benefits program, US Military, Veterans Administration, and the Indian Health Service [136]. Some of these institutions have separate policies protecting individuals from genetic information related discrimination. Of note, GINA protections do not extend to life insurance, disability, and long-term care insurance.

There are several states with laws against genetic information discrimination, which vary significantly in scope and application. These state laws may be more or less stringent compared to GINA, and some also extend to life, disability, and long-term care insurance. GINA provides a baseline protection, which states can model after, but it does not preempt more stringent state laws. A list of state laws related to genetic information discrimination may be located at https://www.genome.gov/policyethics/legdatabase/pubsearch.cfm.

The Affordable Care Act (ACA) of 2010

The ACA changes the prism through which GINA may be viewed. The ACA prohibits discrimination by health insurers (both group AND individual) on the basis of preexisting conditions, including genetic test results, thereby closing the gap in health insurance protection for persons with manifest disease [139]. Also, certain health insurance issuers cannot increase or adjust health insurance premiums based on a "preexisting condition," but have to determine premiums on the basis of age, geographic area, etc. [139].

With regard to employment discrimination, the ACA did not resolve the shortcomings of both GINA and the ADA. While GINA excludes "manifest disease" and ADA has protections in place for those with severe impairment, people with manifest disease who are not yet disabled remain unprotected on the federal level [140].

Unique Considerations

As medicine increasingly relies upon genetic information, this data does not only pertain to a singular patient, but rather to a family or lineage. There are unique considerations when navigating the responsibility to disseminate sensitive genetic information to at-risk relatives, or when questions arise regarding genetically testing minors for adult-onset diseases. Unlike federal and state legislation which help set precedent, there are several ethical matters in the realm of *BRCA* genetic testing that do not come with official statute or law. These matters continually challenge the field, providers, and patients alike.

Duty to Warn

Genetic responsibility toward oneself and others is a highly debated implication of genetic testing for cancer predisposition [24, 25, 141, 142]. Ensuring that family members are made aware of the risk is not always easy, as patients are not legally required to disclose medical information to their relatives [143]. Additionally, HIPAA prohibits healthcare professionals from disclosing patient information to third parties, including relatives, without appropriate consent [135, 143]. To date, with few exceptions, genetic testing information is not exempt from privacy law [143]. In the ethical and medical literature, there is general agreement for a compelling moral argument affirming the importance of sharing genetic information with relatives who may potentially be at risk. Yet, most medical associations hesitate to breach patient confidentiality and warn patients' relatives directly, instead suggesting healthcare providers have a duty to advise patients of the relevance and importance genetic information can have to their relatives [24]. Two primary court cases established the concept of "duty to warn," although these generated different conclusions about what constitutes reasonable disclosure.

1. In Pate v. Threlkel, the Supreme Court of Florida ruled a physician has a duty to warn patients of the genetically transferable nature of the condition for which they are being treated. Though this duty extends to informing the patient's children, the court held that the duty is satisfied by warning the patient of the familial implications of genetic testing [144, 145].
2. Safer v. Estate of Pack. Shortly after Pate, the court in Safer v. Pack held that a physician's duty to warn those known to be at risk of avoidable harm from a genetically transferable condition might not be satisfied by telling only the patient [146].

These disparate outcomes speak to the shades of gray and lack of objective clarity about satisfactory dissemination of health information to those who could medically act.

Although outright refusal to share information is rare, select circumstances seriously challenge healthcare providers, in terms of their responsibility to avoid harm to others. At present, for those who engage in genetic counseling, the *Code of Medical Ethics* of the American Medical Association stresses that pre- and post-test genetic counseling must include implications of genetic information for their patients' biological relatives. However, there is no mandate that the patient act on this information [147].

Next of Kin and Release of Information

Due to the nature of inherited *BRCA1* and *BRCA2* mutations, it is not uncommon for family members to inquire about a deceased relative's positive mutation report for the purposes of their own genetic testing. A positive mutation report is often needed by the testing laboratory to inform their testing process and to ensure the most accurate result, whether positive or negative. Often, the release of this information must be authorized by the "next of kin," or nearest blood relative of a person who has died, including the surviving spouse. Hospitals, private practices, as well as larger academic centers typically have their own documentation requirements, which may be vetted by a legal advisor.

Although not explicitly described in the section prior, the concept of "duty to warn" extends postmortem and is particularly salient given the patient's permission can no longer be sought. Through the process of genetic counseling, patients may be approached during the pre-test consultation regarding to whom they would desire information to be released in the event of their passing. This may be helpful for immediate disclosure purposes, such as a terminal cancer patient on hospice permitting the provider to disclose the result to a spouse and/or family member, or this may fulfill a more long-term need, for relatives who may inquire about test results at any point in time. Genetics is a shared family matter, and the test result of one person may remain tremendously valuable to family members for a number of years post-testing.

Testing Minors for *BRCA1* and *BRCA2* Gene Mutations

Historically, predictive genetic testing in minors for late-onset conditions has been discouraged by professional organizations, if not proscribed, due to the plausible psychological harm to the child, as well as lack of clear medical benefit [40]. The National Society of Genetic Counselors (NSGC), for example, recommends deferring predictive genetic testing of minors for adult-onset conditions whenever possible [39]. This general consensus is applicable to predictive genetic testing to evaluate for HBOC syndrome. Some recent data has emerged suggesting less harm than anticipated, including reports of considerable resiliency and an ability of minors to successfully incorporate risk information into their self-concepts and life plans [40]. Nonetheless, the universal recommendation remains that genetic testing for late-onset inherited conditions be deferred until adulthood [39, 40]. Should testing be considered in opposition to this recommendation, exploration of the medical and psychosocial benefits, and harms should be thoroughly vetted between the healthcare provider, the patient, and his/her immediate family.

In minors, predictive genetic testing for disease risk should focus firstly on the child's medical best interest and whether the intervention will be met with timely medical benefit. If the medical benefits are uncertain or will be deferred to a later time, such as having a *BRCA1* or *BRCA2* mutation, justification for testing is less compelling [40]. Parents, guardians, or children may be motivated beyond medical benefit, instead focusing on childhood testing for possible psychological benefit. Such motivations may be to alleviate uncertainty or to enable earlier adjustment to risk information. The decision to test, when the test is not anticipated to impact medical management in the near term, should be made carefully and on a case-by-case basis. Deferring predictive genetic testing also enables individuals to make decisions about genetic testing for themselves as adults, taking into account their own circumstances, preferences, and beliefs. NSGC strongly recommends that families facing these decisions meet with a certified genetic counselor or medical geneticist to review the clinical and personal implications of testing [39].

In the setting of predictive *BRCA1* and *BRCA2* testing, often the child has experienced cancer in their family, and this history may have impressed upon them a possible sense of urgency or cancer fear. Alternatively, a parent may wish to pursue genetic testing in their child due to their own sense of mortality, in order to ensure children are tested before they pass away. Thorough genetic counseling before predictive testing is essential to ensure that parents, guardians, and maturing minors fully comprehend the limits of genetic knowledge and the potential for psychological harm, as well as to explore whether motivations are sincerely felt to be in the best interest of the child [39]. Because genetic testing is considered an elective procedure, proceeding is fully conditional on the child's assent as well as parental permission [147]. If a child/adolescent does not desire genetic testing, that dissent is the final authority. Additionally, if a minor is young or immature, delaying testing until the minor can actively participate is recommended. Especially in the setting of predictive *BRCA* genetic testing, a healthcare provider may decline involvement given absence of clinical utility.

Conclusion

This chapter summarized some of the main psychosocial considerations from the time a *BRCA1/2*-positive individual seeks genetic counseling to the subsequent journey navigating familial communication and medical decision-making. It is vital for healthcare providers to have an appreciation and understanding that an individual's journey with hereditary cancer may have been imprinted upon them long before arriving for genetic testing, influenced by personal experience with cancer or the testament of their family history. This imprinting understandably drives reasoning and the decisions people make.

In addition, this chapter highlighted both legal and ethical considerations related to hereditary cancer genetic testing. These considerations extend beyond *BRCA1* and *BRCA2* mutation carriers into other hereditary cancer syndromes. They are subject to change and evolution over time, but should provide some foundation for careful thought on these issues.

Acknowledgements The authors wish to thank Kara Milliron and Jill Stopfer for their time and assistance in preparation of this chapter.

References

1. National Society of Genetic Counselors' Definition Task Force, Resta R, Biesecker BB, Bennett RL, Blum S, Hahn SE, et al. A new definition of genetic counseling: national society of genetic counselors' task force report. J Genet Couns. 2006;15(2):77–83.
2. Berliner JL, Fay AM, Cummings SA, Burnett B, Tillmanns T. NSGC practice guideline: risk assessment and genetic counseling for hereditary breast and ovarian cancer. J Genet Couns. 2013;22(2):155–63.
3. Robson ME, Bradbury AR, Arun B, Domchek SM, Ford JM, Hampel HL, et al. American society of clinical oncology policy statement update: genetic and genomic testing for cancer susceptibility. J Clin Oncol. 2015;33(31):3660–7.
4. Domchek SM, Bradbury A, Garber JE, Offit K, Robson ME. Multiplex genetic testing for cancer susceptibility: out on the high wire without a net? J Clin Oncol. 2013;31(10):1267–70.
5. Braithwaite D, Emery J, Walter F, Prevost AT, Sutton S. Psychological impact of genetic counseling for familial cancer: a systematic review and meta-analysis. JNCI J Nat Cancer Inst. 2004;96(2):122–33.
6. Robson ME, Storm CD, Weitzel J, Wollins DS, Offit K, American Society of Clinical Oncology. American society of clinical oncology policy statement update: genetic and genomic testing for cancer susceptibility. J Clin Oncol. 2010;28(5):893–901.
7. Surgeons TASoB. Consensus guideline on hereditary genetic testing for patients with and without breastcancer 2016 [cited 11 Nov 2016]. Available from: https://www.breastsurgeons.org/new_layout/about/statements/PDF_Statements/BRCA_Testing.pdf.
8. Stopfer JE. Genetic counseling and clinical cancer genetics services. Semin Surg Oncol. 2000;18(4):347–57.
9. Nelson HD, Fu R, Goddard K, et al. Risk assessment, genetic counseling, and genetic testing for *BRCA*-related cancer: systematic review to update the U.S. preventive services task force recommendation. Evidence syntheses, no. 101 Rockville (MD): Agency for Healthcare Research and Quality (US); 2013. Available from: https://www.ncbi.nlm.nih.gov/books/NBK179205/. 31 Oct 2016.
10. Cancer ACoSCo. Cancer program standards: ensuring patient-centered care 2016 [cited 11 Nov 2016]. Available from: https://www.facs.org/~/media/files/quality%20programs/cancer/coc/2016%20coc%20standards%20manual_interactive%20pdf.ashx.
11. CIGNA. Genetic testing and counseling program 2016 [cited 11 Nov 2016]. Available from: http://www.cigna.com/healthcare-professionals/resources-for-health-care-professionals/genetic-testing-and-counseling-program.
12. Trepanier AM, Allain DC. Models of service delivery for cancer genetic risk assessment and counseling. J Genet Couns. 2014;23(2):239–53.
13. Patrick-Miller L, Harris D, Stevens E, Egleston B, Fleisher L, Mueller R, et al. Telemedicine: expanding access to cancer genetic services to underserved populations. Cancer Res. 2015;75.
14. Patrick-Miller L, Egleston BL, Daly M, Stevens E, Fetzer D, Forman A, et al. Implementation and outcomes of telephone disclosure of clinical *BRCA1/2* test results. Patient Educ Couns. 2013;93(3):413–9.
15. Mosher CE, Danoff-Burg S. Psychosocial impact of parental cancer in adulthood: a conceptual and empirical review. Clin Psychol Rev. 2005;25(3):365–82.
16. van Oostrom I, Meijers-Heijboer H, Lodder LN, Duivenvoorden HJ, van Gool AR, Seynaeve C, et al. Long-term psychological impact of carrying a *BRCA1/2* mutation and prophylactic surgery: a 5-year follow-up study. J Clin Oncol. 2003;21(20):3867–74.

17. Foster C, Watson M, Eeles R, Eccles D, Ashley S, Davidson R, et al. Predictive genetic testing for *BRCA1/2* in a UK clinical cohort: three-year follow-up. Br J Cancer. 2007;96 (5):718–24.
18. Reichelt JG, Heimdal K, Moller P, Dahl AA. *BRCA1* testing with definitive results: a prospective study of psychological distress in a large clinic-based sample. Fam Cancer. 2004;3(1):21–8.
19. Graves KD, Vegella P, Poggi EA, Peshkin BN, Tong A, Isaacs C, et al. Long-term psychosocial outcomes of *BRCA1/BRCA2* testing: differences across affected status and risk-reducing surgery choice. Cancer Epidemiol Biomarkers Prev. 2012;21(3):445–55.
20. Vodermaier A, Esplen MJ, Maheu C. Can self-esteem, mastery and perceived stigma predict long-term adjustment in women carrying a *BRCA1/2*-mutation? Evidence from a multi-center study. Fam Cancer. 2010;9(3):305–11.
21. Watson M, Foster C, Eeles R, Eccles D, Ashley S, Davidson R, et al. Psychosocial impact of breast/ovarian (*BRCA1/2*) cancer-predictive genetic testing in a UK multi-centre clinical cohort. Br J Cancer. 2004;91(10):1787–94.
22. Caiata-Zufferey M, Pagani O, Cina V, Membrez V, Taborelli M, Unger S, et al. Challenges in managing genetic cancer risk: a long-term qualitative study of unaffected women carrying *BRCA1/BRCA2* mutations. Genet Med. 2015;17(9):726–32.
23. Daly MB, Montgomery S, Bingler R, Ruth K. Communicating genetic test results within the family: is it lost in translation? A survey of relatives in the randomized six-step study. Fam Cancer. 2016:1–10.
24. Offit K, Groeger E, Turner S, Wadsworth EA, Weiser MA. The "duty to warn" a patient's family members about hereditary disease risks. JAMA. 2004;292(12):1469–73.
25. Dugan RB, Wiesner GL, Juengst ET, O'Riordan M, Matthews AL, Robin NH. Duty to warn at-risk relatives for genetic disease: genetic counselors' clinical experience. Am J Med Genet C Semin Med Genet. 2003;119C(1):27–34.
26. Julian-Reynier C, Eisinger F, Chabal F, Lasset C, Nogues C, Stoppa-Lyonnet D, et al. Disclosure to the family of breast/ovarian cancer genetic test results: patient's willingness and associated factors. Am J Med Genet. 2000;94(1):13–8.
27. Koehly LM, Peters JA, Kenen R, Hoskins LM, Ersig AL, Kuhn NR, et al. Characteristics of health information gatherers, disseminators, and blockers within families at risk of hereditary cancer: implications for family health communication interventions. Am J Public Health. 2009;99(12):2203–9.
28. Patenaude AF, Dorval M, DiGianni LS, Schneider KA, Chittenden A, Garber JE. Sharing *BRCA1/2* test results with first-degree relatives: factors predicting who women tell. J Clin Oncol. 2006;24(4):700–6.
29. McGivern B, Everett J, Yager GG, Baumiller RC, Hafertepen A, Saal HM. Family communication about positive *BRCA1* and *BRCA2* genetic test results. Genet Med. 2004;6 (6):503–9.
30. Claes E, Evers-Kiebooms G, Boogaerts A, Decruyenaere M, Denayer L, Legius E. Communication with close and distant relatives in the context of genetic testing for hereditary breast and ovarian cancer in cancer patients. Am J Med Genet A. 2003;116A (1):11–9.
31. MacDonald DJ, Sarna L, van Servellen G, Bastani R, Giger JN, Weitzel JN. Selection of family members for communication of cancer risk and barriers to this communication before and after genetic cancer risk assessment. Genet Med. 2007;9(5):275–82.
32. Peters JA, Kenen R, Hoskins LM, Koehly LM, Graubard B, Loud JT, et al. Unpacking the blockers: understanding perceptions and social constraints of health communication in hereditary breast ovarian cancer (hboc) susceptibility families. J Genet Couns. 2011;20 (5):450–64.
33. Hodgson J, Gaff C. Enhancing family communication about genetics: ethical and professional dilemmas. J Genet Couns. 2013;22(1):16–21.

34. Forrest LE, Delatycki MB, Curnow L, Skene L, Aitken M. Genetic health professionals and the communication of genetic information in families: practice during and after a genetic consultation. Am J Med Genet A. 2010;152A(6):1458–66.
35. BRCA BCf. 2016 [cited 10 Nov 2016]. Available from: basser.org.
36. de Geus E, Eijzenga W, Menko FH, Sijmons RH, de Haes HC, Aalfs CM, et al. Design and feasibility of an intervention to support cancer genetic counselees in informing their at-risk relatives. J Genet Couns. 2016.
37. Gaff C, Hodgson J. A genetic counseling intervention to facilitate family communication about inherited conditions. J Genet Couns. 2014;23(5):814–23.
38. Montgomery SV, Barsevick AM, Egleston BL, Bingler R, Ruth K, Miller SM, et al. Preparing individuals to communicate genetic test results to their relatives: report of a randomized control trial. Fam Cancer. 2013;12(3):537–46.
39. (NSGC) NSoGC. Genetic testing of minors for adult-onset conditions. http://www.nsgc.org/p/bl/et/blogid=47&blogaid=28: 2012 n/a.
40. Ross LF, Saal HM, David KL, Anderson RR, American Academy of Pediatrics, American College of Medical Genetics and Genomics, et al. Technical report: ethical and policy issues in genetic testing and screening of children. Genet Med. 2013;15(3):234–45.
41. Botkin JR, Belmont JW, Berg JS, Berkman BE, Bombard Y, Holm IA, et al. Points to consider: ethical, legal, and psychosocial implications of genetic testing in children and adolescents. Am J Hum Genet. 2015;97(1):6–21.
42. Bradbury AR, Dignam JJ, Ibe CN, Auh SL, Hlubocky FJ, Cummings SA, et al. How often do *BRCA* mutation carriers tell their young children of the family's risk for cancer? A study of parental disclosure of *BRCA* mutations to minors and young adults. J Clin Oncol. 2007;25 (24):3705–11.
43. Bradbury AR, Patrick-Miller L, Egleston BL, Olopade OI, Daly MB, Moore CW, et al. When parents disclose *BRCA1/2* test results: their communication and perceptions of offspring response. Cancer Am Cancer Soc. 2012;118(13):3417–25.
44. Patenaude AF, Tung N, Ryan PD, Ellisen LW, Hewitt L, Schneider KA, et al. Young adult daughters of *BRCA1/2* positive mothers: What do they know about hereditary cancer and how much do they worry? Psychol Oncol. 2013;22(9):2024–31.
45. Castonguay C, Friedman S, Leininger A, McKenna M, Ramos E, Rando R. Talking about *BRCA* in your family tree 2014 [cited 11 Nov 2016]. Available from: http://www.facingourrisk.org/understanding-brca-and-hboc/publications/documents/booklet-talking-about-brca-family.pdf.
46. Tonin P, Weber B, Offit K, Couch F, Rebbeck TR, Neuhausen S, et al. Frequency of recurrent *BRCA1* and *BRCA2* mutations in Ashkenazi Jewish breast cancer families. Nat Med. 1996;2(11):1179–83.
47. Powers J, Stopfer JE. Risk assessment, genetic counseling, and clinical care for hereditary breast cancer. Jognn J Obst Gyn Neo. 2014;43(3):361–73.
48. Grossman SG. The Angelina Jolie effect in Jewish law: Prophylactic mastectomy and oophorectomy in *BRCA* carriers. Rambam Maimonides Med J. 2015;6(4).
49. Ricker C, Lagos V, Feldman N, Hiyama S, Fuentes S, Kumar V, et al. If we build it … will they come?—establishing a cancer genetics services clinic for an underserved predominantly Latina cohort. J Genet Couns. 2006;15(6):505–14.
50. Komenaka IK, Nodora JN, Madlensky L, Winton LM, Heberer MA, Schwab RB, et al. Participation of low-income women in genetic cancer risk assessment and *BRCA1/2* testing: the experience of a safety-net institution. J Community Genet. 2016;7(3):177–83.
51. Sheppard VB, Mays D, LaVeist T, Tercyak KP. Medical mistrust influences black women's level of engagement in *BRCA1/2* genetic counseling and testing. J Natl Med Assoc. 2013;105(1):17–22.
52. Levy DE, Byfield SD, Comstock CB, Garber JE, Syngal S, Crown WH, et al. Underutilization of *BRCA1/2* testing to guide breast cancer treatment: black and Hispanic women particularly at risk. Genet Med. 2011;13(4):349–55.

53. Randall TC, Armstrong K. Health care disparities in hereditary ovarian cancer: are we reaching the underserved population? Curr Treat Options Oncol. 2016;17(8):39.
54. Haroun I, Graham T, Poll A, Sun P, Hill K, Weitzner E, et al. Reasons for risk-reducing mastectomy versus MRI-screening in a cohort of women at high hereditary risk of breast cancer. Breast. 2011;20(3):254–8.
55. Schwartz MD, Kaufman E, Peshkin BN, Isaacs C, Hughes C, DeMarco T, et al. Bilateral prophylactic oophorectomy and ovarian cancer screening following *BRCA1*/*BRCA2* mutation testing. J Clin Oncol. 2003;21(21):4034–41.
56. Singh K, Lester J, Karlan B, Bresee C, Geva T, Gordon O. Impact of family history on choosing risk-reducing surgery among *BRCA* mutation carriers. Am J Obstet Gynecol. 2013;208(4):329 e1–6.
57. Borreani C, Manoukian S, Bianchi E, Brunelli C, Peissel B, Caruso A, et al. The psychological impact of breast and ovarian cancer preventive options in *BRCA1* and *BRCA2* mutation carriers. Clin Genet. 2014;85(1):7–15.
58. Rebbeck TR, Lynch HT, Neuhausen SL, Narod SA, Van't Veer L, Garber JE, et al. Prophylactic oophorectomy in carriers of *BRCA1* or *BRCA2* mutations. N Engl J Med. 2002;346(21):1616–22.
59. Domchek SM, Friebel TM, Neuhausen SL, Wagner T, Evans G, Isaacs C, et al. Mortality after bilateral salpingo-oophorectomy in *BRCA1* and *BRCA2* mutation carriers: a prospective cohort study. Lancet Oncol. 2006;7(3):223–9.
60. Lord SJ, Lei W, Craft P, Cawson JN, Morris I, Walleser S, et al. A systematic review of the effectiveness of magnetic resonance imaging (MRI) as an addition to mammography and ultrasound in screening young women at high risk of breast cancer. Eur J Cancer. 2007;43 (13):1905–17.
61. Warner E, Messersmith H, Causer P, Eisen A, Shumak R, Plewes D. Systematic review: using magnetic resonance imaging to screen women at high risk for breast cancer. Ann Intern Med. 2008;148(9):671–9.
62. Domchek SM, Friebel TM, Singer CF, Evans DG, Lynch HT, Isaacs C, et al. Association of risk-reducing surgery in *BRCA1* or *BRCA2* mutation carriers with cancer risk and mortality. JAMA. 2010;304(9):967–75.
63. Warner E, Plewes DB, Hill KA, Causer PA, Zubovits JT, Jong RA, et al. Surveillance of *BRCA1* and *BRCA2* mutation carriers with magnetic resonance imaging, ultrasound, mammography, and clinical breast examination. JAMA. 2004;292(11):1317–25.
64. Hagen AI, Kvistad KA, Maehle L, Holmen MM, Aase H, Styr B, et al. Sensitivity of MRI versus conventional screening in the diagnosis of *BRCA*-associated breast cancer in a national prospective series. Breast. 2007;16(4):367–74.
65. Sardanelli F, Podo F, D'Agnolo G, Verdecchia A, Santaquilani M, Musumeci R, et al. Multicenter comparative multimodality surveillance of women at genetic-familial high risk for breast cancer (HIBCRIT study): interim results. Radiology. 2007;242(3):698–715.
66. Kurian AW, Sigal BM, Plevritis SK. Survival analysis of cancer risk reduction strategies for *BRCA1/2* mutation carriers. J Clin Oncol. 2010;28(2):222–31.
67. Lerman C, Trock B, Rimer BK, Jepson C, Brody D, Boyce A. Psychological side effects of breast cancer screening. Health Psychol. 1991;10(4):259–67.
68. Cockburn J, Staples M, Hurley SF, De Luise T. Psychological consequences of screening mammography. J Med Screen. 1994;1(1):7–12.
69. Sutton S, Saidi G, Bickler G, Hunter J. Does routine screening for breast cancer raise anxiety? Results from a three wave prospective study in England. J Epidemiol Commun Health. 1995;49(4):413–8.
70. Gilbert FJ, Cordiner CM, Affleck IR, Hood DB, Mathieson D, Walker LG. Breast screening: the psychological sequelae of false-positive recall in women with and without a family history of breast cancer. Eur J Cancer. 1998;34(13):2010–4.

71. O'Neill SM, Rubinstein WS, Sener SF, Weissman SM, Newlin AC, West DK, et al. Psychological impact of recall in high-risk breast MRI screening. Breast Cancer Res Treat. 2009;115(2):365–71.
72. van Dooren S, Seynaeve C, Rijnsburger AJ, Duivenvoorden HJ, Essink-Bot ML, Tilanus-Linthorst MM, et al. Exploring the course of psychological distress around two successive control visits in women at hereditary risk of breast cancer. Eur J Cancer. 2005;41 (10):1416–25.
73. Lodder LN, Frets PG, Trijsburg RW, Meijers-Heijboer EJ, Klijn JG, Seynaeve C, et al. One year follow-up of women opting for presymptomatic testing for *BRCA1* and *BRCA2*: emotional impact of the test outcome and decisions on risk management (surveillance or prophylactic surgery). Breast Cancer Res Treat. 2002;73(2):97–112.
74. Evans DG, Lalloo F, Ashcroft L, Shenton A, Clancy T, Baildam AD, et al. Uptake of risk-reducing surgery in unaffected women at high risk of breast and ovarian cancer is risk, age, and time dependent. Cancer Epidemiol Biomark Prev. 2009;18(8):2318–24.
75. Hoerger M, Scherer LD, Fagerlin A. Affective forecasting and medication decision making in breast-cancer prevention. Health Psychol. 2016;35(6):594–603.
76. Metcalfe KA, Snyder C, Seidel J, Hanna D, Lynch HT, Narod S. The use of preventive measures among healthy women who carry a *BRCA1* or *BRCA2* mutation. Fam Cancer. 2005;4(2):97–103.
77. Metcalfe KA, Birenbaum-Carmeli D, Lubinski J, Gronwald J, Lynch H, Moller P, et al. International variation in rates of uptake of preventive options in *BRCA1* and *BRCA2* mutation carriers. Int J Cancer. 2008;122(9):2017–22.
78. Karavites LC, Allu S, Khan SA, Kaiser K. Awareness of preventive medication among women at high risk for breast cancer and their willingness to consider transdermal or oral tamoxifen: a focus group study. BMC Cancer. 2015;15:878.
79. Peshkin BN, Isaacs C, Finch C, Kent S, Schwartz MD. Tamoxifen as chemoprevention in BRCA1 and BRCA2 mutation carriers with breast cancer: a pilot survey of physicians. J Clin Oncol. 2003;21(23):4322–8.
80. Frost MH, Schaid DJ, Sellers TA, Slezak JM, Arnold PG, Woods JE, et al. Long-term satisfaction and psychological and social function following bilateral prophylactic mastectomy. JAMA. 2000;284(3):319–24.
81. Metcalfe KA, Esplen MJ, Goel V, Narod SA. Psychosocial functioning in women who have undergone bilateral prophylactic mastectomy. Psychooncology. 2004;13(1):14–25.
82. Metcalfe KA, Semple JL, Narod SA. Satisfaction with breast reconstruction in women with bilateral prophylactic mastectomy: a descriptive study. Plast Reconstr Surg. 2004;114 (2):360–6.
83. Bresser PJ, Seynaeve C, Van Gool AR, Brekelmans CT, Meijers-Heijboer H, van Geel AN, et al. Satisfaction with prophylactic mastectomy and breast reconstruction in genetically predisposed women. Plast Reconstr Surg. 2006;117(6):1675–82; Discussion 83–4.
84. Hopwood P, Lee A, Shenton A, Baildam A, Brain A, Lalloo F, et al. Clinical follow-up after bilateral risk reducing ('prophylactic') mastectomy: mental health and body image outcomes. Psychooncology. 2000;9(6):462–72.
85. Bresser PJ, Van Gool AR, Seynaeve C, Duivenvoorden HJ, Niermeijer MF, van Geel AN, et al. Who is prone to high levels of distress after prophylactic mastectomy and/or salpingo-ovariectomy? Ann Oncol. 2007;18(10):1641–5.
86. McGaughey A. Body image after bilateral prophylactic mastectomy: an integrative literature review. J Midwifery Womens Health. 2006;51(6):e45–9.
87. Gopie JP, Mureau MA, Seynaeve C, Ter Kuile MM, Menke-Pluymers MB, Timman R, et al. Body image issues after bilateral prophylactic mastectomy with breast reconstruction in healthy women at risk for hereditary breast cancer. Fam Cancer. 2013;12(3):479–87.
88. den Heijer M, Seynaeve C, Timman R, Duivenvoorden HJ, Vanheusden K, Tilanus-Linthorst M, et al. Body image and psychological distress after prophylactic

mastectomy and breast reconstruction in genetically predisposed women: a prospective long-term follow-up study. Eur J Cancer. 2012;48(9):1263–8.

89. Lloyd SM, Watson M, Oaker G, Sacks N, Querci della Rovere U, Gui G. Understanding the experience of prophylactic bilateral mastectomy: a qualitative study of ten women. Psychooncology. 2000;9(6):473–85.
90. Hatcher MB, Fallowfield L, A'Hern R. The psychosocial impact of bilateral prophylactic mastectomy: prospective study using questionnaires and semistructured interviews. BMJ. 2001;322(7278):76.
91. Bedrosian I, Hu CY, Chang GJ. Population-based study of contralateral prophylactic mastectomy and survival outcomes of breast cancer patients. J Natl Cancer Inst. 2010;102 (6):401–9.
92. Brewster AM, Bedrosian I, Parker PA, Dong W, Peterson SK, Cantor SB, et al. Association between contralateral prophylactic mastectomy and breast cancer outcomes by hormone receptor status. Cancer Am Cancer Soc. 2012;118(22):5637–43.
93. Portschy PR, Kuntz KM, Tuttle TM. Survival outcomes after contralateral prophylactic mastectomy: a decision analysis. J Natl Cancer Inst. 2014;106(8).
94. Schwartz MD, Lerman C, Brogan B, Peshkin BN, Halbert CH, DeMarco T, et al. Impact of BRCA1/BRCA2 counseling and testing on newly diagnosed breast cancer patients. J Clin Oncol. 2004;22(10):1823–9.
95. Evans DG, Lalloo F, Hopwood P, Maurice A, Baildam A, Brain A, et al. Surgical decisions made by 158 women with hereditary breast cancer aged <50 years. Eur J Surg Oncol. 2005;31(10):1112–8.
96. Weitzel JN, McCaffrey SM, Nedelcu R, MacDonald DJ, Blazer KR, Cullinane CA. Effect of genetic cancer risk assessment on surgical decisions at breast cancer diagnosis. Arch Surg. 2003;138(12):1323–8; Discussion 9.
97. Daly MB, Pilarski R, Berry M, Buys SS, Farmer M, Friedman S, et al. Genetic/familial high-risk assessment: breast and ovarian, Version 1.2017 2016. Available from: https://www.nccn.org/professionals/physician_gls/pdf/genetics_screening.pdf. 19 Sept 2016.
98. Kauff ND, Satagopan JM, Robson ME, Scheuer L, Hensley M, Hudis CA, et al. Risk-reducing salpingo-oophorectomy in women with a *BRCA1* or *BRCA2* mutation. N Engl J Med. 2002;346(21):1609–15.
99. Scheuer L, Kauff N, Robson M, Kelly B, Barakat R, Satagopan J, et al. Outcome of preventive surgery and screening for breast and ovarian cancer in *BRCA* mutation carriers. J Clin Oncol. 2002;20(5):1260–8.
100. Claes E, Evers-Kiebooms G, Decruyenaere M, Denayer L, Boogaerts A, Philippe K, et al. Surveillance behavior and prophylactic surgery after predictive testing for hereditary breast/ovarian cancer. Behav Med. 2005;31(3):93–105.
101. Bradbury AR, Ibe CN, Dignam JJ, Cummings SA, Verp M, White M, et al. Uptake and timing of bilateral prophylactic salpingo-oophorectomy among *BRCA1* and *BRCA2* mutation carriers. Genet Med. 2008;10(3):161–6.
102. Madalinska JB, Hollenstein J, Bleiker E, van Beurden M, Valdimarsdottir HB, Massuger LF, et al. Quality-of-life effects of prophylactic salpingo-oophorectomy versus gynecologic screening among women at increased risk of hereditary ovarian cancer. J Clin Oncol. 2005;23(28):6890–8.
103. Campfield Bonadies D, Moyer A, Matloff ET. What i wish i'd known before surgery: BRCA carriers' perspectives after bilateral salipingo-oophorectomy. Fam Cancer. 2011;10(1): 79–85.
104. Rubinstein WS. Surgical management of *BRCA1* and *BRCA2* carriers: bitter choices slightly sweetened. J Clin Oncol. 2005;23(31):7772–4.
105. Rebbeck TR, Friebel T, Wagner T, Lynch HT, Garber JE, Daly MB, et al. Effect of short-term hormone replacement therapy on breast cancer risk reduction after bilateral prophylactic oophorectomy in *BRCA1* and *BRCA2* mutation carriers: the PROSE study group. J Clin Oncol. 2005;23(31):7804–10.

106. Stromsvik N, Raheim M, Gjengedal E. Cancer worry among Norwegian male *BRCA1/2* mutation carriers. Fam Cancer. 2011;10(3):597–603.
107. Hallowell N, Arden-Jones A, Eeles R, Foster C, Lucassen A, Moynihan C, et al. Guilt, blame and responsibility: men's understanding of their role in the transmission of *BRCA1/2* mutations within their family. Sociol Health Illn. 2006;28(7):969–88.
108. Graves KD, Gatammah R, Peshkin BN, Krieger A, Gell C, Valdimarsdottir HB, et al. *BRCA1/2* genetic testing uptake and psychosocial outcomes in men. Fam Cancer. 2011;10 (2):213–23.
109. Liede A, Metcalfe K, Hanna D, Hoodfar E, Snyder C, Durham C, et al. Evaluation of the needs of male carriers of mutations in *BRCA1* or *BRCA2* who have undergone genetic counseling. Am J Hum Genet. 2000;67(6):1494–504.
110. McAllister MF, Evans DG, Ormiston W, Daly P. Men in breast cancer families: a preliminary qualitative study of awareness and experience. J Med Genet. 1998;35(9): 739–44.
111. Andriole GL, Crawford ED, Grubb RL 3rd, Buys SS, Chia D, Church TR, et al. Mortality results from a randomized prostate-cancer screening trial. N Engl J Med. 2009;360 (13):1310–9.
112. Hayes JH, Barry MJ. Screening for prostate cancer with the prostate-specific antigen test: a review of current evidence. JAMA. 2014;311(11):1143–9.
113. Schroder FH, Hugosson J, Roobol MJ, Tammela TL, Zappa M, Nelen V, et al. Screening and prostate cancer mortality: results of the European randomised study of screening for prostate cancer (ERSPC) at 13 years of follow-up. Lancet. 2014;384(9959):2027–35.
114. Bancroft EK, Page EC, Castro E, Lilja H, Vickers A, Sjoberg D, et al. Targeted prostate cancer screening in *BRCA1* and *BRCA2* mutation carriers: results from the initial screening round of the IMPACT study. Eur Urol. 2014;66(3):489–99.
115. Mitra A, Fisher C, Foster CS, Jameson C, Barbachanno Y, Bartlett J, et al. Prostate cancer in male *BRCA1* and *BRCA2* mutation carriers has a more aggressive phenotype. Br J Cancer. 2008;98(2):502–7.
116. Mitra AV, Bancroft EK, Barbachano Y, Page EC, Foster CS, Jameson C, et al. Targeted prostate cancer screening in men with mutations in *BRCA1* and *BRCA2* detects aggressive prostate cancer: preliminary analysis of the results of the IMPACT study. BJU Int. 2011;107 (1):28–39.
117. Tai YC, Domchek S, Parmigiani G, Chen S. Breast cancer risk among male *BRCA1* and *BRCA2* mutation carriers. J Natl Cancer Inst. 2007;99(23):1811–4.
118. Giordano SH, Cohen DS, Buzdar AU, Perkins G, Hortobagyi GN. Breast carcinoma in men: a population-based study. Cancer Am Cancer Soc. 2004;101(1):51–7.
119. Patterson SK, Helvie MA, Aziz K, Nees AV. Outcome of men presenting with clinical breast problems: the role of mammography and ultrasound. Breast J. 2006;12(5):418–23.
120. France L, Michie S, Barrett-Lee P, Brain K, Harper P, Gray J. Male cancer: a qualitative study of male breast cancer. Breast. 2000;9(6):343–8.
121. Iredale R, Brain K, Williams B, France E, Gray J. The experiences of men with breast cancer in the United Kingdom. Eur J Cancer. 2006;42(3):334–41.
122. Robinson JD, Metoyer KP Jr, Bhayani N. Breast cancer in men: a need for psychological intervention. J Clin Psychol Med Settings. 2008;15(2):134–9.
123. Bradbury AR, Patrick-Miller L, Egleston B, Sands C, Feigon M, Pawlowski K, et al. Parent perceptions of offspring responses to parental communication of *BRCA1/2* test results. J Clin Oncol. 2009;27(15).
124. Bradbury AR, Patrick-Miller L, Schwartz L, Egleston B, Sands CB, Chung WK, et al. Psychosocial Adjustment in school-age girls with a family history of breast cancer. Pediatrics. 2015;136(5):927–37.
125. Bradbury AR, Patrick-Miller L, Egleston B, Schwartz L, Sands CB, Chung W, et al. Psychosocial adjustment, cancer worry and perceived risk in 6–13 year old girls from breast cancer families. Cancer Res. 2015;75.

126. Bradbury AR, Patrick-Miller L, Schwartz LA, Egleston BL, Henry-Moss D, Domchek SM, et al. Psychosocial adjustment and perceived risk among adolescent girls from families with *BRCA1/2* or breast cancer history. J Clin Oncol. 2016;34(28):3409–16.
127. Werner-Lin A, Hoskins LM, Doyle MH, Greene MH. 'Cancer doesn't have an age': Genetic testing and cancer risk management in *BRCA1/2* mutation-positive women aged 18–24. Health London. 2012;16(6):636–54.
128. Werner-Lin A. Beating the biological clock: the compressed family life cycle of young women with *BRCA* gene alterations. Soc Work Health Care. 2008;47(4):416–37.
129. Prince AER, Roche MI. Genetic information, non-discrimination, and privacy protections in genetic counseling practice. J Genet Couns. 2014;23(6):891–902.
130. Hall MA, McEwen JE, Barton JC, Walker AP, Howe EG, Reiss JA, et al. Concerns in a primary care population about genetic discrimination by insurers. Genet Med. 2005;7 (5):311–6.
131. Pollitz K, Peshkin BN, Bangit E, Lucia K. Genetic discrimination in health insurance: current legal protections and industry practices. Inquiry. 2007;44(3):350–68.
132. Hall MA, Rich SS. Patients' fear of genetic discrimination by health insurers: the impact of legal protections. Genet Med. 2000;2(4):214–21.
133. Peterson EA, Milliron KJ, Lewis KE, Goold SD, Merajver SD. Health insurance and discrimination concerns and *BRCA1/2* testing in a clinic population. Cancer Epidem Biomar. 2002;11(1):79–87.
134. Hadley DW, Jenkins J, Dimond E, Nakahara K, Grogan L, Liewehr DJ, et al. Genetic counseling and testing in families with hereditary nonpolyposis colorectal cancer. Arch Intern Med. 2003;163(5):573–82.
135. United States. Health Insurance Portability and Accountability Act of 1996. Public law 104–191. US Statut Large. 1996;110:1936–2103.
136. Genetic Information Nondiscrimination Act H.R. 493110–233. 2008.
137. United States. Americans with Disabilities Act of 1990. Public law no. 101–336. US Statut Large. 1990;104:327–78.
138. Leiker M. The Americans with disabilities amendment act–are you ready for the changes? WMJ. 2008;107(8):393–4.
139. Patient Protection and Affordable Care Act. Public law no. 111–148. 2010.
140. Green RC, Lautenbach D, McGuire AL. GINA, genetic discrimination, and genomic medicine. N Engl J Med. 2015;372(5):397–9.
141. Rhodes R. Genetic links, family ties, and social bonds: rights and responsibilities in the face of genetic knowledge. J Med Philos. 1998;23(1):10–30.
142. Falk MJ, Dugan RB, O'Riordan MA, Matthews AL, Robin NH. Medical geneticists' duty to warn at-risk relatives for genetic disease. Am J Med Genet A. 2003;120A(3):374–80.
143. Agnese DM, Pollock RE. Breast cancer genetic counseling: a surgeon's perspective. Front Surg. 2016;3:4.
144. Florida. District Court of Appeal FD. Pate v. Threlkel. Wests South Report. 1994;640:183–6.
145. Florida. Supreme Court. Pate v. Threlkel. Wests South Rep. 1995;661:278–82.
146. New Jersey. Superior Court, Appellate Division. Safer v. estate of pack. Atl Rep. 1996;677:1188–93.
147. Committee on Bioethics, American Academy of Pediatrics. Informed consent, parental permission, and assent in pediatric practice. Pediatrics. 1995;95(2):314–7.

Index

A.B. Chagpar (ed.), *Managing BRCA Mutation Carriers*,
DOI 10.1007/978-3-319-59198-8

MIX
Papier aus verantwortungsvollen Quellen
Paper from responsible sources
FSC® C105338

FSC
www.fsc.org

If you have any concerns about our products,
you can contact us on
ProductSafety@springernature.com

In case Publisher is established outside the EU,
the EU authorized representative is:
Springer Nature Customer Service Center GmbH
Europaplatz 3, 69115 Heidelberg, Germany

Printed by Libri Plureos GmbH
in Hamburg, Germany